AF684191

PRINCIPIA PRACTICA

The Logic of Practice

Arnold A. Johanson

University Press of America,® Inc.
Lanham • New York • Oxford

Copyright © 2000 by

University Press of America,® Inc.
4720 Boston Way
Lanham, Maryland 20706

12 Hid's Copse Rd.
Cumnor Hill, Oxford OX2 9JJ

All rights reserved

Printed in the United States of America
British Library Cataloging in Publication Information Available

Library of Congress Cataloging-in-Publication Data

Johanson, Arnold A.
Principia practica : the logic of practice / Arnold A. Johanson.
p. cm.
Includes bibliographical references and index.
l. Practice (Philosophy) I. Title.
B831.3.J64 1999 170—dc21 99-049178 CIP

ISBN 0-7618-1530-9 (cloth: alk. ppr.)

∞™ The paper used in this publication meets the minimum requirements of American National Standard for Information Sciences—Permanence of Paper for Printed Library Materials, ANSI Z39.48—1984

To Stig Kanger and Philip Blair Rice

Contents

PART TWO. A Logical Basis for Practical Philosophy

PART THREE. A Defense of a Practical Philosophy

Preface

There is little doubt that Western practical philosophy has lost its way. How, where, and when did this happen and who is responsible? It is my theory that it began with the Greeks. (Recall the warning of Virgil: "Beware of Greeks bearing gifts!") The scene of the crime is ancient Athens where Socrates, that venerable Greek gadfly, challenges his fellow Athenians to answer some simple questions, questions like "What is goodness?", "What is justice?", and "What is piety?" Challenging all comers, Socrates knows full well that they cannot answer, and that not even he knows the answers—but he enjoys himself, first flattering them, pretending that he hopes to learn from them, and then crushing their pride as they each go down to defeat admitting they know nothing. After years of suffering him, the Athenians decide that, on the whole, Athens would be a happier place without him. Socrates gets the message and drinks the hemlock. Philosophy is a hazardous occupation!

The failure of the Greeks, including Socrates, to answer the simple questions posed by him has not stopped philosophers down the ages from asking the same questions; consequently, questions of this kind remain the fashion in ethics even unto the present day. Now I do not pretend to know if such questions have any answers, but what I do know is that they are not just hard questions. They are the wrong questions!

You ask, if these are the wrong questions, what are the right questions? The question is fair enough. I think I know some of the right questions, and I think I know some of the answers. Some of the right questions are questions like "What is the function of society?" and "What is important in life?" I suspect, however, that the main reason that the "What is?" questions favored by the Greeks remain the fashion today is that moral philosophers have stuck steadfast to a logical framework more suited to mathematics and science than to subjects like ethics and politics.

During the past century there have been attempts to build a new logic more suited to practical philosophy. Pioneers in this effort have been Mally [1926] and von Wright [1951]. Somewhat reminiscent of early attempts to found mathematics on set theory, their theories are vitiated by paradoxes. In my opinion the best way out of these paradoxes is to follow a path outlined by Mill over one hundred and fifty years ago. Up to now, Mill's suggestion has not been taken seriously. (I refer here to a suggestion of Mill in his *A SYSTEM OF LOGIC* and not to the popular doctrine of utilitarianism for which he is justly famous.) The present work carries forth Mill's idea but in a way that Mill could not have anticipated. The mathematical tools, necessary for carrying carry out Mill's idea, did not exist in Mill's time. This may suggest that only those with a technical facility can read this work, but that is not so: the bulk of this work is designed to be read by those without a technical background in mathematics and logic.

Though there are already a great number of works on practical philosophy, I am driven to introduce a new work in this field, not only because of a profound dissatisfaction with current practical philosophy, but because of a concern for the state of the world. Perhaps one can do little about the state of the world except to point out folly wherever it occurs, and cry out against cruelty, but I believe that something can be done about the present condition of practical philosophy. It is possible, moreover, that if practical philosophy were put on a sound intellectual basis, this could, in itself, have a positive effect on the state of the world.

It is roughly twenty five hundred years since Socrates went down the wrong road and the West followed him. It is time to ask for new directions. Twenty five hundred years are enough!

Acknowledgments

During his life Stig Kanger was an inspiration to me. I not only profited from studying his work, but I appreciated his hearty laugh, his kindness, and his encouragement of my own efforts. Over the years J.J.C.Smart has been my mentor, both in extensive correspondence and in personal contact. Richard Robin of Holyoke College has been a constant source of encouragement. And I have profited from John L. McKenney's careful reading of my work and his trenchant comments. Finally, there are two others that I want to acknowledge. One is my teacher Philip Blair Rice who introduced me to the subject and woke me from my skeptical slumbers. The other is my wife Mary, who fought to keep my syntax in order and without whom none of this would have mattered.

Introductory

1 Prologue

The modern world faces a paradox. On one hand, the failure of the old paradigms of morality to fit contemporary problems produces a disastrous breakdown in the common faith in ideals and principles; on the other, fear of a loss of ideals and principles inspires a fanatical adherence to the old paradigms. The fact that one can through rational thought demonstrate the failure of the old paradigms is viewed in sharply different ways in the two opposing camps. To fanatical adherents of the old paradigms it is a *reductio ad absurdum* of rational thought; to skeptics it is too often a demonstration of the absurdity of ideals and principles.

This work is written in the belief that one cannot accept either of these extreme views. As John Dewey tells us the destruction of old ideals and principles does not necessarily mean an abandonment of ethics, but is a necessary preliminary to building a new ethic to fit new conditions. I do not wish to argue here at length for either rationality or for ideals and principles, but I wish to remind the reader that cynicism about ethics fosters a narrowness of vision whose devastating effects are nowadays found both in the individual and society (see Chapter One).

This work has two goals: one, to do what one can to put practical philosophy on a sound intellectual basis; the other, to construct, on this basis, normative systems both for the governance of society and the guidance of individuals. Can these goals be achieved? It is perhaps not too difficult to construct a practical philosophy suitable for these times since the literature of the subject is vast and one can draw on a legacy of the work of many fine thinkers wiser than oneself. The really difficult part is to insure that the philosophy which is to be constructed has a sound intellectual basis. It must be admitted that the assumption that it is possible to construct practical philosophy on a sound intellectual basis flies in the face of all experience. I am reminded of the remark that Dr. Johnson made when he learned of a man who, having been unhappy in his first marriage, married immediately after his wife died: it was the triumph of hope over experience! In my case, my hopes are fueled by a faith in reason and a conviction that internecine conflict between warring factions can only be stopped, if at all, by an ethic based on reason.

Perhaps one saving grace is that I do not propose to talk about *values*. According to the eminent twentieth century philosopher, Karl Popper, most talk about "values" tends to be hot air. The quote is interesting so let me give it in full. In his autobiography, *UNENDED QUEST*, he writes as follows:

> ...few scientists, and few philosophers with scientific training, care to write about values. The reason is that most talk about values is so much hot air. So many of us fear that we too will only produce hot air, or if not that, something not easily distinguished from it.

I tend to agree with Popper. The nebulousness of concepts like "value" has made it difficult to develop a foundation for practical philosophy and this very difficulty has made the history of practical philosophy one of continuous wrangling. A certain amount of conflict in any subject is healthy, but the wrangling in practical philosophy has been singularly unproductive. The lack of a foundation for practical philosophy disturbed John Stuart Mill [*UTILITARIANISM*, paragraph one] so much that he wrote, almost in despair:

> From the dawn of philosophy, the question concerning the *summum* b*onum*, or what is the same thing, concerning the foundations of morality, has been accounted the main problem in speculative thought, has occupied the most gifted intellects, and divided them

> into sects and schools, carrying on a vigorous warfare against one another. After more than two thousand years the same discussions continue, philosophers are still ranged under the same contending banners, and neither thinkers nor mankind at large seem nearer to being unanimous on the subject, than when the youthful Socrates listened to the old Protagoras, and asserted (if Plato's dialogue be grounded on real conversation) the theory of utilitarianism against the popular morality of the so-called sophist.

Despite Mill's best efforts and the efforts of generations of philosophers since Mill, little has changed. In fact, two synergistic forces in the modern world combine to threaten the whole enterprise of moral philosophy even further. One is the devastating effect of science and rationalism on traditional morality; the other is the rise among philosophers of the view expressed by Shakespeare's Hamlet, "There is nothing either good or bad but thinking makes it so," a view that has come to be known as "noncognitivism." With few exceptions, the response of philosophers to these threats has been ineffective. In fact, the usual works of modern academic moral philosophy—which if they do not degenerate into hot air or clever academic exercises—either wallow in existential despair or consist in moralistic sermons in academic dress. Sad to say, works that are intended to inspire are, hardly without exception, without a logical foundation, and works whose intent is give practical philosophy a logical foundation are not only uninspiring, which is perhaps forgivable, but fall short of their object.

Any work on practical philosophy, if it is to be worth its salt, must have something to say not only about theories of ethics and politics but about how to live one's life. Having something to say about how to live is what it traditionally meant to "have a philosophy." Moreover, the saving grace of the great moral philosophers is that they have something to say on matters of concern to most of humanity. One finds, for example, that a significant part of works like Aristotle's *ETHICS* (and his *POLITICS*), Plato's *REPUBLIC*, Spinoza's *ETHICS*, and Kant's *GROUNDWORK* is concerned with purely practical matters. It is in the spirit of this tradition that the present work contains a philosophy of life (see Chapters One, Six, and Seven), a philosophy of religion and art (see Chapter Seven), an ethics (see Chapters One, Five, and Six), and a political theory (see Chapters Four and Eight).

The work is divided into *three parts.* (See the table of contents.) ***Part One*** is essentially an informal introduction to the ideas of this work. The material in ***Part One*** is presented without technical language and it is my hope and intention that it can be read and understood by nonspecialists. Except for some of the material in Chapter Two, the ideas in the first eight chapters that make up ***Part One*** are perhaps not as revolutionary as those in the later chapters, but it is in defense of the ideas in ***Part One*** that the second and third parts of the book (***Part Two*** and ***Part Three)*** were conceived.

2 A New Framework

After these preliminary remarks, it is time to offer a brief explanation of the practical philosophy which this work contains. Before proceeding further, however, let us be clear on what is meant here by "practical philosophy." (I have not been able to find this term in dictionaries or encyclopedias.) Practical philosophy in a narrow sense is the study of ethics and politics; but, in the broader sense used in this work, ***practical philosophy is the study of normative systems***. And what is a normative system? Roughly speaking, a normative system is a system of imperatives. Examples include such systems as the *TORAH, THE NEW TESTAMENT,* and *THE CONSTITUTION OF THE UNITED STATES OF AMERICA.* The normative system in the *TORAH* has as its primary purpose that of promoting harmony in a community, while the normative system in the *NEW TESTAMENT* has as its primary purpose the development of personal enlightenment and spiritual growth. These systems contain imperatives like "Love thy neighbor as thyself!" and "Take no thought for your life, what ye shall eat, or what ye shall drink; nor yet for your body, what ye shall put on!"

Clearly, it is impossible to understand normative systems, as ordinarily constituted, without making sense of imperatives. Some years ago I published an article in which I attempted to do just that (see Johanson [1988]). When I later discovered that John Stuart Mill [1898ed.] thought along similar lines, I was naturally delighted to get the great man's *imprimatur*. (How he would react to my attempts to fill in the bare bones of his idea I, of course, do not know.) Mill begins with two points: one, science is of use in solving practical problems; the other, the language used in science differs from the language of practice. Mill clarifies the second point as follows: "on one hand, the proper language of practice is the language of sentences in the imperative mood; on the other; the language of science is the language of sentences in the indicative mood." Mill naturally concludes that there must be a way of going back and forth between these two languages. Mill's problem, in a nutshell, is to develop a theory that allows one to apply the logic of science to the solution of practical problems. This problem is solved in this work by the introduction of two interrelated measures: one, a clarification of the notion of "normative system"; the other, the introduction of the notion of "requirement" as a substitute for "truth." A few prefatory remarks to illustrate these ideas may not be amiss, but before make them, I should like to quote an observation by Bertrand Russell on the difficulties encountered by a writer when he or she strays from classical language in an attempt to express ideas that have not yet had time to acquire "pleasant literary clothes."

Russell [1940] writes:

> The question of what can and what cannot be expressed in philosophy without the use of a crabbed and difficult vocabulary and syntax is a difficult one, which has more influence than might be thought upon the actual content of a writer's opinions. Broadly speaking, old conceptions have acquired pleasant literary clothes, whereas new ones still appear uncouth. An aesthetic bias in favour of good literary form is therefore likely to be associated with conservatism. [Russell here considers some examples.] In such ways those who insist upon elegant literary form are compelled to lag behind—often far behind—the best thought of their time. Per contra, conservatives have a great aesthetic advantage over innovators, for ideas, unlike animals become more beautiful as they grow older.

I try in the major part of this work to avoid "crabbed and difficult vocabulary and syntax," but the point of this work will be lost if I do not at the outset explain, at least heuristically, some technical terms. With this in mind, I hope less technically advantaged readers will bear with me. (I direct more technically advantaged readers to the note at the end of this chapter.)

Requirement is a notion that applies to imperatives. As an illustration consider the following: Suppose a car on a motor way is approaching a red traffic signal. In such circumstances one can say that *it is required that it shall be the case that presently the car comes to a stop*. Here the notion of requirement is applied to the imperative *"It* ***shall*** *be the case that presently the car comes to a stop!"* The phrase *"it is required that"* appended to *"it* ***shall*** *be the case that presently the car comes to a stop"* turns ***a sentence in the imperative mood*** into ***a sentence in the indicative mood***: *"It is required that it shall be the case* *that presently the car comes to a stop*." The phrase **"it is required that"** is an operator that is applied to a **command** (*It **shall** be the case that presently the car comes to a stop*!) rather than to an **activity**. (The theory neither asserts nor denies, however, that there is a property "required"—or indeed any other "value"—which may be attributed to an action or an activity—such as "to stop the car"). Without a notion of ***requirement*** or something like it, it would not be possible to develop a logic of imperatives. The notion of requirement is needed as a substitute for the notion of "truth."

Now we began by saying that in order to apply science to practical problems we need a way to go from the language of imperatives to the language of indicatives, ***and back again***. The operator "it is required that" takes us from imperatives to indicatives so we need an operator that goes in the opposite direction. Such an operator is already well known; this is the operator "it **shall** be the case that" called the "**imperative operator**." If we take the imperative operator "it **shall** be the case that" and append it to—for example—the sentence "presently the car comes to a stop," we get the imperative "*it **shall** be the case that presently the car comes to a stop*!" Adopting as our second operator, the imperative operator, we obtain the two operators we have sought (one in each direction). To a mathematician a natural question arises: are these operators "adjoints"? The reader will naturally ask, "What is **adjointness**?" That is a question that is not addressed in these introductory remarks (see Chapter Two, however). Suffice it to say here that adjointness lies at the foundation of almost all (if not all) of mathematics and logic. If I am not mistaken, adjointness is as fundamental in the theory of normative systems as it is in most mathematical or logical theories. In the body of this work (see ***Part Two***) it is shown that if one assumes this one property (of adjointness) all the formulas in the logic of imperatives flow in a natural way.

A practical logic based on the above theory has many advantages. I mention three of them here. First of all, unlike traditional logic, such a logic allows premises such as Kant's categorical imperative, which have no place in traditional logic. Secondly, out of the theory there emerges a language that enables us to avoid some troubling meta-ethical questions—particularly, questions about the truth or falsehood of ethical judgments. As a consequence, some epistemological obstacles to establishing proposed solutions to problems in the field are removed. The present work, though not directly supporting Bertrand Russell's views on ethics and politics, may be regarded as an answer to Russell's critics who claimed that since he did not believe that ethical judgments were capable of truth or falsehood he had no business making pronouncements on the conduct of life and politics. Finally, and most important, the present theory offers a theoretical basis for basing rules or commands (expressed by sentences in the imperative mood) on empirical data (expressed by sentences in the indicative mood). *This enables us to approach many problems in practical philosophy in the same piecemeal fashion that has been so effective in science.*

3 Chapter Summary

We have already noted that this work has three parts. The first part states a practical philosophy, the second part builds a logical foundation on which this practical philosophy can be supported, and the third part presents a defense of the practical philosophy based on the foundation laid down in the second part. I begin this section with a summary of the chapters in the first part.

The first three chapters of ***Part One*** form a prolegomenon to the moral and political philosophy presented in the next five chapters. The latter chapters therefore constitute, in a sense, the heart of this work. In these chapters, two normative systems are constructed: one for politics, the other for ethics. The former is contained in Chapters Four and Five; the latter in Chapters Six and Seven. The fourth chapter argues for a normative system for politics based on democratic principles and in the fifth chapter a social ethic is derived from this system. I argue for democratic principles on the basis of what may be called *the paradigm of* (*societal*) *functionality.* Functionality, as a principle of social consstruction, hardly seems to me controversial—what I have to say in its behalf will be found in Chapter Four. The next two chapters (Chapter Six, *What Are We To Do With Our Lives?*; Chapter Seven, *Beauty, Mysticism, and Religion*) contain a philosophy of life founded basically on what I call *the paradigm of growth.* This philosophy attempts to do justice not only to the classic insights of the Greeks, who believed in a philosophy of full functionality or ***aretē*** but to the more modern insights of Whitehead, Maslow, and Dewey. My arguments for the principle of growth are to be found scattered throughout Chapters One, Two, Three, Six, and Seven, but the nub of these arguments is to be found in Chapter Two. The last chapter of ***Part One*** (Chapter Eight, *Several Suggestions for the Salvation of the Planet*) deals with what I have to say on the problems that humanity faces on a global scale. This chapter is an application in a global context of the democratic principles of Chapter Four and the humanistic principles of Chapter Six.

It remains to discuss the first three chapters in ***Part One***. The first chapter (*The Essence of Ethics*) is an anticipation of some of the conclusions of Chapter Six. I might have called it "If I could preach just once." The work begins with this chapter for two reasons: one, it is accessible; the other, it gives the reader an idea of where the practical part of this work is going. Nonetheless this chapter may be skipped by those whose interest in ethics and politics is purely theoretical. The real meat of the book begins with the second chapter. In this chapter (called "*A Brief Tour of Western Moral Philosophy*") the reader is introduced to the main ideas of this work. These ideas are presented through a critical history of the development of Western moral philosophy. The chapter reaches perhaps its highest point in the second section where the underlying logic of practical philosophy is discussed. This section culminates in an informal presentation of the methods (or logic) of this work, which we have already discussed briefly.

Practical philosophy cannot be founded totally on logic, however. Despite Hume's plea to philosophers to reject every system of ethics "not founded on fact and observation," modern moral philosophers, in their overweening concern for method, have neglected science and consequently have paid scant attention to the work of scientists such as Maslow or Lorenz. The correction of this oversight is begun in the first section of Chapter Two and continued in Chapter Three (*The Springs of Human Action*). In this part of the work, I depend on science and, in particular, on the fundamental work of psychologists such as Hebb and Maslow and ethologists such as Lorenz. Here the reader is introduced to (or perhaps reminded of) ideas—such as a social hierarchy and a hierarchy of higher needs—which play an important role in succeeding chapters. The inspiration here comes from sources as diverse as William James, John Dewey, Donald Hebb, Abraham Maslow, Alfred North Whitehead, Bertrand Russell, Jean Paul Sartre, and Konrad Lorenz.

Having discussed all the chapters in ***Part One***, I turn now to ***Part Two***. In ***Part Two*** the theory of normative systems is given a formal presentation. ***Part Two*** contains two chapters: Chapters Nine and Ten. Chapter Nine (*The Language of Requirement*) develops a theory of norms (or normative statements) based on the notion of requirement (and the dual concept of "admissibility"). *Simple examples of norms, in the sense of this work, are statements that set forth requirements or admissibilities.* (See for example, the principle of functionality and the principle of the consent of the governed in Chapter Four, ***Part One***.) The theory of norms is, of course, important since normative statements are the raw material out of which ethics and politics and normative systems of every stripe are made. Chapter Ten (*Imperative Logic*) generalizes the theory introduced in the first chapter of ***Part One*** and has a brief discussion of the formal semantics of requirement. A large part of Chapter Ten is devoted to dyadic norms. Though this chapter is self-contained, it is openly dependent in some respects on the mathematical theory of categories and for that reason may be difficult for some readers.

After reading the first chapter of ***Part Two***, some readers may wish to go immediately to ***Part Three.*** *For an understanding of* ***Part Three****, it is not necessary to have read the second chapter of* ***Part Two****.*

Part Three (A Defense of a Practical Philosophy) contains only one brief chapter (Chapter Eleven, *Arguments for Ethical and Political Principles: A Formal Analysis*). After some preliminaries, the chapter does two things: first it proposes a formal structure for each of the *two normative systems*—one for ethics, the other for politics—that were introduced informally in ***Part One***; secondly, it shows how this formal structure may be used to back up, by formal and logical demonstration, the arguments for the ethical and political principles in ***Part One***. The demonstrations in these systems depends on two sources: the facts of the case as determined by empirical observation; the other, the axioms of the given system. The axioms (as in any formal theory) are of two kinds: one, the logical axioms; the other, the proper axioms. The proper axioms also divide into two kinds: one, requirements that are imposed on human beings by raw fact; the other, requirements that are imposed on the world (and other human beings) by human beings. Among the latter are the requirements implicit in the paradigms on which the two normative systems are partly based; namely—in the case of the societal system—the paradigm of (societal) functionality and—in the case of the system for the guidance of individuals—the paradigm of (individual) growth. It follows that the reasonableness of the practical philosophy in these pages depends ultimately on the reasonableness of these two paradigms. Though I believe that valid arguments may be given for the principles on which these paradigms are designed, this does not mean that I believe they may be established beyond the shadow of a doubt—but in this respect they do not differ from principles in the reigning paradigms in mathematics and physics. For example, principles such as the law of excluded middle, the axiom of infinity, and the axiom of choice that characterize the reigning mathematical paradigm are challenged by those who support the intuitionistic paradigm of Brouwer. I shall indeed be satisfied if the paradigms in this work are as defensible as (say) the current paradigms in physics.

This completes my summary of what is in the book. Those readers who are anxious to learn immediately what this work teaches about a particular matter may wish to skip some or all of the first few chapters. Those who wish to know what I say about ethics may (after perusing Chapter One) find it profitable to turn to Chapter Six (*What Are We To Do With Our Lives?*). People who are oriented toward what are generally called "matters of ultimate importance" could begin with Chapters One and Six and follow it up with a reading of Chapter Seven. Others, who are more interested in political theory, may turn first to Chapter Four and follow it up with Chapter Eight. Moralists might want to start with Chapter Five (*Social Ethics*). Those chiefly interested in the empirical background of this work should turn to the first section of Chapter Two and follow it up with a reading of Chapter Three. I suggest philosophers start with Chapter Two. Logicians might consider reading Chapter Two first and then turn to the chapters (Nine, Ten, and Eleven) in PARTS TWO and THREE.

Note (to specialists):

Though I shall not in the body of this work slavishly stick to the **underlying language** of this work—for in keeping with my intention **not** to write only for specialists, I make some concessions to ordinary speech—***any statement not formulated in that language must be considered heuristic.*** In other words, statements not in the formal language must receive their sense by being translated into the formal language, much as Russell in his theory of descriptions translated apparently meaningless sentences (such as those containing empty phrases like "the king of France") into sentences that have meaning in the formal language. One could use Wittgenstein's ladder analogy here, and say that any statement in traditional literary form (and not in the formal language) is to be used as a ladder to reach the corresponding more precise, but less intuitively understandable, assertion.

Part One

A PRACTICAL PHILOSOPHY STATED

Chapter One

The Essence of Ethics

As a rough definition, let us suppose that ethics, as an inquiry, is a search for a pattern by which to live one's life. To some extent our pattern of life is chosen for us by the circumstances of our birth and our mental and physical inheritance, but there is at least a grain of truth in the dictum of Sartre [1946] that men and women make themselves. The result of ethical inquiry may be viewed as a series of guidelines, somewhat like the Hebrew Decalogue, for making oneself. As a way of introducing my own views on how to choose the guidelines, I begin by anticipating some of the things I say in the next chapter about the views of Plato and Aristotle.

As explained in Chapter Two, Plato thinks that every earthly thing is flawed because it does not live up to the model for it laid up in heaven, the timeless and changeless ***form.*** Aristotle, Plato's student, takes over Plato's idea and makes it his own. According to Aristotle, the form of a thing is its ***essence*** and the more nearly a thing partakes of its essence the more nearly perfect it is. Moreover, Aristotle thinks that the essence of each thing is captured by giving its definition; for example, a person may be defined as a "rational animal." Hence, for Aristotle, a "virtuous" person is one who closely resembles the model thinker-cum-animal. The physiologist and psychologist today have a similar concept since a normal or "healthy" person may be defined as one who is not only, physiologically speaking, a reasonably perfect animal, but one who is *rational* as well.

It might be thought that the ideas of health and normality have nothing to do with ethics. But even Kant, in agreement with Aristotle, thinks that rationality is important in ethics—in fact the great Kant thinks it is the essence of ethics. It is true that the *authorities* on rationality—I refer here to psychologists and psychiatrists—are not usually thought of as experts on ethics: yet they are often asked in criminal cases to judge whether a defendant is capable of distinguishing "right" from "wrong." Can one not conclude that to be moral (or immoral) is above all to be rational? People even hold God to account (as in the book of *JOB*) because he (or she) is supposed to be rational.

The psychologist Maslow [1971] goes further and identifies morality with psychological health. Now, many philosophers discount this theory because, they say, it confuses facts with values. This may be true (see the discussion of meta-ethics in the next chapter) and yet there is no dispute that medicine is a *normative* science. Even philosophers do not often hesitate to follow a physician's prescriptions.

As we said, broadly speaking, ethics, as an inquiry, is a search for a pattern by which to live one's life. In certain areas of life, and in certain circumstances, the problem of how to live one's life has a straightforward solution. In the case of physiological health there is a large establishment of people—physiologists, biochemists, clinicians, etc.—whose prescriptions, by and large, guide rational people. In this field it is the needs of the body that are important. For some reason, however, it is not thought to be a matter of ethics—except when it comes to deliberate self-murder—how well people treat their own bodies. It is only when we come to the arena of social relations that tradition holds that personal conduct is of ethical concern. In this field it is social needs that are important. On the basis of social needs (and the needs of society) there have arisen institutions that determine what behavior is advisable and what is not. In this realm, there are two kinds of prescriptions: one, laws sanctioned by religion and encoded in books such as the *HOLY BIBLE*; the other, laws passed by civic authorities and encoded in what are called *STATUTE BOOKS*. In ancient times no such distinction was made since civic rulers, like the Pharaoh in Egypt, were considered gods or representatives of the gods. This realm resembles the field of medicine in that there is a battery of experts, including psychologists and lawyers, whose prescriptions, by and large, guide rational people. The medical people do not entirely abandon you even here since your social behavior impinges on both your physiological and psychological health. In Chapter Five, I argue that what is considered ethical in this domain is whatever contributes to the "social health" of society (Cp. Leslie Stephen [1882]).

Ethics in its most developed form, however, is to be found in none of the areas of life we have discussed so far. On the contrary, it is to be found far beyond the range of law and medicine in what may be called the "spiritual realm." This realm arises, like the realms we have already discussed, from human needs.

Maslow divides needs into three kinds: lower, intermediate, and higher. The lower needs are the physiological needs like the needs for food, water, sex, shelter, etc. The intermediate needs are the social needs, the needs for love, safety, power, and esteem. Beyond the physiological and social needs lie motivations that are more dimly apprehended including those that lead into purely human pursuits such as art, science, and religion. The higher needs are more difficult to describe, but they include (for example) the needs for knowledge and beauty. Physiological health occurs when the lower needs are satisfied; social health occurs when the intermediate needs are satisfied. I shall argue here, and in Chapter Six, that to be ethical, in the highest sense, is to completely satisfy the higher needs.

Maslow notes that before higher needs become important, the lower needs must be satisfied. For the person who is starving, food is the only important thing. According to Maslow [1954, p. 214]:

> Just as a tree needs sunshine and water and food, so do people need love, safety, and other basic need gratifications that come only from without. But once these external satisfiers are obtained, once these inner deficiencies are satiated by outside satisfiers, the true problem of individual human development begins.

As human beings develop, they grow both in power and understanding. Immature people, like a child in a tantrum, use power in destructive ways. Maturity, in part, is the ability to make rational use of power. We may postulate that people have a need not only to develop resources but to make constructive use of those resources. Now we may also postulate that people have within them a limited potential to develop resources and, moreover, a native propensity to develop in a limited number of directions. Some may have the natural propensity, under a suitable environment, to be musicians, others to be mathematicians, and still others to be cooks, teachers, novelists, gardeners, nurses, home-makers, priests, politicians, vagabonds, etc. This matter is more thoroughly explored in Chapters Two, Three, Six, and Seven.

In later chapters (see Chapters Two and Three), I define *growth* in an adult human being as the tendency to "become less occupied with the fulfillment of the lower needs and, correlatively, to be more and more directed by the higher needs." People are not satisfied simply to fill their bellies, nor are they completely satisfied to have power, riches, and even love, but need something more. The higher needs are needed to make sense out of life and to inform us what is truly important. The need to make sense out of life and the need to know what is truly important are ***meta-needs***. (See Chapter Three.)

What do we live for? Though even Einstein [1931] finds part of the answer to that question is provided by the fulfillment of social needs—the need (as he says) to accommodate to those on "whose smile and well-being our own happiness depends..."—ultimately the social needs do not completely answer the question. Bertrand Russell [1967], after mentioning that he has sought love and the ecstasy and the "prefiguring vision of heaven " it can bring, writes: "With equal passion I have sought knowledge. I have wished to understand the hearts of men. I have wished to know why the stars shine. And I have tried to apprehend the Pythagorean power by which number holds sway above the flux." In a similar vein, H. G. Wells [1935] writes: "Man does not live by bread alone; he eats that he may learn and adventure creatively,..." For ordinary people—as well as for the likes of Einstein, Russell, and Wells—a profession or trade of some kind, whether as an artist, scientist, or even a stone-mason, provides something to live for. But even professional life does not yield a complete answer to the question. People, if they have satisfied their bodily and social needs, and have attained something like a profession or calling in life, need something more. I call this need, a need for **transcendence**. Transcendence, for those who have reached a certain measure of growth, is the most important of the higher needs. There is a hint of transcendence in the best of science and art. Moreover, the need for transcendence is the source of religion at its deepest level. Spinoza described the fulfillment of this need as "the intellectual love of God." Aristotle called it "divine activity," and found it in contemplation. "Transcendence" may be defined as that which frees us not only from the here and now but from imprisonment in narrow self-interest. The need for transcendence is sometimes expressed as a yearning to break out of the confinement of the finite into the boundless realm of the infinite. If the assertions of mystics are correct, this realm can be found only by looking inward. For example, Emerson [1841], the American transcendentalist, asserts that within each human being "is the soul of the whole; the wise silence; the universal beauty, to which every part and particle is equally related; the eternal ONE" which "abolishes time and space." Moreover, Emerson makes the claim that each human being is "the facade of a temple wherein all wisdom and all good abide." Using a similar metaphor, Russell [1903] in his *FREE MAN'S WORSHIP*, writes: "Haunting foreshadowings of the temple appear in the realm of the imagination, in music, in architecture, in the untroubled kingdom of reason, and in the golden sunset magic of lyrics..." It is apparent that, in all the animal kingdom, only human beings have an idea of things unseen, beyond the here and now. At the very least, human beings have an idea of a vast unending future. Moreover, in the imagination of human beings there are no limits. We humans are impatient of *all* boundaries—if there is a limit we want to surpass it. In short: human beings yearn for the infinite. One may ask, why this "divine discontent"?

The infinite like love, is mysterious. Sidney Harris, the late popular columnist, writes, "Love is the celebration of a mystery." The lure of mystery—or what transcends understanding—is the source not only of romance but that which is deepest in science. According to Einstein, "The most beautiful thing we can experience is the mysterious." Some people are touched by the romance of science; for others science is destructive of romance. We all want, however, to know all the things that can be known through science and empirical observation: What is beyond this wall? What is beyond the next hill? What is beyond the sea? What is in the interior of the earth? What lies inside the atom? What is it like on the moon and the other planets? What is beyond the solar system? What is beyond the stars?

But the deepest riddle of all—the mystery of existence—lies beyond the ken of science. Einstein [1931] puts it this way: "Strange is our situation here on earth. Each of us comes for a short time, not knowing why, yet sometimes seeming to divine a purpose." It is apparent to many of us that the cosmos, if it has a soul, does not care for human life. But, by and large, people tend to rebel against this thought and think that not only does the cosmos have a soul, but it cares for human life. For the believer a human life (not to mention a sparrow) has cosmic significance. ("What is a man, that thou art mindful of him...?") Even atheists, however, have a need to believe that there is something that **is** of cosmic significance—if not a single human life, then something transcendent.

People wish not only to escape the boundaries of the known world, but, unless I am mistaken, wish to escape from the "prison of self."

Freedom from the "prison of self" may be called "**self-transcendence.**" Blackwell [1985] in *THE SPINOZISTIC ETHICS OF BERTRAND RUSSELL* aptly characterizes what I call self-transcendence as "impersonal self-enlargement." According to Blackwell, impersonal self-enlargement itself involves what has been called a "Spinozistic development of the understanding toward a reduction of separateness." Accordingly, I interpret transcendence to mean, in part, that the following maxim is obeyed: ***Do not think primarily of yourself, but let your first thought be for your family and community , and be mindful therefore of the species to which you belong and the planet on which you live.***

One caveat, however: self-transcendence is not to be understood as self-abnegation. One of the higher needs (see Chapter Six) is to be dedicated to something to which one is idiosyncratically suited. People like Einstein, Wells, and Russell, in developing their own talents, are contributing to the whole. If you do not develop and express what is in you, you are liable to be stunted spiritually. Nietzsche was correct to warn of *ressentiment.*

Some may question the reality of the higher needs, especially that of self-transcendence. These, it may be thought, are simply cultural artefacts. But skepticism of this kind leaves unexplained culture itself. To be sure, the need for self-transcendence comes in part from social needs like love, companionship, and tribal identity. The latter needs, however, can just as often lead to hate and destruction. But the human need for self-transcendence asks for something more. Even captains of industry, after they have apparently gained the whole world, seem to want to go beyond their own narrow selves, beyond their friends and relatives, and even beyond their own tribe, to give something to the world.

As explained more fully in Chapter Three, Whitehead relates morality to importance and importance to perspective. The view I defend in this work is similar: morality demands that people do not remain imprisoned in a particular perspective but adhere to what is truly important. With apologies to Kant, let us define morality as *treating, insofar as feasible, all living things as ends in themselves and not merely as means.* It is easy to see that transcendence implies morality in this sense since not to treat living things as ends in themselves is a failure to keep oneself in perspective and to take the interests of living things other than oneself seriously.

It may be fairly asked, if all people have a need for self-transcendence, why is it that morality is so rare? To some extent this begs the question; for it is only in ***developed*** people that the need for self-transcendence assumes any importance in their lives. Moreover, there is a sense in which this question is somewhat like asking why people get sick if people have a need for robust health. Or like asking why is it that though people have a need to relax, so few people relax, especially in America. It seems that disease is almost more universal than health.

Health is difficult to achieve and no more so than in the realm of ethics. One is reminded of the saying of Spinoza: "...all excellent things are as difficult as they are rare." But what is merely difficult for one person may be impossible for another. One does not expect fair play from a toddler, nor does one expect headhunters from Borneo to control (without outside intervention) their tendency toward taking the heads of their fellows. This is not because either the toddler or headhunter is intrinsically inferior, but because neither has had the opportunities for development of (say) the average college graduate. I have always been haunted by the following lines from the poem, *THE MAN WITH THE HOE*, by Edwin Markham [1899]:

> Slave of the wheel of labor, what to him
> Are Plato and the swing of Pleiades?

Most human beings develop some modicum of physical strength, but many fail, for many reasons, to undergo a comparable development in either the social or spiritual spheres.

To sum up the argument so far: the ethical paradigm that I urge on my readers agrees with that of Aristotle and Maslow who claim that to be ethical (or "virtuous") means to be a healthy human being, especially in the realm of spirit. Physical health makes the body whole, social health makes the community whole, and spiritual health, if human beings can aspire to it, will (I claim) make the planet whole. In the cases of physical and social health the criteria are fairly clear. Babies are healthy if they undergo physiological development and grow physically and mentally; children are healthy if they not only grow physically and mentally, but grow socially. The criteria for adult health are deeper and more obscure. I claim that adults are healthy if they not only have undergone physical, mental and social development, but are growing spiritually. The thrust of what I have been saying is that spiritual growth occurs when the lower and intermediate needs recede in importance and the higher needs, including the need for transcendence, take over. The higher needs, as described in later

Chapters (especially Chapter Six) may be arranged in a hierarchy as follows: first, beauty; second, creation and discovery; third, autonomy; fourth, "identity"; and, finally, "transcendence." We have seen that Maslow claims that the higher needs only come into play when the lower needs are to some extent satisfied; similarly, I claim that each need in this list of higher needs tends to become more important as the needs that precede it are satisfied.

Because of the human need for transcendence, I believe that to be healthy in the realm of the spirit means to be faithful to the deepest (most important) human ideals revealed in what may be called the "mystic vision." (See Chapter Seven: *Beauty, Mysticism, and Religion.*) In this vision, according to Emerson and countless others, is revealed "that Unity, that Over-soul, within which each man's particular being is contained and made one with all other." Modern people tend to accept—as an ideal if not in practice—the call of the mystic vision, typical of prophets from Buddha to Christ, for love, forgiveness, and a breakdown in separateness. This is a call for self-transcendence.

What reasons can be given for a requirement to satisfy higher needs such as self-transcendence? I discuss this here in a preliminary way although the main apology for the requirement to satisfy the higher needs (and thereby undergo spiritual growth) is to be found in Chapter Six.

For those who have been touched by and accept as persuasive the message of the mystic vision, no argument is necessary. For those who are not so persuaded, at least three grounds can be given for self-transcendence: one, the ground of physiological health and vitality; two, the ground of aesthetic sensitivity; finally, the ground of the importance of the ***social bond*** or what Russell [1930] calls a "profound instinctive union" with other people. I consider each in order. There are two ways to health and vitality: first, slough off an excessive concern for achievement and success, take time for relaxation and meditation, eat right, and do moderate exercise; second, "widen your horizon" (as Russell [1979] puts it) and develop interests which are outside your own skin. For most people the prospects for following these precepts are not large, especially when things go "wrong." Nonetheless the prospects are not hopeless; for there is—as we already stated—at least a grain of truth in the dictum of Sartre that each of us makes himself or herself. Moreover techniques for effecting such changes are available (see Chapter Six).

Aesthetic sensitivity, the next ground to be considered, is perhaps a more powerful motive for self-transcendence. Unless we are very fortunate, when we look out our windows we see a steady erosion of the aesthetic quality of the environment. The wild places are being spoiled, forests are disappearing, the number of species is rapidly decreasing, and the world is rife with poverty, slums, murder, drug-running, pollution, disease, and wars. There is therefore no doubt that the combination of industrial development, overpopulation, and poverty is resulting in a world in which aesthetic quality is eroding. The major cause of this erosion is not hard to find: it is a narrowness of outlook and a consequent lack of awareness. Everywhere one finds a short-sightedness that makes people take advantage of every opportunity, heedless of the consequences. The experts are agreed that if we are to stop this erosion, we must expand our awareness and end our narrowness of outlook. I take examples from two different areas to illustrate this point: one, from the eco-system; the other, from the area of population control.

In discussing deforestation and the consequent imminent collapse of the world's ecosystem, E. O. Wilson [1992] makes the claim that the deforesters are acting against their own economic interests. It is Wilson's position that if the deforesters were sufficiently aware of their own long-term interests they would turn to alternatives such as "extraction farming"—that is to say, harvesting the natural products of the forest without harming it. As the forests dwindle, the adequacy of the food supply and habitat of many species has become moot. Even zoo directors, who might be expected to be sympathetic to the plight of wildlife, exacerbate the problem. According to Schaller (See Ward [1993]): *In an astonishingly shortsighted display of greed, zoo directors all around the world struggled to outbid one another for* ***pandas,*** *heedless of the damage they might do for the species.* The situation is similar in river and water management. According to Kevin Coyle, the president of *AMERICAN RIVERS* (See Stevens [1993]), "the four horsemen of river destruction" are dams, diversion of water, alteration of channels, and land development. Much of this destruction is a result of the short-sightedness of the Army Corps of Engineers; for in an attempt to control flooding, the Corps has built many dams and turned many of America's twisting rivers into straight canals. Stevens reports that the Corps now hopes to rectify some of the damage done to the ecosystem by restoring the twists and turns in some rivers—for example in the Kissimee River in Florida.

Consider now population control. According to Weisskopf (See Perutz [1992]), the Northern hemisphere of the earth has presented the Southern hemisphere with the problem of overpopulation by "the introduction of death control without birth control." The short-sightedness of the Food and Drug Administration in Washington, together with the narrowness of the religious right the world over, has stifled the development and use of cheap and effective methods of birth control. Perutz [ibid.], in his review of the autobiographies of Carl Djerasi and Etienne-Emile Baulieu, inventors of pills effective in population control, writes: *Both their autobiographies express bitterness that religious and political pressures are preventing the introduction of these pills into some of the countries that need them most.*

Finally, let us turn to the most important of the reasons for self-transcendence. This is the reason based on the social bond. There is little doubt that for most people the social bond is the source of what matters most deeply. The fact is that we are, as Aristotle says, social animals. The greatest punishment short of torture or execution that can be meted out to us is to be separated from others. (Or as William James [1890, p.293] points out, to *"remain absolutely unnoticed"* by others.) Psychological experiments show unequivocably that expanding one's social bonds brings dividends, while a contraction of social bonds is depressing; for, on one hand, acting in an expansive and generous manner has the psychological effect of making one cheerful; on the other, being selfish is depressing (Cf. Myers [1987, 1989]). Moreover, it is a common observation that human beings, like all animals, tend to reciprocate the treatment they receive from others (Ecclesiastes 11:1-2, Spinoza *ETHICS* part 3, prop. 43, Spencer [1892], Russell [1988]). This fact may be illustrated from history: it is well known that, on one hand, one of the causes of World War II was the repressive provisions of the Treaty of Versailles while, on the other hand, the generous treatment of its former enemies in World War II by the United States (and its allies) made former enemies into friends. Hence, just as enmity produces enmity, friendship begets friendship.

Russell [1930] puts the case for the development of an identification with what he calls "the stream of life" in the following way:

> To be happy in this world, especially when youth is past, it is necessary to feel oneself not merely an isolated individual whose day will soon be over, but part of the stream of life flowing from the first germ to the remote and unknown future...The happy man feels himself not really separate from those who will come after him. It is in such profound instinctive union with the stream of life that the

greatest joy is to be found.

Russell's argument for an identification with the stream of life can—if extended slightly—be seen to be an argument for bonding with the whole of humanity. In other words it can be construed as an argument for making one's community—or what Clifford [1886] calls "the tribal self"—the whole world. This may be seen as follows.

First of all, if you have an interest in anyone in your family who has a child, you have, in normal circumstances, an interest in common with the parents and grandparents of that child. More generally, we may say that if you have an interest in the descendants of persons in your family, your interest in common descendants has the potential of forming a bond between you and all the ancestors of these descendants. But as H.G.Wells [1908] observes:

> We are all going to mingle our blood again. We cannot keep ourselves apart; the worst enemies will some day come to the Peace of Verona. All the Montagues and Capulets are doomed to intermarry. A time will come in less than fifty generations when all the population of the world will have my blood, and I and my worst enemy will be unable to say which child is his or mine.

Short of genocide on a scale surpassing that of Hitler, no one can prevent this. Consequently the world is becoming so interconnected that the totality of ancestors of the descendants of the members of a given family or community is likely to include a significant representation from every ethnic group and every large population on the face of the globe. Hence, if we were sufficiently imaginative, an interest in common descendants would be capable of bonding together almost all the people now living on earth. Furthermore there is strong reason to believe that we are all blood relatives; for there is scientific evidence that tends to show that all members of the human species are descendants of one person—a woman in Africa. We may conclude that if you form a social bond with people because they are your blood relatives, or because they are potentially the ancestors of a blood relative, you must (if you are consistent) form a social bond with all (or nearly all) people on earth.

Nonetheless, there are many who are so in the grip of passion, such as revenge or lust for power, that they wish to destroy those who belong to another group no matter at what personal cost. Some, like Sampson, are out for revenge. The ficitional character, Satan, in Milton's

PARADISE LOST, having lost favor with God, cares for nothing but destruction. A George Bush gets so engrossed in the conquest of his foe that the destruction of a 100,000 enemies does not weigh on his conscience. Those who either have given up on happiness—and may even be suicidal—or those, like Bush, who are lucky in their choice of enemies, may be incapable of believing the assertion that a style of life that is hurtful to their enemies is self-defeating.

People of this kind, who are merely limited in their sympathies, pose a great difficulty for the moralist. But ***sociopaths,*** i.e. people who have no capacity at all for forming social bonds, pose even more of a difficulty. The extent to which sociopaths can learn to extend their sympathies may be moot. Nonetheless I think it can be asserted that sociopaths miss what matters most in life, and as a consequence, perhaps—just perhaps—they can be appealed to on the ground that a radically different lifestyle—one of healing rather than destruction—is capable of yielding important rewards of which they are now unaware. Is it not sometimes possible, even for those with a deep and abiding hatred of enemies, to have a rebirth of spirit and to find that making people unhappy and causing destruction yields less of a reason for being than making new friends and bringing happiness to others? There is some testimony to support this possibility from those who, after a life devoted to prodigality or hate, have adopted "Christian love," and gone on to serve humanity. Such a conversion need not, however, involve the adoption of the creed of any church.

I conclude that a person who eschews narrow partisanship and is thoughtful, kind, and considerate of others—whatever their race, sex, or geographical origin—has the power to stimulate the growth of the deepest source of significance: human bonds. Those bonds may result in the greatest joy, or in the greatest sorrow, but except for those of us who are the victim of some sort of pathology, they are, after all, what matters to us all.

To sum up: Ethics, as an inquiry, is a search for an ethical paradigm. An ethical paradigm is appropriate, in my view, to the extent that it promotes what is important to human beings. Moreover, what is important to human beings—by and large—is the satisfaction of human needs. There are three levels of human needs: the bodily level, the social level, and the level of the human spirit. Determining human requirements at the first level is the business of medicine, determining human requirements at the second level is the business of politics, and determining human requirements at the third level is, if I mistake not, the business of ethics. I contend that *spiritual growth* is what is required to satisfy human needs at the highest level and, thereby, promote what is most important to human beings; consequently, I conclude that the essence of ethics, as a guide to human behavior, is ***spiritual growth*** (cp. Dewey [1920]). This point of view is buttressed by what follows in the rest of this work, especially Chapters Six and Seven.

Chapter Two

A Brief Tour of Western Moral Philosophy

The main ideas of this work are introduced in this chapter through a critical history of Western moral philosophy. For convenience I divide this history into two sections: the first deals largely with content, the next with method. Later sections deal with the topics introduced. The numbers in braces refer to notes at the end of the chapter.

1 From Plato to Whitehead and Russell

The concept of *paradigm* has its roots in Plato and his theory of ideas. Modern thinkers, especially those of a realist persuasion, are accustomed to thinking of ideas and theories as representing the world of objects, and when an idea or theory does not fit the world faithfully, they tend to blame the idea or theory. As natural as such a view may seem, the great philosopher Plato takes a different view. Instead of blaming the idea for not taking a proper measure of the world, Plato blames the world for not measuring up to the idea. True reality, the Platonic doctrine asserts, consists in ***ideas***—fixed, immutable, and unimprovable. An object, according to Plato, is merely an imperfect copy of the perfect idea. It follows that, for Plato, an idea is a *paradigm*, i.e. a pattern to be emulated. It is understandable, therefore, that for Plato the basic idea is "the good."

In the modern world paradigms come and go: Newton's paradigm is replaced by Einstein's; the paradigm of class consciousness is replaced by the democratic paradigm, etc. Such paradigms, of course, are systems of ideas or theories rather than simple ideas. Popper [1981, p. 159]), who believes that systems of ideas form a world of their own, points out that just as honey is the product of bees, such systems are the product of human beings. Accordingly, a paradigm, unlike a Platonic idea, need not be conceived as eternal and immutable. It is Plato who began the search for a paradigm in moral philosophy {1}. Plato's *REPUBLIC* stands as an inspiration for all socio-political ideologies or paradigms, including More's *UTOPIA* and Wells's *A MODERN UTOPIA*. Plato's pupil, Aristotle [1096-7], continues the search for a paradigm, but criticizes Plato's search for an all embracing "good." Accordingly, Aristotle is the first to develop two paradigms: one, a guide to the constitution of the ideal human being; the other, a guide to the constitution of the ideal "polis"—the former being the subject of his *ETHICS*, the latter, his *POLITICS*.

The ideal human being for Aristotle—as for Plato—is a person who fills to perfection all the qualities of being human {2} {3}. Such a person has *aretē*—a Greek word associated with the idea of "fulfillment of function." Aristotle's opinion of what constitutes being human is well known: he defines "man" as a rational animal and notes further that "man is a social animal." Aristotle concludes that the ideal human being has the perfections of both a thinker and a social animal. This means that both knowledge and social skills are *important*. Of course, people are but imperfect copies of this ideal; for they—for physical, moral, or mental reasons—fall short of perfection in at least three ways: one, they fail to cope with their physical environment in a rational way; two, they fail to function well in society; three, they fail either to have intellectual awareness or function well as a thinker. Aristotle says that those who are able to cope with their environment and to function well in society have *phronesis* or practical wisdom {4}. Moreover, those who have intellectual awareness and function well as a thinker—the other aspect of excellence—are said to have *sophia* or higher wisdom. Quite obviously, *sophia* is held by Aristotle in higher esteem than mere prudence or practical wisdom.

According to Aristotle, the activities of all human beings have but one aim, the aim of *eudaimonia*. *Eudaimonia* is usually translated as "being happy," but, for Aristotle, being happy and being the ideal human being are one and the same, i.e. to be happy one must fill to perfection all the qualities of being human. Prudence (*phronesis*) and higher wisdom (*sophia*) are the means by which one achieves this aim. Hence, according to Aristotle, happiness is activity in strict accordance with *phronesis* and *sophia* {5}.

In the *POLITICS* Aristotle states that it is imperative that the constitution of the state "...be one...so ordered that any person whatsoever may act and live happily" [1324a24-25]. The doctrine that acts or rules are to be judged on the basis of well-being or happiness is called "utilitarianism." For the utilitarian it is well-being and happiness that are *important.*

There are actually two forms of utilitarianism, one in which the utilitarian criterion applies to acts, the other in which it applies to rules. Naturally enough, the former is called *act utilitarianism*, the latter, *rule utilitarianism.* For example, Aristotle—at least in his *POLITICS*—is a rule utilitarian; Bentham, on the other hand, is philosophically an act utilitarian, but his effect on legislation has been such that he is considered a rule utilitarian.

Let us define a ***social ethic*** (or ***socio-political ethic***) as an ethic derived from a socio-political scheme or paradigm. In ancient Greece, both morals and politics were thought to belong to the sphere of socio-political wisdom. According to the Greek scholar, Sinclair [1962]:

> For Aristotle, as for Plato, the subject of political philosophy, or ***politikē***, embraced the whole of human behavior, the conduct of the individual equally with the behavior of the group. Ethics was, therefore, part of politics; we might also say that politics was part of ethics. *It was the aim of political philosophy to establish standards of social behavior* [emphasis added].

Aristotle [1094b15] says: "The question of the morally fine and just—...this is what political science attempts to answer..." But what is it that is "morally fine and just"? Aristotle [1129b20-23] answers this in part by saying: "in one sense we call *just* anything that tends to produce or conserve the happiness of a political organization [emphasis added]." In modern times a defense of rule utilitarianism is given by Taylor [1975]. Echoing Plato and Aristotle, he writes: "...moral rules...may be conceived as devices invented by men [and women] to ensure the carrying on of civilized life." It should be clear that rule utilitarianism belongs to socio-political ethics.

The modern rival to utilitarianism is the ethical system of Kant. The great Kant is, I think, the cleverest mountebank among the first rank of intellectuals. (Nietzsche [1889] is much less complimentary to Kant than I, calling him "the most deformed intellectual cripple there has ever been" and Russell (more gently) [1942] classifies him as one of the four misfortunes of philosophy.) The underlying intent of Kant's moral philosophy, if I am not mistaken, is not only to add to the human conceit of being above and beyond the animal kingdom, but to show that the received rules of society can be established by pure thought without any appeal to the animal side of human nature. By the cleverness of his argument, and his obvious sincerity, he convinces not only himself, but perhaps half of the academic world. I shall try to show, however, that his system does not differ in essential respects from rule utilitarianism.

Explicitly Kant's ethic is based on the concept of "good will," but the "good will" of which Kant writes is not a will expressing something like Christian charity, but the will—which may be rather grim—to do one's duty." But what is "duty"? According to Kant, one does one's duty when one acts in accordance with his famous ***categorical imperative***:

> *Act only on that maxim by which you can at the same time will that it shall become a universal law.*

Despite the Judeo-Christian doctrine (as found in *GENESIS*) that God did not want human beings to have a *knowledge* of morality—let alone *reason* about it—Kant says that reason is given to human beings, presumably by God, to produce a good will; that is to say, the function of reason is to produce a good will. But what is the function of a "good will"? This Kant cannot tell because—he says—a "good will" is "good in itself." One might have thought that the function of the "good will" is to make people "good" or to increase the amount of "good" in the world, but this Kant does not say because he realizes that "good will" sometimes fails in its object. For Kant morality is something that—like music—is not in itself of use, but nonetheless something many people cherish. But surely Kantians cannot deny that morality is useful. Even music has social uses that are not immediately apparent. What is more important, however, is that—as I intend to show—Kant's attempt to derive morality from pure thought must be accounted a failure.

Kant's system is supposed to differ from rule utilitarianism in being "deontological," i.e. independent of consequences. This distinction was questioned long ago by John Stuart Mill [1871]. To see that Mill's doubts are justified, consider one of Kant's illustrations, say, the duty to pay one's debts. Kant notes quite correctly that if people were not to pay their debts, the institution of lending would grind to a halt. The natural *consequence* of this would be destruction of the whole banking system. Kant claims it is impossible for one, as a rational being, to will such a consequence, but this is simply not so. If I were a communist, I *would* will this consequence! One must conclude that one of the hidden axioms of Kant's system is that it is impossible for a rational being to will the destruction of the prevailing social system. But I think it is precisely the prevailing social norms, derived from a combination of capitalism and Christianity, that Kant wishes to establish.

Kant's ethic is different from rule utilitarianism in rejecting happiness as a criterion, but this is less so than appears on the surface. Kant not only acknowledges the importance of ends, but even admits that "the natural end which all men seek is their own happiness" [ibid., p.98]. Moreover, he claims that the categorical imperative implies that those who do their duty do not treat people merely as means, and on this basis concludes that those who do their duty do not carry out an action involving others whose end the others do not share. From this we may infer that according to Kant, one has a duty to seek happiness for oneself and others. {6}

Another similarity between Kantian ethics and rule utilitarianism is that both insist on general rules. A consequence of insisting on general rules is that sentiment has to be downgraded, and this is as true of rule utilitarianism as it is of Kant's ethic. For this reason there is actually very little, if any, practical distinction between the Kantian system and rule utilitarianism. Even though rule utilitarianism is based on an end—namely that of happiness—Olson [1967] concludes that rule utilitarianism is "a modified form of deontological ethics."

Both Kant's ethics and rule utilitarianism have some disturbing side effects. In both Kant's system and rule utilitarianism, rules are imposed despite their effect on individual cases. The adherents of rule utilitarianism—which include those nineteenth century liberals and modern conservatives who believe in *laissez faire* capitalism and free-trade (and excludes twentieth century bleeding-heart liberals like myself)—resemble Kant in that they are concerned with rationality rather than sentiment and, moreover, they are concerned with consequences in the large rather than the small. Kant [1785] states that moral principles must "expect nothing from the natural inclinations of man, but everything from the supremacy of the law and the reverence due to it." A true utilitarian or a true Kantian will not shrink from his or her duty merely because some unfortunates have to be sacrificed for the welfare of the whole. "Goodness" of whatever stripe can be rather harsh on the individual. In Kant's case these effects are ordained by a "good will" and in rule utilitarianism they are ordained for the sake of the "greater good." A willingness to sacrifice a few unfortunates for the sake of the whole sometimes makes for strange bedfellows. For example, Bertrand Russell, the great pacifist, puts himself in the same bed as the warmakers in at least two instances: in one Russell [1916a] (like Mill [1873]) defends the genocidal policy of the United States government directed against the original natives of North America on the ground that the greater good is served when the forceful displacement and brutal treatment of an indigenous population brings to a benighted land the blessings of civilization; in the other, he defends his proposal to threaten the use of the atomic bomb on the USSR on the ground that the evil of the loss of a few million lives is a small price to pay for the greater good of the prevention of an all out atomic war (Cf. Clarke [1976]).

Kant's system is sometimes called "ethical formalism." In this chapter I accept this term but broaden its meaning. I define *ethical formalism* as any ethic in which the natural inclinations of human beings are relegated to minor importance compared to the supremacy of rules. Since an ethic derived from a socio-political scheme or paradigm is apt to relegate the natural inclinations of human beings to a minor role, there are perhaps few socio-political ethics that may not be classified as ethical formalism. So far we have identified two varieties of ethical formalism: Kantian ethics and rule utilitarianism. In addition I would add the ethics of Plato. Almost the whole of Plato's *REPUBLIC* is concerned with the definition of "justice" and his definition—"having and doing what is a man's own and belonging to him"—leaves little room for personal judgment. In contrast to Kant, however, Plato takes great pains to show that the virtuous person is personally happy {7}.

So far we have only considered ethics in which the underlying motive seems to be socio-political. Herbert Spencer [1879, p.326] divides ethics into "two great divisions, personal and social {8}." This division is accepted by Russell [1954] who writes: *Ethical beliefs, throughout recorded history, have had two very different sources, one political, the other concerned with personal religious and moral convictions* {9}. He goes on to say: *One of the problems that an ethical system must consider is the relation of these two kinds of moral system, and the delimitation of their respective spheres.*

Act utilitarianism is peculiar because it seems to straddle both spheres. The underlying goal is socio-political—the health, wealth, and happiness of humanity—but this goal is said to be based on moral conviction. According to Smart [1967], the leading adherent of act utilitarianism: *In putting forward act utilitarianism as a* ***normative system****, we express an attitude of generalized benevolence and appeal to a similar attitude in our audience* [emphasis added]. Like ethical formalism, act utilitarianism has some harsh consequences. As paradoxical as it may sound, a true act utilitarian cannot allow sentiment to stand in the way of principle. It is true that the act utilitarian wishes to see caring developed, but he or she notes that unless caring is judiciously applied the general human happiness is liable to suffer. For example, punishment of criminals is sometimes justifiable on the ground of the greater good. As an extreme example—and extreme examples are the ultimate test of a theory—according to act utilitarianism, murder is "correct" in those cases in which almost everybody would be happier if some one person were to perish in a totally discreet way, unbeknownst to others. (Let us suppose this person, though a nuisance, is not a criminal.) I think most act utilitarians would themselves shrink from carrying out all the consequences of their own doctrine {10}. There are nonetheless those who do not shrink.

(Consider, for example, the members of conspiracy groups such as the Ku Klux Klan or the CIA. Do they not think they are working for the general happiness of the greater number?)

Russell [ibid.] writes, "...I do not think that ethics can be *wholly* social." A clear example of an ethic that is not wholly social and in which the natural inclinations of human beings are taken into account is that of Epicurus—a younger contemporary of Aristotle. For Epicureans freedom from pain and the cultivation of simple pleasures such as friendship are the guide to life. A perhaps less clear example may be found in Aristotle; for despite Aristotle's assertion that the study of "the good" is the science of politics, one may extract a personal ethic from the Aristotelian doctrine of "virtues" in the *ETHICS* {11}. In fact, **personal ethics** may be defined as that part of ethics that deals with "virtues," i.e. the characteristics of the ideal person.

Different philosophers have their own conception of the ideal person, of course, and there is no reason to suppose that one can prove that one conception of the ideal person is the correct one. On the other hand, there is not as much divergence on "virtue" among the great philosophers as might be supposed. Let us try to see why this is the case.

It is natural to suppose that for an object to be ideal it must be an "exemplary" specimen of its kind. But "exemplary" means different things for different kinds of objects. For an item that is of little use to be exemplary it need only be a typical object of its kind. This, however, is not sufficient for an item that has a modicum of utility; for example, an ordinary stone is exemplary if it is a typical stone, but a diamond, which is useful for its glitter, is exemplary only if it is untypical. It may be supposed that a human being is ideal (and therefore exemplary) only if he or she is of some use, whence the ideal human not only has the properties of being human, but, like a diamond, is "useful" and rare. This raises at least two questions: one, useful to whom?; the other, has a human being any use and if so what?

We need not, at this point, explore such matters in detail, but merely assume that human beings ***are*** of some use, if not to themselves, at least to others (see Chapter Six). Instead, let us resume consideration of how usefulness is related to the ideal. To take an example, consider the usefulness of a knife: a knife has a use because it has a function; namely, to cut. Moreover, the usefulness of a knife is increased the more its *power* to cut is increased. One may suppose that in a similar way, the usefulness in human beings is related to human faculties and may be measured by a human being's power to function. I conclude that what is ideal in a human being is related to his or her power. As we have noted, in classical Greece the ideal person is said to have *aretē,* which means "power to function." The corresponding Latin word is *virtus*, which originally meant maleness or strength and power; hence *aretē* is usually translated as "virtue." Thus, as Nietzsche [1886] noted, in classical times people are said to have virtue if they are powerful. There are at least three kinds of power, however: influence, capability, and self-overcoming. The Stoics dropped as conditions for virtue the first two kinds of power and adopted the third; as a consequence the term "virtue" no longer connotes all kinds of ability, but only such as show power over what are considered the baser parts of human nature, among which are avarice, fear, and anger. In this work, however, insofar as I use the term "virtue" at all, I mean any characteristic of the ideal human being. Such characteristics vary, of course, according to one's philosophy of the ideal human being.

Let us define a ***natural*** ethic as a personal ethic founded on the needs, attitudes, or native propensities of human beings. Of course, the classic example of such an ethic is the ethic of Aristotle (in the *ETHICS*). For Aristotle, human beings are motivated by a search for happiness which turns out to be nothing but the full development of one's powers as a social and rational animal. The Aristotelian person is ideal to the extent he or she is successful in this pursuit—which Aristotle says can only be determined after a person is dead! Another example of a natural ethic is the ethics of Epicurus. The Epicurean criteria for the ideal are simpler, however. According to Epicurus there is only one power that needs to be developed: the power to overcome fears and to obtain peace of mind. The ethic of Stoicism is similar, the chief difference being that while the Epicurean retires from the world and enjoys simple pleasures, the Stoic is indifferent to both pleasure and pain and does not let even public duty destroy his or her equanimity. Nonetheless, Epicurus, by all accounts (Cf. Russell [1945]) was a true Stoic; for he maintained that a person could be happy on the rack and, as proof of this, endured his last illness proclaiming joy in his heart.

In the West it is in Stoicism (or in some forms of Christianity) that one finds the highest possible integration of a personal ethic with a socio-political ethic. Such an ethic seems to be derived from mysticism. The great philosophers like Plato, as well the great religious figures like Buddha and Christ were influenced by mysticism. Moreover, Socrates exemplifies the mystic tendency and we find the same tendency in the Stoics who followed in his footsteps. Insofar as their morality is based on an attitude of generalized benevolence, the morality of mystics is personal, but insofar as the mystics actually identify with the whole of society, their morality is social.

Of the great philosophers, probably nobody exhibits the mystic tendency more than Alfred North Whitehead. In *PROCESS AND REALITY* [1929] he writes:

> Morality of outlook is inseparably conjoined with generality of outlook. The antithesis between the general good and the individual interest can be abolished only when the individual is such that its interest is the general good, thus exemplifying the loss of minor intensities in order to find them again with finer composition in a wider sweep of interest.

According to the view of Spencer [1892, p.261], the mystics are the forerunners of a human being that through the process of evolution will eventually become common. He writes:

> The ultimate man will be one whose private requirements coincide with public ones. He will be that manner of man who, in spontaneously fulfilling his own nature, incidentally performs the function of a social unit; and yet is only enabled so to fulfill his own nature, by all others doing the like. {12}

Of course such an evolution may not be devoutly desired: no one wishes to emulate the ants who are better adapted to social life than are human beings. On the other hand, it is clear that there is a reciprocal relationship between the development of society and the development of the individual. As Maslow [1954] points out, those individuals whose development is at a maximum—for example, Einstein—are most likely to have an identification with the whole of humankind (as well as with creation as a whole) {13}. Moreover, those societies that are the most highly developed are those that are most likely to promote the free development of the individual. Herbert Spencer may be overly sanguine about the prospect of human development, but I think it is clear that the human species may be able to transcend some of the features, inherited from an animal past, that keep down its prospects for a brighter tomorrow. Let us examine briefly what is meant by this.

Lower animals operate largely by reflex and instinct, but the human animal, through the development and use of the brain, particularly the cerebral cortex, is able to chart its own course and to override the primitive mammalian (and reptilian) brain—the so-called "old Adam" which it has as part of its biological inheritance. Thus the possession of a brain superior to that of other species makes it possible—but not necessary—for human beings to rely more on rationality than on brute strength; therefore human beings are able to develop societies based on law, constitutional government, and equality, and slough off patterns of behavior that are more appropriate to lower animals. In other words, people have become human by "transcending their animal nature" and developing their potential for what Aristotle calls "something divine within" {14}.

Recently moral philosophy has returned to Aristotle and his theory of virtues. Instrumental in this revival have been two philosophers: Elizabeth Anscombe [1958] and Alasdair MacIntyre [1981]. Modern virtue theory, however, is subordinate to politics; for according to modern virtue theory the ideal democrat has different virtues than (say) the ideal aristocrat. In particular, benevolence is a virtue for the democrat, but too much benevolence is a positive hindrance to the aristocrat. It seems to me, however, that virtue theory cannot advance substantially beyond Aristotle, unless it dumps its antiquated psychology. (It will presently be clear what I mean by this.)

Ethics takes a psychological turn in the seventeenth century with the publication of the ethics of Hobbes. The quintessential natural ethic in modern times, however, is that of Spinoza [1677] who combines a natural ethic like that of Aristotle with an outlook similar to the Stoics and the Epicureans. His emphasis on transcending limitations and developing one's powers, leads Nietzsche to exclaim (Cf. Kaufmann [1968]): "I am utterly amazed, utterly enchanted! I have a *precursor*, and what a precursor! [emphasis in original]." For Spinoza, error or imperfection consists in being limited and, moreover, the only unlimited being is God. God is defined as the sum total of all beings (and things) and hence is not only infinite but perfect. Spinoza resembles Aristotle in emphasizing the development of one's powers, but his conception of the possibilities that lie dormant in human beings is much more developed. He is the first major philosopher for whom transcending limitations, outgrowing narrowness, and growth toward perfection and knowledge is the major *desideratum* in an exemplary life. One can say that Spinoza's paradigm for ethics is based on *growth*. Moreover, like the Stoics and Epicureans, the underlying motive of Spinoza's philosophy is to get rid of the passions—at least those which destroy peace of mind. According to Spinoza, such passions result from inadequate ideas. Ideas are inadequate insofar as they depart from ideas in the mind of God. "Joy" is defined as "the passion by which the mind passes to greater perfection."

Before we move on to Nietzsche, there is one other philosopher who deserves our attention. This is Humboldt—not the famous Alexander von Humboldt but his brother Wilhelm. Humboldt [1854] wrote a work (published posthumously, but written more than a half a century earlier under the impact of the French Revolution) in which he explicitly (perhaps for the first time) takes up the theme of growth as a guide to life. According to Humboldt:

> The true end of Man, or that which is prescribed by the eternal and immutable dictates of reason, and not suggested by vague and transient desires, ***is the highest and most harmonious development of his powers to a complete and consistent whole*** [emphasis added].

John Stuart Mill [1859] quotes Humboldt extensively in his famous work *ON LIBERTY.* It is interesting that the currents of German censorship were such that Humboldt could not get this work published during his own lifetime.

For Nietzsche [ibid.], transcending limitations is the *sine qua non* for the development of the ideal person, which he called the "superman." As already indicated, Nietzsche acknowledges Spinoza as a pioneer in this point of view. Nietzsche is not as refined a type as

Spinoza, however. Consider, for example, the impressions they each give on how they regard their fellow human beings—especially women! It is true that both Spinoza and Nietzsche downgrade pity, but the difference is that Spinoza offers us *reasons* for helping others—one looks in vain for anything similar in Nietzsche. Spinoza—like Kant and the Stoics—wants people to help others out of reason rather than pity, but he adds: "For he who is moved neither by reason nor pity to be of service of others [in the relief of suffering] is properly called inhuman; for...he seems to be unlike a man." {15}

The prototype of Nietzsche's superman is perhaps to be found in the self-reliant hero idealized by Emerson [1841]. Nietzsche [1974 ed.] indicates his indebtedness to Emerson in the first edition of *DIE FRÖLICHE WISSENSCHAFT.* It is, after all, Emerson [1860], not Nietzsche, who first asserts "Life is a search after power..." But Nietzsche never aspired to the grandeur and magnanimity of Emerson, the author of *THE OVER-SOUL.*

Nietzsche's most eloquent work is reputed to be *ALSO SPRACH ZARATHUSTRA*, in which Nietzsche [1885] enunciates the dictum "Become what thou art!" According to this dictum, Nietzsche subscribes to the doctrine of self-realization introduced by such authors as Fichte [1794], Hegel [1837], and Emerson [ibid.]. The doctrine of self-realization has been a popular one throughout nineteenth and twentieth century ethics, even though its central premise apparently rests on what is a philosophical mistake: for self-realization seems to imply the existence of a self that exists prior to its realization. Some of the philosophers who have raised doubts about this doctrine include William James, John Dewey, and John Paul Sartre. According to William James [1890, p.288], "The problem [a person has at a particular moment] is less what act he shall now do, than ***what being he shall now resolve to become*** [emphasis added]". Dewey [1898] echoes this. He writes, "You have no self until you find what may be called its realization." The existentialists agree with James and Dewey and, in the ringing phrase of Sartre [1946], make the claim "Man makes himself." Instead of "self-realization," Dewey [1902] prefers to speak of growth. In *RECONSTRUCTION IN PHILOSOPHY,* Dewey [1920] declares: *Growth itself is the only moral end.* Russell [1916b] takes up this theme in his work *PRINCIPLES OF SOCIAL RECONSTRUCTION.* The first chapter of that work is entitled: *The Principle of Growth.*

Russell describes growth in the following terms:

> The impulses and desires of men and women, ***in so far as they are of real importance in their lives,*** are not detached one from another, but proceed from a central principle of growth, an instinctive urgency leading them in a certain direction, as trees seek the light. [emphasis added] {16}.

(Note the use of the word "importance" here. For what follows this is worth bearing in mind.) The existentialists, who deny the self has an essence, nevertheless recognize the importance of becoming one's own person when they speak of "authenticity" (Cf. Heidegger [1949], May [1953]). The psychotherapist, Carl Rogers [1961], also recognizes this when he urges people to ask of themselves: "Am I living in a way which is deeply satisfying to me, and which truly expresses me?"

The great neurologist Goldstein [1939] believes that growth is the primary—if not the only—drive in an organism. He calls it "self-actualization." His concept of growth, somewhat similar to Russell's, is as follows:

> The organism has definite potentialities, and because it has them has the need to actualize or realize them. Driven by such needs, we are experiencing ourselves as active personalities not, however, passively impelled by drives experienced as conflicting with the personality.

Inspired by Goldstein, Maslow builds a psychology on the notion of *self-actualization.* Maslow [1954, p.115] aligns himself with other psychologists—among whom he lists Goldstein, Jung, Adler, Angyal, Horney, Fromm, May, Rogers—"who postulate some positive growth tendency in the organism which, from within, drives it to fuller development." According to Maslow [ibid., p.296]: *Self-actualization* [is] *the coming to full development and actuality of the potentialities of the organism...it is not acquired from without but is rather an unfolding from within of what is, in a subtle sense, already there.* The preceding quotations from Russell, Goldstein, and Maslow seem to imply that talents and interests spring from a mysterious source which, "*in a subtle sense, is already there.*" The notion of growth as described by these writers is therefore subject to all the criticisms of the notion of self-realization by James, Dewey, and Sartre. As Dewey makes plain, however, those who believe in the concept of growth have no need to assume that each person has a mysterious essence lying in wait to be actualized. The nub of the concept of growth is that each person has his or her own distinctive set of potentialities that are most *important* in his or her life and which, if not realized, result in frustration and therefore the pattern of development for one person need not be the pattern of development for all.

The importance of the concept of *importance* in ethics has, I believe, been profound and almost unrecognized {17}. We have noted that some philosophers think knowledge is important, others pleasure, some growth, and others "functionality" or power, but the notion of importance itself receives relatively little attention {18}. The major exception to this is Whitehead [1938] {19}. Some philosophers think the concepts of interest or meaning, which are related to the concept of importance, are important, but I believe one of the major deficiencies in the great ethical systems is a preoccupation with the concepts of "value" and "duty" (or "good" and "ought") and a relative neglect of such concepts as requirement, importance, meaning, significance, or "what matters."

Whitehead [ibid.], commenting on common usage, complains that the word, "importance," has been reduced to "a silly little pomposity." As everyone knows, the adjective "important" is sometimes applied to persons or groups of persons as well as to events or other factors in peoples's lives. In this sense of "importance," important persons are people who are significant in a person's life, such as a spouse, sibling, or parent, or they are people of note. Aristotle's "magnificent man" [*ETHICS* 1122a18-b1] is important in the latter sense. In this work I use the word "important" as a word applied to aspects of life rather to than to persons or groups; therefore, in referring to the attainment of importance, I am not talking about the attainment of the importance of persons or societies but of the attainment of what is truly important to a person.

Except for value, *meaning* is most often cited as being important in life. For the sake of clarity, however, I generally avoid the terms "meaning" and "value" as synonyms for "importance." It is easy to see why. The clarity of the term "meaning," for instance, is vitiated by having too many meanings. Morris [1964] points out that the term "meaning" has two poles: "that which something signifies and the value or significance of what is signified." These two poles are, of course, quite distinct. A piece of music or a poem, e.g. Coleridge's *KUBLAI KHAN,* may be *interesting*, despite the fact that it doesn't *signify* anything. Macbeth said that life "is a tale told by an idiot,...signifying nothing," but even if Macbeth is correct and life *signifies* nothing, it might still turn out that life is significant in the sense of being important or exciting an interest. Accordingly, for the sake of clarity, I ordinarily restrict the term "meaning" to the former of the two poles and use the term "significance" when referring to the latter. Thus, according to me, the word "life" is meaningful, but it is bad grammar to say life itself is meaningful. Instead of talking of the meaning of life, I talk of its interest, significance, imperativeness, or importance {20}. But what about *value*? It is true that if we accept the famous definition of Perry [1926] that a ***value*** is any object of any interest, those things that have any interest, significance, or importance have "value." Nonetheless, since I agree with Popper [1976] that most talk of values is so much hot air, I intend to avoid the term "value" in this work.

Whitehead relates importance to interest, feeling, significance, perspective, selection, intellectual freedom, purpose, unity, novelty, harmony, intensity, vividness, matter-of-fact, and many other things, but not specifically to excitement and pleasure {21}. "Excitement" is a synonym for "interest." The importance of pleasure (and excitement) is twofold: one, it is important as an end-in-itself; the other, pleasure along with pain and absence of pleasure are important as motives for action. (How pleasure--as well as excitement--has come to be an end-in-itself is discussed in the next chapter, Chapter Three: The Springs of Human Action.) Though it is possible, for many reasons, to be critical of pleasure as a motive for action, absence of pleasure may make it difficult to see any significance in life. To paraphrase Kant we may say: pleasures without importance are empty; life without pleasures, dull. We return to this point when we come to discuss the methods of practical philosophy in Section 2.

Though living may with some justice be described—like the wit, Tom Lehrer, does—as "sliding down the razor blade of life," one may ask if it is possible to avoid letting adversity and the external events of life determine the significance of life {22}. Stillman [1993] writes poignantly about the plight of those in nursing homes: "Have you ever been in one of these places? It is like being in hell, with ancient whitened corpses tied to wheelchairs, gazing into space." Nonetheless, there are many who, despite illness or other adversities, remain interested and engaged in the events that surround them. Darwin, for example, led a life of great purpose despite being ill all his life. Moreover, as the Stoics taught, tragedy can heighten rather than destroy the importance of living. The last days of Socrates, for example, were filled with significance. Of course, what is important to a man like Socrates may not be significant to a person of the common persuasion; for Socrates developed beyond the ordinary.

This brings us back to the notion of human development or growth.

In order to clarify the notion of human development, it is, I think, necessary to make use of some concepts of modern psychology, particularly the humanistic psychology of Maslow.

The most important concept of humanistic psychology is that of a "hierarchy of needs." In this hierarchy, needs are arranged in the order of appearance in human development. The imperatives at the two lower levels are shared by nonhuman animals, but the imperatives at the highest level are almost entirely human. Thus at the bottom of this hierarchy are the bodily needs: these are the needs for air, water, food, etc. At the intermediate level are the psychological needs for love, social acceptance, safety, power, self-esteem, etc. At the highest level is the need for what Maslow, following Goldstein, calls "self-actualization."

The higher needs become important when the lower needs are satisfied; moreover, in order for the needs at the higher level to be felt at all, it is often necessary for the lower needs to be so satisfied that they are of little concern. The higher needs are more difficult to achieve if the lower needs are not met; on the other hand, those who have satisfied all their social needs generally find it easy to satisfy needs at the lowest level, and what is more important, those who live at the highest level of spiritual development find that the needs at the lower levels recede in importance. For social ethics and politics it is the intermediate needs that are important, but for personal ethics the lower and intermediate needs recede in importance and the higher needs come to the fore.

Maslow's hierarchy is incomplete; for he pays little attention to the ordering of higher needs. As suggested in the previous chapter, I propose the following as a list of higher needs in which each need in the list becomes more important as the needs that precede it tend to be satisfied: first, beauty; second, creation and discovery; third, autonomy; fourth, "identity"; and, finally, at the apex, what may be called "transcendence." Transcendence is what frees us not only from the here and now but from the "prison of self." As indicated in Chapter One, self-transcendence may, at least in part, be characterized by what Blackwell [1985] in *THE SPINOZISTIC ETHICS OF BERTRAND RUSSELL* calls "impersonal self-enlargement." Impersonal self-enlargement itself involves what has been called a "Spinozistic development of the understanding toward a reduction of separateness."

The concept of growth or self-actualization has been criticized by some out of a concern that those who advocate it are pandering to egoism. The foregoing hierarchy shows that this is a complete misunderstanding. William James [1890 I, p. 324] points out that "in a thousand ways the primitive sphere of the egoistic emotions may enlarge and change its boundaries." Historically, the concept of self-realization was introduced as a way to show that the will of the true self is identical to the will of God. According to Fichte, one realizes oneself when one identifies one's own will with the infinite will of the universe. "Growth," as I understand the term, is the tendency of a human being to become less occupied with the fulfillment of the lower needs and more and more directed by the higher needs. Since at the summit of the higher needs is transcendence, which, as we said, tends to free us from the "prison of self," I conclude that growth is not at all pointed toward aggrandizement of a personal self and that the charge of egoism is false.

Another objection to an ethic of "growth" or self-actualization is that the notion of growth is too vague to supply a guide to the conduct of life. According to this objection, developmental concepts such as that of Maslow and others are too vague or mystical to serve as an adequate basis for an ethic. Even Russell [ibid.] seems to admit such vagueness when he writes: "...the full growth which may be hoped for cannot be defined or demonstrated; it is subtle and complex, it can only be felt by a delicate intuition and dimly apprehended by imagination and respect." This somewhat mystical vagueness in talking about growth or free development is at least partially overcome in this work by giving an outline of the guiding principles of a paradigm for human development (or growth) established on the basis of the hierarchy of higher needs. (See Chapter Six: *What Are We To Do With Our Lives*?).

The existentialists (such as Heidegger and Sartre) with their emphasis on "being," remind us that a major problem faced by all of us, whether we are conscious of it or not, is whether to continue to be. One reason for living is fear of the alternative. Hamlet, in his soliloquy on suicide, says, "Conscience does make cowards of us all." But if we are not to live simply because we fear the alternative or simply out of animal instinct there have to be less trivial reasons for continuing to be. The deepest purposes of a human being are those that supply the rationale, if any, of a person's life. If it is the case that there are degrees of importance, it is because some purposes are deeper—if not more immediate—than others. The purposes that lie deepest are, by definition, those that do not lie close to the surface and therefore are difficult to discern. I suggest that the depth of a purpose (and the importance of its realization) may be measured by the extent to which it supplies a reason for living.

As we have seen, the notion of importance plays a key role in arranging needs in a hierarchy. Can one ask, pace Popper, "What is importance?" I postpone further discussion of the ordering of the items in the hierarchy—particularly the higher needs—until Chapter Three (The Springs of Human Action); but in Section 3 of this chapter I consider the meaning of importance.

So ends the first part of our survey of Western moral philosophy. The first part has focussed on content rather than method and for this reason I have neglected many philosophers that are important in the history of ethics. One apparent oversight is Plato's great teacher Socrates. This oversight is rectified, I trust, in the next section.

2 From Socrates to Mill, Kanger, and Chisholm

In the preface to this work I indicate my belief that Western philosophy lost its way when it became beguiled by the dialectical method of Socrates. I am not unique in this belief. Popper [1950, p.512] raises the doubt whether questions of the kind Socrates always asks—questions like "What is the good?" and "What is justice?"—are not a colossal waste of time. Wittgenstein [1953] is a pioneer in questioning the assumption that there is something in common, rather than simply a family resemblance, to all instances of the use of a word. Socrates, himself—as reported in Plato's *APOLOGY*—in his last year of life, when asked about his apparent claim to have knowledge of human and political goodness, sums up his search for knowledge in these domains as follows: "...the truth is that I have no knowledge of the kind." As we have seen, Aristotle concludes that for a practical ethics one need not have a generic concept of "the good."

In modern times, Hume [1739] strikes a blow against the entire enterprise of defining words like "good" and "ought." He argues that there is a demarcation between the language of practice containing words like "ought" and the language of matter of fact containing words like "is," and moreover sentences in the language of practice cannot be established solely by an appeal to facts. Philosophers such as Searle [1964] have attempted to show that the gap, suggested by Hume, between the practical and the theoretical (or "ought" and "is") does not exist. Such attempts are not, in my opinion, successful (Cf. Johanson [1973]). The gap does exist: nonetheless, I think it can be bridged.

Hume [1751] himself did not abandon the effort to connect ethics with nature. Hume even attempts to define, if not "good," at least a roughly equivalent term, in natural terms. He says "virtue" is "*whatever mental action or activity gives to the spectator the pleasing sentiment of approbation.*" After making the claim that utility pleases, Hume draws the conclusion that "Whatever is valuable in any kind classes itself under the division of *useful* or *agreeable*... [all emphasis in the original]." Other philosophers, however, are more skeptical of definitions in ethics. Bentham [1932 ed.], who claims that all qualities, relations, and classes are fictions, thinks that sentences containing words like "duty" and "right" are, in Russell's sense, incomplete symbols and therefore, best defined by translating the sentences in which they occur into other sentences in which all the words refer to real entities (Cf. Monro [1967], McKenney [1958]). At the turn of the century, Sidgwick [1901] and Moore [1903] conclude that all effort to define "good" have been failures and, moreover, a definition of "good" is in principle impossible. Moore and Sidgwick come to very different conclusions, however. Sidgwick is in the tradition of Hume, and believes that it is intuitively clear that "pleasure alone is good as an end," but Moore—abetted by Russell [1910]—divorces ethics from nature. It is Moore's theory that to say something is good is to assert that it has an indefinable property of a kind unknown in natural science. Santayana [1913], protesting against such a separation of ethics from nature, answers with a devastating critique of the early views of Moore and Russell in *WINDS OF DOCTRINE*. Meanwhile, the Swedish philosopher Hägerstrom [1911] decides that sentences of the form "x is good" are not capable of truth or falsehood, a view not unlike that which Shakespeare attributes to Hamlet. This view comes to be known as "noncognitivism." Later Russell [1927], acknowledging his debt to Santayana, defects from the camp of the absolutists and becomes a noncognitivist. Russell is soon to be followed by other philosophers including Carnap [1935], Ayer [1936], and Hare [1952].

As a result of these developments two opposing doctrines regarding methods in ethics now prevail: one, the intuitionistic doctrine that good and bad are absolutes that are somehow intuited correctly by people with a sufficiently developed moral sense; the other, the noncognitivist's doctrine that assertions of good and bad express the attitude of the speaker and are therefore, as P. B. Rice [1955] used to say disdainfully, mere "ejaculations of emotion." At the present time there is a standoff between the noncognitivists like Hare and their opponents on the nature of ethical language.

Neither of these views is completely satisfactory. In the last subsection, we noted that what makes something ideal varies according to whether or not it is useful. Thus a stone, if it not to be used for some purpose, is a good stone if it is a standard or average stone, but a diamond is good to the degree it has some sort of utility. Moreover, a good knife is one *that is sharp and cuts well* {23}. In ethics, however, one is not concerned with the evaluation of inanimate objects such as stones, diamonds, or knives, but of such things as life (especially human life), activities, motives, and states of mind. As Hume notes, when the latter are useful people tend to find them pleasing. But are there (or are there not?) useless motives or states of mind that please as well? No one seems to say that *activities* are good in themselves, but it is otherwise with human life, motives, and states of mind. (In the discussion that follows I confine my attention to states of mind {24}.) Most writers agree with Moore—and presumably with Hume, Spencer, and Russell—that a state of mind, unlike activities in general, may be good even though it serves no utilitarian purpose. I do not agree, however. On the contrary, we have already indicated that a good state of mind (or pleasure in general) serves a distinct biological purpose: the purpose of enabling a person to function well (either as an individual or as a member of society). Now the states of mind that Moore says are valuable are described by him as pleasures and it may not seem that pleasures serve an immediate purpose. But a life without pleasure not only tends to be mercifully short, but such a life, if prolonged, leads to a loss of the sort of zest without which people do not function well, either out in the world or domestically. What is more, pleasure is good only to the extent it does not interfere with a person's ability to function {25}. It will of course be replied that pleasure is *intrinsically* good, even if it serves as an inducement to self-destruction, but this only shows the futility of arguing about the usage of the words "good" and "bad." There is no doubt that people desire pleasure for its own sake, but the notion of an intrinsic end (or "value") only muddies the waters. As John Dewey [ibid.] is fond of pointing out, there are no intrinsic ends. If our purpose is to decide which sorts of things are to be approved of and which are to be condemned then the words "good" and "bad" might as well be abandoned to the metaphysicians—as Bentham long ago claimed—and let us get on with the proper business of ethics which (roughly speaking) is to decide on the general principles of the conduct of life. Popper [1950; 1976, p.62], reacting to the failure of philosophers to agree on the definition of ethical terms, enunciates a principle which may be formulated as follows: *let nothing depend on the meaning of words and, what is more, let it be that the attempt to answer "what is" questions is abandoned in philosophy.* I tend to agree with Popper in this point of view—at least as far as words like "good," "ought," and "justice" are concerned—and hence in this work, I follow Popper's dictum and abandon the Socratic quest for the meaning of terms of this kind. (I return briefly to this subject, however, in Section 5: *Remarks on meta-ethics.* See also Chapter Three where I discuss "ends.")

Nonetheless, we still have to deal with the gap, noted by Hume, between the language of practice and the language of fact and science. If ethics is not to be completely divorced from nature there must be some way to bridge that gap. So far as I know, the first clue on how one might do this is to be found in the writings of John Stuart Mill [1898 ed.]. Mill, writing on "The logic of practice, or art, including morality and policy" says "the imperative mood is characteristic of art." (In much of the rest of the chapter, Mill seems to forget this insight, but that is neither here nor there.) Mill calls for the development of a logic of practice that does not ignore sentences in the imperative mood.

Up to Mill's time, the most noteworthy attempt to achieve a logic of practice was that of Aristotle. Since Aristotle was also a pioneer in the development of logic, it is not surprising that he had something to say about applications of logic to practical reasoning. Though the study of logic on the theoretical side has led through Leibniz, Boole and others to modern formal logic, practical logic has been relatively neglected.

Aristotle's distinctive contribution to practical logic is his construction of a practical syllogism on the model of its theoretical counterpart. Like a regular syllogism, a practical syllogism has a major premise, a minor premise, and a conclusion. The major premise states some end to be accomplished (or something prized) and the minor premise expresses an activity which either accomplishes or frustrates the end. Curiously enough, though the major and minor premises are regarded as statements, the conclusion is said to be an activity.

If Aristotle had employed sentences in the imperative mood he might have given the following as an example of a practical inference. Let the major premise be "Be prudent" and the minor premise be "Those who eat sweet food are imprudent"; then "Don't eat sweet food" may be inferred as a conclusion.

It seems that almost all philosophers believe that the language of practice is the same as the language of science, i.e. the language of declarative sentences. This is true despite the use of sentences in the imperative mood by philosophers such as Kant. Nonetheless, the fact that imperatives have been relatively neglected has not prevented the development of a concerted effort in the twentieth century to construct a practical logic.

The most important development in this field is the invention of standard ***deontic logic*** by von Wright [1951]. Standard deontic logic is attractive because of its simple logical structure. It was soon found that this logic, despite its charms, is not completely satisfactory, however. It is riddled with paradoxes and makes no concessions to noncognitivists. Many varieties of deontic logic have been invented during the last half century in an effort to improve it, but I think that the doubts of noncognitivists and others over deontic logic cannot be allayed without the use of imperatives and unless the insights of Mill are taken into account. If this is true, further development hinges on the development of a logic of imperatives.

A major obstacle to building a logic of imperatives, is that sentences in the imperative mood do not have truth values. But are imperatives any different in this respect than statements in traditional ethics or deontic logic? Kanger [1957] suggests, as a concession to noncognitivists, that one might abandon the word "truth" as applied to ethical judgments, and use the word "correct" instead. Castaneda [1983] classifies imperatives as either "legitimate" or "illegitimate." This is reasonable. Not all imperatives are on a par—some commands make sense and others do not—and hence one may partition them into (say) "sheep" and "goats." It is Kanger's claim that even if imperatives (or even deontic statements such as "One ought to do x in situation y") do not now have the property of truth, it is still possible to invent a property, analogous to truth, which would serve in its place. On such a basis Kanger [ibid.], anticipating Kripke [1963], develops a scientific semantics for modal logic in general and deontic logic in particular {26}.

In the present theory, the function of Kanger's notion of *correctness* is taken over by a concept of *requirement*. Requirement is a concept very familiar in discussions of practical matters. Kohler [1938] uses the concept of requirement as the *sine qua non* of value. Chisholm [1964] speaks of requirement in terms of one condition requiring another. But in the sense I speak of it here the notion of requirement applies only to imperatives. Moreover, the concept of requirement here presupposes that it is possible to speak of one imperative implying another. There is nothing remarkable in this; for it is ordinarily assumed that from the sign "No smoking!" one may infer the warning "Don't smoke a cigar!"

Mill's idea is, basically, that sentences in the imperative mood (or optative mood) can be mapped into sentences in the indicative mood and *vice versa*. He does not put it that way, but that is what it amounts to. Mill's own words are as follows:

> Now, the imperative mood is characteristic of art, as distinguished from science. Whatever speaks in rules or precepts, not in assertions respecting matters of fact, is art...
>
> The Method, therefore, of Ethics can be no other than that of Art, or Practice, in general..

Then Mill makes the claim:

> Now the reasons of a maxim or policy, or of any rule of art, can be no other than the theorems of the corresponding science.
>
> The relation in which rules of art stand to doctrines of science may be thus characterised. The art proposes itself an end to be attained, defines the end, and hands it over to science. The science receives it, considers it as a phenomenon or effect to be studied, and having investigated its causes and conditions, sends it back to art with a theorem of the combination of circumstances by which it could be produced.

The procedure Mill has in mind involves the use of two transitions between art and science that may be diagrammed as follows:

$$\text{ART} \quad \begin{matrix} \rightarrow \\ \\ \leftarrow \end{matrix} \quad \text{SCIENCE}$$

Mathematics has long dealt with this kind of situation; for in mathematics one often finds that a problem which is intractable in one domain is easily solved in another. For example, every school child is taught that the logarithm may be used to transform a difficult problem of multiplication into an easy problem of addition. In algebraic topology, operators of this kind are routinely used to transform intractable geometric problems into tractable algebraic ones. Such operators occur so often in mathematics that Eilenberg and MacLane [1942] give them a special name. The name they give is "functor," which philosophers will recognize as the name Carnap uses to refer to a function symbol.

Like the exponential and logarithmic operators in arithmetic, the most useful and interesting operators in mathematics occur in pairs. Moreover, in important cases, the operators are functors that have a property, described below, called "adjointness." Adjointness in this context was conceived in the late fifties by Kan [1958]; it is one of the most far-reaching ideas in twentieth century mathematics. Mill's idea suggests the possiblity that there are pairs of adjoint functors between the categories represented, on one hand, by statements of fact and, on the other, by imperatives or optatives; furthermore, it suggests that adjoint functors might be important—as they prove to be in all branches of mathematics—for the foundations of practical philosophy (Cf. Lawvere [1969]).

This suggestion raises at least two questions: one, just what is an "adjoint functor"?; the other, are there adjoint functors in practical philosophy, and if so what are they?

Without getting into highly technical matters these two questions can be answered at this point only by example and what better example can we give than the mappings or operations that are the chief object of study here? These operations are "letting" and "requirement." If x is a statement, "Letting x be true" is the imperative (or optative) that says "It *shall* be the case that x is true." We have already mentioned "requirement." The present logic of imperatives is based on the assumption that **requirement is adjoint to letting**. Roughly, this means that if x is a statement and u is an imperative, letting x entails u if and only if x entails ***it is required*** that u. For example, in traffic theory, "letting the light be red implies the command to stop" is true if and only if the light is red implies ***it is required*** to stop. (This omits niceties—the logical details are to be found in ***Part Two***, Chapter Nine.) Another pair of adjoints are the related operators: admissibility and sustaining. For example, in traffic theory, "the command to go is admissible implies that the traffic light is green" is true if and only if the command to go implies the optative "sustain that the light is green!" If requirement is taken as basic, admissibility may be defined in terms of requirement; for it is admissible that it *shall* be the case that x be true if and only if it is not required that it *shall* be the case that x be false. Similarly, sustaining and letting are also interdefinable; for letting x is the same as not sustaining the negation of x.

Now the operations of letting, sustaining, requirement, and admissibility supply us not only with a language for doing ethics and politics—as well as other practical subjects, e.g. the theory of law and computing (Cf. Brown & Carmo [1996])—but a logic as well. What one wants is not only to set up standards of how to conduct personal and public affairs, but a system of logical relations between these standards. Out of the theory it is easy to derive a theory of norms which has many advantages over previous theories including the theory of rights of Kanger [1966] and Lindahl [1977].

Ethics and politics cannot, however, be obtained entirely out of logic or reason. Hume [ibid., p.8] wrote in this regard: *It is full time that they* [i.e. philosophers] *reject every system of ethics, however subtle and ingenious, which is not founded on fact and observation.* Despite Kant, norms cannot be established simply by an appeal to reason. *EXCEPT IN VERY TRIVIAL CASES, WHAT IS REQUIRED AND WHAT IS ADMISSIBLE (IN THE ULTIMATE SENSE) CAN ONLY BE ESTABLISHED BY* ***EMPIRICAL OBSERVATION*** *ON THE NEEDS OF HUMAN BEINGS.* I believe with Santayana that ethics cannot be reasonably divorced from nature. This means that the ultimate data of ethics—in Spencer's sense—belong at least partly to science. Everyone knows that science is of use in the solution of practical problems. As we noted in the last section, the science of choice in ethics is psychology. In other areas such as politics, other sciences of behavior are equally useful. The purpose of Mill's practical logic is to clarify the logic behind this use.

Science is no panacea, however; ***I make no claim that merely by reducing problems of practice to problems of science all practical problems can be solved. But*** apart from changes in social consciences such as those which have led governments—as well as many households—away from supporting slavery and racial intolerance, ***it is mainly the piecemeal approach of science which has made possible whatever advances have been made in the reduction of suffering***.

The idea of using operators such as "requirement" and "letting" to bridge the gap between the normative and nonnormative, which is already adumbrated by Mill, is, if I am not mistaken, the correct one. Fortunately, the tools necessary for bringing this idea to fruition, unavailable to Mill, are now at hand. These have come from mathematicians rather than philosophers, however.

There are three mathematicians chiefly responsible for these tools: Eilenberg and MacLane [1945], and Kan [1958]. The first two invented the theory of categories, and the last, the theory of adjoint functors. The use of the functors, "requirement" and "admissibility" and their adjoints, enables one to exploit that relationship. Operators constructed out of these functors form modal operators, analogous to truth, that satisfy a Kanger-Kripke type of semantics. Using these operators one is able to develop a theory of norms which in formal structure resembles, at least superficially, Kanger's typology of "rights." In this work it is shown that this theory of norms can be used profitably in laying a foundation for normative systems (See Chapter Eleven, ***Part Three***).

In this work, normative systems for ethics and politics are based on the logic of requirement; moreover, I assume that requirements are partly social and partly individual. Individual requirements are especially important in personal ethics. Requirements for a person vary not only with the conditions and circumstances of a person at a given time, but the stage of development of that person; hence, for a personal ethic to make sense, it must take into account the state of a person at a given time. For example, it makes no sense to talk of higher things when a person's children are starving. Moreover, it may not make sense to speak of ethics at all when social structure breaks down and living conditions are sufficiently hard. In particular, children born in chaos, who are shown no examples of decent behavior, have no chance to develop into moral human beings (Cf. Turnbull [1972]). In these circumstances one is reduced to a Hobbesian world of a war of all against all. A certain degree of human development or growth is necessary before even a rudimentary ethic makes sense. Similar remarks apply to politics; for a theory of politics that applies to human civilization at one stage of development can hardly be applied to one in early primitive stages. (I return to this topic in Chapter Three: The Springs of Human Action.)

That is not to say that even when human development reaches a relatively advanced stage that ethics and politics become a straightforward matter. Ethics and politics in all known civilizations are based on the prevailing mores, usually expressed in terms of imperatives, like the *TEN COMMANDMENTS.* In order to criticize the prevailing mores, therefore, one needs to examine the use (and possible abuse) of imperatives in human discourse.

Kant [1785, p. 84] classifies imperatives into three kinds which he calls *technical, pragmatic,* and *moral.* In the first Kant includes rules of skill (like prescriptions in medicine), in the second counsels of prudence (like "Look both ways before crossing the street!"), and in the last "laws of morality." A modern person would probably expand the classification of technical imperatives to include not only rules of skill but lists of instructions like traffic rules, the rules of a game (e.g. baseball), computer programs, and algorithms in mathematics. Technical imperatives arise in the context of a problem, and I claim that when there is a problem to be solved it is *prima facie* the case that ***it is admissible that it shall be the case that one does whatever is needed to solve the problem.***

Note, however, that there are always requirements that constrain how human beings solve problems, and hence not every solution to a problem may be "admissible." Some constraints are merely technical, others are in the form of counsels of prudence or moral laws. When does an imperative express a rational constraint and when does an imperative merely express a *tabu*. I would say that an imperative u is irrational (i.e. it is not admissible that u) if, by and large, to satisfy u interferes with the complete fulfillment of human needs, whether these needs are physiological, social, or spiritual. This view of admissibility is in harmony with the views of John Dewey [ibid.] who urged that philosophers abandon their homage to the outmoded views of the past and turn toward "open [scientific] inquiry in an attempt to discover the patterns of behavior that most adequately meet the current problems of society, while at the same time providing for the all-around *growth* of every mature and maturing member of society."{27} {28}

This brings to a close our survey of the development of the philosophy of practice in Western countries.

3 What is importance?

It is not easy to give a necessary and sufficient condition for importance, but I think one can say that if something is important there is a rationale for it that can be explained either in terms of its significance or in terms of requirements. Consider, for example, the importance of keeping a promise. Keeping a promise is important only if there is *reason* to keep it, and this reason, if it exists, stems from the ***requirements*** of social life. Moreover, the things, like joy, that are *significant* are important because they supply a reason for living. In a world without purpose or need it is not clear that there would be any requirements that could supply a rationale for anything. The relation of importance to requirement is complex, however, because at different times and places and different stages of life some needs or purposes are more pressing than others. One may ask how this can be the case if requirements of this kind are the reason for the activities of an organism and therefore determine their importance? The answer is obvious; some purposes are deeper than others and the reasonableness of a given activity depends on the state as well as the stage of maturity of the organism. This matter is explored further in the next chapter.

In the case of the higher needs it is easy to give a necessary condition for activity inspired by them to be of importance; namely, the condition that either all lower needs are satisfied or it is possible to neglect them. It is for this reason that Christ in *THE SERMON ON THE MOUNT* says "Take no thought for the morrow." Those who seek to fulfill the higher needs are often like that saint of the Stoics, Socrates, who had an indifference to heat and cold, adopted a simplicity in matters of dress and food, and showed complete independence of bodily comfort. The practice of renunciation among monks and nuns—whether Buddhist, Christian, or other—is designed to achieve this state {29}.

I suggest that the deepest purposes correspond to the higher needs and the deepest purpose of all is the search for transcendence. Plato, in his *SYMPOSIUM*, anticipates Aristotle's conception of what is of *ultimate importance* in life, in that, for Plato, the highest form of intellectual life is the contemplation of the forms, particularly the form of beauty—which, in his description, is hardly distinguishable from "the good." (Cp. Philebus 64e) Whether or not such a pure form exists I would not presume to say (or deny!). Nonetheless, though the dialogue is not explicit on the point, I can see that it might be said that the *SYMPOSIUM* anticipates my thesis that the very human purpose of penetrating beyond the here and now into the infinite lies at the very top of the list of things that supply a reason for living; that is to say, *make living important.*

4 Science and practical inquiry

Kuhn [1970], who points out that science has two phases, normal and revolutionary, fails to note—unless I am mistaken—that the same phases may be discerned in art. Most scientists, according to Kuhn, operate within so-called normal science, but the same thing may be said of the technical experts in art. The technical expert, whether in science or art, has a paradigm in mind to which he tries to fit the world in Procrustean fashion. For example, technical experts in science use the Newtonian paradigm for the solution of problems in launching satellites, and similarly Supreme Court Justices in the United States use the United States Constitution as the paradigm by which to judge the admissibility of State or Federal laws. Sometimes the old paradigm becomes unwieldy, however. When this happens the proper response is to engage in *inquiry*. This means you criticize the old paradigm and, if necessary, invent a new one. For example, in science Einstein proposes to replace the Newtonian paradigm with General Relativity; moreover, in practical inquiry, Jefferson proposes to amend the Constitution of the United States by adding the Bill of Rights.

Let us call the use of the old paradigm *technics*. Technics and inquiry are dual subjects in the following sense: on one hand, technics fits the world to a paradigm; on the other, inquiry fits a paradigm to the world.

There is a large difference between the paradigms of science and those of practical inquiry, however. This difference is somewhat like the difference between a myth and a computer manual. On one hand, a scientific paradigm is used as a basis for a scientific theory, conceived as a system of laws of how things behave, rather than how one would like them to behave; on the other, a practical paradigm forms a guiding principle for the construction of a normative system whose purpose is to set forth norms and imperatives for the regulation of the operations (and/or activities) of human beings (or data processors). (For a precise account of what is meant in these pages by the terms "norm" and "normative system" the reader is referred to Chapter Nine, ***Part Two***.) Normative systems are in some ways more complex than scientific theories. Complexity does not necessarily translate into greater difficulty, however, since the simplest theories are the notoriously difficult mathematical ones. There are at least two reasons for the complexity of normative systems. One is that, in general, normative systems presuppose science. The parts of science which are especially important for normative systems are the behavioral sciences such as psychology, ethology, sociology, and political science. The other is that in addition to having to have the usual logical apparatus of a scientific theory, a normative system has to have special features for handling imperatives or norms.

Admittedly there are limits to the use of science in the field of practical philosophy. The most substantial difficulty is that any proposed list of *design principles* for a normative system—which are the principles needed to justify the introduction of imperatives or norms—is subject to dispute (Cf. Chapters Four, Six, and Eleven). Nevertheless I think that there are good reasons, on which rational persons can agree, for accepting some rather than others (Cf. Rawls [1993]). Another difficulty, is that it is not always possible to substantiate the empirical generalizations that are sometimes needed for the establishment of imperatives or norms. There are at least three reasons for this: the first, and perhaps the most important, is that emotion, emotive language, and preconceptions about desired results often interfere with objectivity in research in the behavioral sciences; secondly, moral limitations sometimes preclude a totally objective scientific investigation on biological subjects; finally, any investigation into behavioral patterns and attitudes tends to change what is being investigated. Moreover, even if all the above shortcomings are overcome, there remains the following limitation to the scientific method in practical philosophy: it is not likely that there is a normative system that is complete in the sense that every problem expressible in the system can be solved without adding new axioms. The latter need not be completely discouraging, however, since according to the landmark result of Gödel [1931] not even a formal system adequate for arithmetic is complete.

There are two questions about completeness that demand consideration, however: one, in what way is practical inquiry incomplete?; the other, does it matter that practical inquiry is incomplete?

The first question is answered by Aristotle [1103b26]: so much in practical inquiry depends on particular circumstances (and one may add, particular people) that general rules cannot always be established. The classic problem that tests the rule is that of Sartre in which a man is undecided whether to join the resistance in France during World War II or stay home to help his aging mother. In my opinion, there is no possible algorithm in practical inquiry that yields a definitive solution of such a problem.

But do we really desire that each moral problem be solved by an application of a general algorithm? Human society has already had enough of the authoritarianism which results from the conviction that the "right" choice of action is known with certainty (Cf. Popper [1950, p. 511, note 18]). A complete system, if accepted, would likely tend to have a dampening effect on human freedom and creativity. In practical philosophy, what is needed is some way of deciding—with as much certainty as the subject matter allows—on the general principles that are required in modern life; that is to say, relative to modern conditions and considering the present state of human evolution, a method of deciding which general imperatives in ethics and politics are such that their negations are inadmissible. This is exactly the purpose of practical inquiry into both socio-political affairs and the conduct of life.

To sum up: By ***practical inquiry*** we mean an inquiry that aims at the criticism of normative systems and the construction of a new system of norms and imperatives that better fit the needs and aspirations of human beings. Science, by contrast, is an inquiry that aims at the establishment of facts and the construction of scientific theories to fit the facts. Practical inquiry is not a science because it aims at the construction of normative systems and the establishment of norms or imperatives rather than the establishment of scientific theories and the establishment of facts. On the other hand, the present work shows how to reduce many of the problems in practical inquiry to science. It is true that much stands in the way of progress including the religious veneration accorded to ancient paradigms. Yet, despite the great danger to civilization of the misuse of technology, I am persuaded that, given the right choice of paradigms, and the use of the piecemeal approach of science, practical inquiry can help not only to ward off almost certain disaster, but to improve the quality of life for all on this planet.

5 Remarks on meta-ethics

For realists who enter the field of moral philosophy the question of whether moral judgments are capable of truth or falsehood represents a major stumbling block. Moreover, Moore undermines all attempts from Aristotle to Bentham and Mill to connect ethics to nature by his doctrine that those who attempted to base ethics on natural properties such as pleasure are committing a fallacy—which he dubs the "naturalistic fallacy." Perhaps the prime example of a realist who foundered over this issue is Bertrand Russell. Russell [1944, p.723] writes:

> ...as for me it is necessary to acknowledge that I see no property, analogous to "truth," that belongs or does not belong to an ethical judgment. This, it must be admitted, puts ethics in a very different category from that of science.

Russell's contention, if true, imposes a profound difficulty on those who like Russell himself wish to persuade others of the soundness of their ethical judgments. (Note, however, that Russell [1954] reverses himself once more.) Practical inquiry as based on imperative logic is a way of avoiding this difficulty without departing from realism.

Practical inquiry is agnostic on the question of whether sentences containing such words as "ought" and "good" (or "value") are capable of truth or falsehood. That is not to deny that ***evaluative activity*** occurs. But the questions of the existence of *values* or the truth or falsity of obligation or evaluative statements may be left to the metaphysicians. What is claimed is simply that sentences like "If x is true then it is required that it *shall* be the case that y" or "It is admissible that it *shall* be the case that x only if y" are capable of truth or falsehood. Of course this means that one must have some procedure for deciding when it is that sentences of these (and related) types are true. In many cases—especially in ethics and politics—it may not be so obvious how one determines requirement or admissibility. There is a profound difference in this regard between mathematics and the fields of ethics or politics that perhaps bears mentioning at this point: members of a school of mathematicians are seldom seen shooting members of a rival school. The situation is not completely hopeless, however. In practical inquiry, as in geometry or physics, one must start with axioms. According to Russell [1969], "an ethical opinion can only be defended by an ethical axiom,..." But this need not destroy the objectivity of the subject any more than beginning with axioms destroys the objectivity of (say) Euclidean geometry.

Roughly speaking, it is the "ideal" that is the subject of practical philosophy. The perennial argument is whether the ideal is simply the heart's desire or is the ideal based on something more objective and absolute. In some contexts the ideal is called "justice"; in other contexts the ideal is called "the good." Much ink, as well as blood, has been spilled over the meaning of these terms. The game of defining these terms has been going on since the time of Socrates if not already in the Garden of Eden or prehistoric times. Plato, in his dialogue, *THE REPUBLIC*, hit on the right solution although he did little to make this clear. Plato, instead of assuming one knows the meaning of "justice," develops his paradigm for justice without reference to the meaning of the term. Only later does he say—as he intimates at the beginning—*"there it is,* written large in the model." If this procedure is correct, and I believe it is, we do not need to define the word "ideal"—or any of its many synonyms—in order to develop appropiate paradigms for ethical or political systems.

One often hears that we need a new *paradigm* for the modern age. A paradigm in this sense may be considered a model from which one may derive the principles for the guidance of modern people in ethics and politics. There are two constraints that such a model must satisfy if it is to have any chance of acceptance. One is what is often called the tide of history. The other is the force of rational thought. Rational thought can do very little except to try to modify the cultural forms that arise from the historical tide. There are two factors in the historical tide that tend to generate dissatisfaction with the old paradigms. One is the force of circumstance: as time goes on the circumstances of life change and people adapt—or fail to adapt—to the changes. The other factor is the attraction of novelty and dissatisfaction with old things. Generally speaking, the young desire change and the old wish things to remain as they are, or better yet, return to the way they used to be. The fight against change is a losing battle, however; for change is inevitable. There are three possible courses of action: one, try to maintain tradition insofar as possible; two, let things drift; three, try to modify things toward some conception of the ideal.

What is needed in practical inquiry is the development of normative systems of two kinds: one for social structure, the other for personal development. It is not hard to see that corresponding to a normative system for social structure there is a socio-political ethic and corresponding to a normative system for personal development there is a personal ethic. (For example, Plato's *REPUBLIC* contains implicitly a socio-political ethic and Aristotle's *ETHICS* contains a personal ethic.) Neither of the two normative systems can be judged satisfactory unless the two ethics that are extracted from the systems are *complementary* to each other; that is to say, (1) the rules of behavior derived from either ethic are compatible, (2) the social ethic, based on the demands of society, promotes the free development of the individual in accordance with the normative system for individual development, and (3) the personal ethic, based on the demands of the individual, promotes the full development of society in accordance with the normative system for the governance of society. I return to this point in Chapter Eleven (***Part Three***).

Any societal paradigm must allow for the fact that not all citizens are ideal, but an ideal society is difficult to achieve without ideal citizens. To ask which comes first—the ideal society or the ideal citizen—is a chicken and egg question. Note that a society that promotes the free development of its citizens is self-defeating and unstable, unless youths, who have known nothing but freedom, tend to feel a loyalty to their own heritage. All known societies try to *mold* their citizens so that the effects of free development are yet untested. The difficulties that stand in the way of the construction of a society that promotes free development is a matter to which we return (see Section 4.4 in Chapter Four and the whole of Chapter Eight).

Despite my professed agnosticism about the truth of statements involving predicates such as "good" or "ought," I have been accused of covertly assuming the truth of such statements as "Freedom (for the individual) is good," and "Society ought to be ruled by law." But I protest that my admittedly positive attitude toward freedom does not commit me to assuming the former statement, nor does my somewhat anti-establishment attitude commit me to denying the latter. The truth or the usefulness of a theory is independent of the attitude of the propounder of the theory. In the field of mathematics, for example, it is absurd to suppose that one's attitude toward numbers has a bearing on the validity of theorems. I do not deny that I have an emotional attachment to the furthering of equilibrium in society and enhancing significance in individual lives and wish that others have the same feelings, but even if human societies "ought" to perish and all human beings "deserve" to live empty lives—or even if, like Milton's Satan, I chose "evil" as my "good"—that would not invalidate practical inquiry any more than the existence of non-Euclidean geometry invalidates Euclidean geometry.

6 Concluding remarks

To whom can human beings look for guidance? Many people still swear their allegiance to religious prophets claiming to speak for God (or the gods), but, increasingly, people are hearkening to secular prophets. In the West people are guided in part by the paradigm of democratic liberalism, whose chief prophets are, on one hand, men like Locke, Jefferson, Madison and Hamilton, and, on the other, men like Adam Smith, Ricardo, and John Stuart Mill. Although many people in the East have until recently been guided by Karl Marx, nowadays the liberal paradigm of free trade—with or without an accompanying obeisance to the democratic paradigm—is becoming ascendant over the whole world. The paradigm of democratic liberalism has spawned a new paradigm, however, a paradigm to end all paradigms. This is the paradigm of innovation and discovery. This new paradigm brings with it a pattern of hostility to the established order of things. The prophets of this new paradigm include Plato, Francis Bacon, Thomas Jefferson, H.G.Wells, Alfred North Whitehead, Karl Popper, and John Dewey. Plato [Republic, 530c] encourages astronomers to "let the heavens alone" and "study problems"; Francis Bacon promotes the modern inductive method and is a pioneer in the attempt at a logical systematization of scientific procedure (Cf. Russell [1945]); Jefferson [1787] says "a little rebellion now and then is a good thing"; Wells [1935] preaches a "scientific world commonweal"; Whitehead [1927] claims that the "rules and canons [of society] require revision in the light of reason"; Karl Popper [1950] advocates "piecewise social engineering"; and John Dewey [1938] teaches that science can take over morals as well as the hard sciences such as physics, chemistry, and biology. Of all the prophets of the new paradigm, John Dewey is the most revolutionary. According to the pragmatic doctrine of Dewey [1920]: "Natural science loses its divorce from humanity; it becomes humanistic in quality."

As I made clear in Section 4, I cannot go all the way with Dewey: I think there are significant differences between science and practical inquiry. Moreover, as Russell [1945, p. 828] points out, the paradigm of innovation and discovery, if not coupled with a certain amount of restraint and humility, can have serious drawbacks {30}. According to Russell, "man, formerly too humble, begins to think of himself as almost a God." These concerns are, of course, not new. The ancient Athenians became fearful of the devastating effect of rational analysis as practiced by Socrates and his disciples. As a consequence, Plato [Republic, 539] warns against the withering effect of rational analysis on the piety and morals of youth. It is for this reason also that Descartes decides that he would not allow his method of radical scepticism to affect his adherence to traditional standards of behavior. This fear of the new paradigm is not altogether groundless; for as noted in the foreword to this work, scientific rationalism can have the effect of making some people give up on the very idea of ideals and principles. What one needs is a Hegelian synthesis uniting the admissible imperatives of the paradigm of innovation and discovery with those imperatives from the older paradigms that humanity can ill afford to lose. The plan of this work is to build such a synthesis, formulate it within a new logical framework, and show how it may be defended. The reader hardly needs to be reminded that no such synthesis can be expected to be the final word; however, I hope that others will find something useful in my proposals and bring to fruition what I begin here.

Notes

1. Normative systems in this work are "paradigms" in the sense that they are *models*, but—unlike the *ideas* of Plato—they are subject to criticism.

2. The word "human" (or "humane") is still used—despite current cynicism—as a term of praise.

3. Aristotle's conception of the ideal human reminds one of the definition of "good" proposed by the axiologist Robert S. Hartman [1967]: to be a good x is to fulfill all the qualities of being an x. Since Plato claims that nothing, except perhaps God, (and filth?) fulfills to perfection its class concept, it follows that almost nothing is "good." Is it not therefore preferable to think in comparative terms since humans, though they may improve, are unlikely to achieve absolute "goodness"?

4. Aristotle relies not only on the concept of the "fully functional human being", but he is guided by the notion that "as human beings, reason and mind are the end to which our *growth* tends" [1334b, 14] [emphasis added] (Cf. Sinclair [1962]).

5. Since Aristotle [1098a18-20] notes that the achievement of perfection can never be completed in less than a lifetime, it would perhaps be more fitting

to say that the ***way*** to happiness is activity in accordance with *phronesis* and *sophia.*

6. On one reading Kant is a true utilitarian. He writes in *THE CRITIQUE OF PURE REASON* [1787, 383] that a necessary idea. which must be placed at the foundation, not only of the first plan of the constitution of a state, but of all its laws, is a "constitution of *the greatest possible human freedom according to laws by which the liberty of every individual can consist with the liberty of every other* (not of the greatest happiness, for this follows necessarily from the former) [emphasis in original]."

7. Kant thinks that the person with "good will" is probably unhappy on earth and therefore should look for his or her reward in heaven.

8. Aside from the distinction between socio-political ethics and personal ethics noticed by Spencer, there is another supposed division of ethics that has attracted some attention in recent times. It is said that there are two major paradigms for ethics, one, "deontological," the other, "teleological." The *teleological* paradigm says that consequences are the determining factor in ethics; the *deontological* paradigm denies the importance of consequences. The teleological paradigm is obviously adhered to in utilitarianism but used covertly in other systems. The deontological paradigm is said to be followed in socio-political theories of the social contract type such as that of Rawls (Cf. Taylor [1975, p.112]). As I have already had occasion to note, I think the usefulness of this classification is questionable. To my mind, the only systems that might conform to the definition of deontological are those systems whose only justification is a Kierkegaardian "leap of faith." I think, however, the phrases "deontological requirements" and "teleological requirements" are useful. The word "deontological," taken literally, means "based on duty" (for the adjective "deontological" comes from the Greek root "deont-" which means "that which is binding"); on the other hand, the word "teleological" means "based on ends" (for the adjective "teleological" comes from the Greek root "telos" meaning "end"); consequently, I think it reasonable to define "deontological requirements" as "moral rules (or "duties") imposed by society" and "teleological requirements" as "ends chosen by the individual."

9. Bergson [1935] divides the *sources* of morality and religion into two: one, he calls social; the other, he calls mystical.

10. Jack Smart tells me he would be a "backslider."

11. Of course, Aristotle not only develops a normative system for the guidance of the individual in his *ETHICS* but a normative system for the governance of society in his *POLITICS.* Were a provision of one of these systems to conflict with another, however, there is little doubt that Aristotle would favor the normative system for the governance of society; for in his *ETHICS* Aristotle [1094b5-10] says: "for while it is desirable to secure what is good in the case of an individual, to do so in the case of a people or a state is something finer and more sublime." Aristotle may not have kept the distinction between the two systems clearly in his own mind since the underlying rationales in each are similar—the provisions in his personal ethic being based on the concept of individual happiness and the provisions of his politics being

based on the happiness of the whole society (provided one excludes slaves!). Aristotle deserves praise for sharply separating practical inquiry (guided by prudence) from theoretical inquiry (guided by *sophia* or the part of the rational soul which contemplates eternal things).

12. Spencer can be interpreted as saying that being moral means keeping a balance between deontological and telelogical requirements, where the deontological requirements (or "duties") come from society, and the teleological requirements are those ends chosen by each person.

13. In discussing his personal tendency toward pacificism, Einstein [1930] writes: "War is low and despicable, and I had rather be smitten to shreds than participate in such doings."

14. Aristotle, however, did not realize the potentiality of people to grow in such a way as to transcend the narrowness of the tribal ways, and therefore did not realize that people in coming ages were to become more democratic and less apt to tolerate slavery, imperialism, or inequality.

15. Nietzsche's [1886, p. 5] final opinion of Spinoza is characteristically harsh and unflattering. He writes:

> ...still more so, the hocus-pocus in mathematical form, by means of which Spinoza, as it were, clad his philosophy in mail and mask—in fact the "love of *his* wisdom," to translate the term fairly and squarely—in order to strike terror at once into the heart of the assailant who should dare to cast a glance on that invincible maiden, that Pallas Athene:—how much of personal timidity and vulnerability does this masquerade of a sickly recluse betray! [emphasis in original]

16. The ancient Hebrews compared a righteous person to a tree planted by a river. Russell, however, may be unconsciously echoing John Stuart Mill [1859] who writes: *Human nature is not a machine to be built after a model, and set to do exactly the work prescribed for it, but a tree, which requires to grow on all sides, according to the tendency of the inward forces which make it a living thing.*

17. With apologies to Russell!

18. If only the great philosophers had agreed on what is important in life, we might now agree on how to approach ethics!

19. It is true, however, that other authors emphasize the importance of enhancing "meaning" and significance in life which, of course, relate to importance. Dewey [1922], for example, emphasizes the importance of attending to *present significance* in the conduct of life rather than to fixed far off goals.

The first draft of this work contained no quotations from Whitehead. In my youth, however, I had read Whitehead and probably I was unconsciously influenced by him when I sat down to write this chapter.

The following quote (see Whitehead [1911]) may be of interest to those who find Whitehead obscure: "It is a safe rule to apply that, when a mathematical or philosophical author writes with a misty profundity he is writing nonsense."

20. In Chapter Three, I use the phrase "What does it all mean?" in a context in which "What does it all signify?" does not seem appropriate. I myself could not find a way to make the same point if I avoid the word "mean."

21. The three factors (excitement, tension, and pleasure) that Wundt (Cf. Lazerson [1975]) uses as variables in his three-dimensional theory of emotion each betoken that something is important. This matter is further explored in Chapter Three.

22.Schopenhauer [1928 ed., p.306] compares life to "a path of red-hot coals with a few cool places here and there."

23. Compare Rawls [1971]. For Rawls a stone is good if it has all the properties it is rational to want in a stone; however, it is not clear to me why it is any more rational to want a stone to be typical than it is to want it to be atypical.

24. The question of whether the goodness of a motive (such as good will) depends on its usefulness was already discussed in connection with Kant. Whether *life*, in and of itself, is useful seems problematic, but one *could* evaluate a life according its usefulness.

25. Some things are pleasurable because they are useful and others are useful because they are pleasurable; still others are pleasurable, but they have drawbacks that tend to destroy their usefulness.

26. Kanger [1966, 1972] also develops a theory of norms which he calls a typology of "rights." This theory has been extended by Lindahl [1977]. While I have found inspiration in Kanger's theory, the theory may be criticized on the ground that it does not seem to flow inevitably from a natural set of assumptions. (See Ch .9, ***Part Two.***)

27. John Dewey is the philosopher who has most emphasized the role of problems in determining human requirements and admissibilities. Dewey, however, was hung up on declarative sentences and therefore did not, so far as I know, consider imperatives. If he had done so, I think it is reasonable to say that his evaluation of whether it is correct to say of an imperative u "it admissible that u" would have been (in the words of Dewey) "on the ground of its ability to resolve the *problem* presented by the conditions under investigation" (and therefore terminate, at least temporarily the process of inquiry). See Dewey [1939, p. 47] [emphasis in original].

28. The quotation in the text is from an unpublished class hand-out entitled "A Brief Exposition of John Dewey's Ethical Theory" by Dr. John L.Mckenney.

29. The primary method of achieving the capacity for renunciation is that of meditation (see Chapter Seven: Mysticism, Beauty, and Religion).

30. According to Wittgenstein (cf. Monk [1990, p. 485]):

> It isn't absurd, e.g., to believe that the age of science and technology is the beginning of the end for humanity; that the idea of great progress is a delusion, along with the idea that the truth will ultimately be known; that there is nothing good or desirable about scientific knowledge and that mankind in seeking it, is falling into a trap.

Chapter Three

The Springs of Human Action

1 Introduction

Psychologists, in general, have been loath to recognize human ends which go beyond the needs of the body; for this reason the eminent psychologist, (Hebb [1953]), was prompted to write: "From what one can gather from the standard textbooks, one would think that all he [i.e. a human being] is really concerned with is to fill his belly and propagate the species." Of course, even in those days, a few psychologists besides Hebb—such as Murray [1938], Harlow [1953], and Maslow [1954]—understood that human beings tend to satisfy requirements other than those that stem from obvious physiological needs. Murray, for example, lists twenty (or more) such requirements and, moreover, the experiments of Harlow [ibid.] show that higher animals, other than *homo sapiens*, meet requirements that go beyond the merely physical or social.

We have already noted in previous chapters that human requirements (or needs) may be arranged in a hierarchy. The idea that needs form a hierarchy, ranging from low, through intermediate, to higher, is nowadays attributed to Maslow, but the classification of needs into three types has a long history. For example, William James [1890] writes of three constituents of the phenomenal self—the material self, the social self, and the spiritual self—each with its own characteristic interests or needs. Examples of higher (or spiritual) needs have been given by Hebb [1953] who writes as follows:

> I propose that Man has an inherent need for work; that he is so made as persistently to seek a certain degree of discomfort, and persistently invite the risk of pain for the sake of that risk; and that a need for mental effort, including philosophic and religious thought, is built right into his nature from the first, as much as swimming is built into a fish.

The physiological needs are relatively unproblematic and hence we confine our attention in this chapter to the intermediate and higher needs. We begin with social needs such as status, power, love, and safety, which are intermediate in Maslow's hierarchy; then turn to the higher needs. The final section (Section 5) points up how the springs of human action are relevant to ethics and politics.

2 Status

Ethologists tell us that a social group of mammals organizes itself through competition between each pair of animals to determine their relative status (Cf. Lorenz [1966]). The result of the relative rankings is called a pecking order or dominance hierarchy. By and large, in groups of lower animals, status among males is achieved by a test of physical strength and only males of high status have what is called "biological fitness," i.e. the ability to mate and leave a large number of offspring. Among some animals the roles of the sexes may to some extent be reversed—the female hyena, for example, being larger and more powerful than the male. As higher animals evolved the competition for status (which we may call the status game) has also undergone evolution and in humans it has become very elaborate. In human society, though high status males still have most of the advantages, the status game is open to both males and females and it is possible for a female to outrank most males either in her own right or through association with a high ranking male.

Status among animals must have originated in the struggle for possession, particularly the possession of females. Status in human society depends partly on a person's role, bankers rating rather high, tradesmen relatively low, menial laborers lower still. Those who have no recognized role in society, or who belong to a despised minority, are rated at the bottom. To have status at all a person has to *be* somebody, either a notable, or someone who contributes something to society, and hence the status system contributes to the specialization of roles without which civilized society is not possible. The specialization of roles has the result that everyone in a community has his or her own responsibilities.

As Wilhelm von Humboldt [1854, p. 186] long ago observed:

> Man is more disposed to domination than freedom; and a structure of domination not only gladdens the eye of the master who rears and protects it, but even the meanest underworkers are uplifted by the thought that they are members of a majestic whole, which rises above the life and strength of single generations.

The person at the absolute top of the pecking order is often regarded with a combination of fear and awe, and at times is worshipped as a god. There are basically two ways to achieve top status in human society: one, through talent and success in certain lines of work; the other, through inherited status. If one gets to the top through works, success is ensured not so much by physical strength as it is through such means as political adeptness, business acumen, physical attractiveness, artistry, and what one may call "razzle dazzle." Billy Graham gets to the top by oratory; Babe Ruth, by spectacular feats; Shirley Temple, by delighting her audiences; Hetty Green, by business acumen; a Ross Perot buys his way to the top; and Fidel Castro shoots his way to the top. Very few, e.g. Einstein, get to the top through intellectual accomplishment.

The curious thing is that people—as well as members of related species—tend to give hero status to individuals who have nothing but the power—or in some cases fan fare or razzle dazzle—to recommend them. To see this one need only take note of the rogues gallery of those who have been regarded as heroes from Alexander to Saddam Hussein. Clearly, power generates genuine admiration, e.g. consider the almost unbelievable phenomenon—such as occurred in the Patty Hearst case—in which people, taken hostage, tend to identify with and make heroes of their captors. Most heads of states make judicious use of razzle dazzle in the form of the thunderous trappings of pomp and circumstance; they also make effective use of the solemn blessings of religious figures to silence any doubts about the legitimacy of their position. According to Jane Goodall [1971] razzle dazzle is an effective tool, even among the chimpanzees. She reports the amusing case of a chimp who became the leader of a troop of chimps in the Gombe Reserve by making a lot of noise with garbage cans.

Inherited status, the easy way to get to the top, is—from the Darwinian standpoint—rather an anomaly since it allows weaker persons to achieve high status and leave many offspring. Inherited status does have some advantages, however. For example, it tends to bypass what otherwise might be a fratricidal struggle for power and hence tends to preserve the unity of the community. Strangely enough, achieving a top place in the dominance hierarchy by heredity (rather than by works) appeals not only to those who benefit by it, but to the rank and file. People are natural believers in grace and aristocracy, and this makes the development of democracy difficult and its preservation tenuous. Status, possession, and power have an intimate connection. Generally speaking, those who have status are awarded property and those who have property are awarded status. When people have wealth, they can convert property into power because they can reward others for doing their bidding. In order for this to work, however, there has to be a criminal justice system that protects property "rights." The "justice" meted out by a system of this kind is—as Thrasymachus said (Cf. Plato's *REPUBLIC)*—the interest of the stronger; for it imposes a moral code which outlaws taking away the excess property despite the great need of others. Though modern societies have ameliorated this a bit, in most societies the function of the law is to enforce the status quo. Those who are at the top do not willingly give up their privileges and accept healthequality. If democracy is to be recommended, it is not on the ground that it best fits human nature as it now seems to be constituted. Wealth is a virtual Gyges ring, unresistable to all but saints. Wealth comes to people of high status either by appropriation, as in the case of a tyrant, or by having it handed to them, as in the cases of royalty, heroes, beauties, and entertaining personalities. A hero or celebrity is showered with gifts, is a shoo-in for high office, and (if male) is besieged by women eager to have babies fathered by a hero, etc. In addition to the relative freedom to exercise one's powers and satisfy one's tastes that position and property convey, the existence of people of lower status affords two opposing satisfactions: on one hand, the pleasure of *noblesse oblige*; on the other, the delight in lording over or tormenting those of a lower class. The former response is that of philanthropy; the latter response is that of tyranny. Tyranny is not confined to human beings but found among other social animals, e.g. wild dogs. Not only tyranny, but philanthropy, would disappear in a society of equals.

In spite of its undoubted rewards, there are burdens connected with status. First of all, the prospect of losing status is a nightmare of which not even people of relatively low status are free. For example, a beauty may rather die than lose her looks. Moreover, not everyone enjoys the attention they receive as the result of celebrity. For example, a Pope cannot ordinarily wander around Rome alone. Einstein said that if he had life to live over again, he would become a plumber. It is perhaps partly for these reasons that people sometimes shy away from the struggle for high status.

Low status people suffer doubly, however: once, through the treatment they receive; the other, through the way they feel about themselves. That in itself is a powerful reason for engaging in the struggle for status. There is an additional powerful reason: the pleasure of the struggle. For those of stout heart, not burdened by anxiety or fear, a struggle for power not only offers the allure of the possibility of power and high status, but the chance for excitement and avoidance of boredom. On the other hand, not everybody enjoys the struggle. Moreover, if fraticidal quarrels are not to break society apart, people have to develop civility and find ways of moderating the status game. There have to be losers who accept their lot in life or at least do not revolt overtly. But we may ask: why do not the losers more often either isolate themselves or join with others in revolt? (Cp. Hume [1777])

One obvious reason is that low-status people, owing to the treatment they receive and the way they feel about themselves, tend to suffer from ill-health (both mentally and physically) and hence have neither the energy nor the resources with which to revolt. Among lower animals the benefit of the struggle for status is, as we noted, biological fitness. This benefit still operates to some extent among human beings; for, as we already noted, high status males get a choice of females, and a female who achieves her status by physical attractiveness and charm has a choice of males. But lower status people seem nowadays to have more biological fitness than higher status ones, and hence there have to be other reasons why people, even though not particularly successful, do not revolt or drop out of what is called the "rat race." Besides the factor already mentioned, one factor is that people are not so much motivated to gain status as not to lose the relative status they already have—or as the orientals have it, not to lose face. But perhaps the most important reasons, besides economic necessity and the desire for security, are the following three: First, to admit defeat and accept one's place in life is both boring and depressing. Second, being a social animal, human beings, as well as apes, find quarrels, however fierce or even violent, easier to take than loneliness, and what is more, though they would have preferred to dominate, they accept the near certainty of defeat rather than the total certainty of the loss of companions. Third, people have a tendency, shared with their simian cousins, toward hero worship. As part of this hero worship, a lower ranked person has a tendency to identify with those at a higher level of status. As an example of the latter, the crowd identifies with the hero and takes pride in the hero's exploits as though they were its own. The hero's interests (and even tastes) are generally adopted as those of the entire group, and for this reason the claim such as that of Louis XIV, "L'etat c'est moi!," becomes plausible. The fact that hero worship and the adoption of the behavior of those at a higher level of status are tendencies that humans share with apes indicates that they are not just based on culture but have a biological foundation.

People at the top of the pecking order fulfill at least two functions: one, the leadership function; the other, the ceremonial function. As leaders, high status people function, like the brain of an organism, as decision makers; as ceremonial figures, they excite veneration, enchantment, and worship and unite the realm into a whole. As long as the people venerate the leadership, trust in and identification with the leadership gives the rank and file individual a feeling of participation in community affairs. In the cases in which the internal struggle for power is not too fractious, a leadership emerges that is able to govern with the consent of the governed and make decisions that are binding on the whole group. People, adopting the purposes and interests of the leadership as their own, unite behind them and present a single front against a hostile environment.

3 Power, love, and safety

According to anthropologists, the size of a primitive band of humans is limited to about 500 persons owing to the tendency of a larger band to fracture. A large group of human beings tends to split like a single cell in mitosis (Cf. Morris, [1969]). As a result of such splitting, the social landscape is entirely altered; for instead of one more or less harmonious society, there is now a group of tribes. When this happens, a new status game emerges in which the units of competition are no longer individuals; the units of competition are, on the contrary, groups. It is a status game of the second order.

Less powerful or less aggressive tribes may attempt to escape from struggle by migration, but migration tends to be limited when hunting and gathering or nomadic life is replaced by agriculture. The struggle may be initiated by population pressure as well as the desire for plunder and excitement, but it is often continued largely owing to the grievances and losses which inevitably arise out of the struggle. The struggle is made all the more fierce not only by the desire for revenge, but because of that peculiar attachment to territory called "love of the homeland." In the struggle for hegemony among the tribes, entire tribes may be wiped out or enslaved. Eventually, by conquest or alliance, a supertribe—or a *state*—is formed.

The process of formation of a state recapitulates to some extent the process of the organization of a society out of individuals, but now the process is not moderated by sentiment. Sometimes, like the Iroquois, tribes who are threatened by one or more other tribes, have sufficient wit to form an alliance, but most states are formed by conquest. In the case of conquest, the members of the conquered tribe are given low status, and may be treated as slaves. The amalgamation of two tribes is less difficult if they are culturally the same, but there is bound to be difficulty when there is a clash of cultures. The culture that finally emerges will probably have elements of both cultures, and it may be that the tribe that is weaker militarily will have the more enduring culture. It often happens that the conquering people adopt a great deal of the culture of the conquered.

Once states are formed, the leaders desire more power, and turn to the conquest of other states. The status game then begins all over again. Instead of pitting tribes of (say) five hundred individuals against each other, however, it is now a struggle between city states (or heads of states) containing thousands or hundreds of thousands. As a result, nation states are formed usually headed by a king. Of course nation states compete against each other and eventually form empires. An empire, once secure in its borders, tends to develop an international force for settling quarrels among subject peoples (i.e. the peoples whom Mark Twain [1901] jokingly called "the peoples that sit in darkness") and produce a kind of peace such as *pax Romana, pax Brittanica*, or *pax Americana.* The clash of empires competing for world hegemony, which we call a world war, is a status game to end all status games.

Wealth has been sought in all ages for the increase in status and power it brings. Two things, which are interrelated, have altered the nature and scale of this struggle. One is the increasing globalization of the struggle owing to the changes wrought by new inventions. First there was the invention of the horse and carriage and the merchant ship; then motorized transport, followed by air travel, data processing, and instant communication. The second factor is the decline of the power of the state, relative to the power of commerce. As a result states have become pawns in the struggle for status among multinational corporations and international financiers. So far multinational corporations and international finance have met with only token resistance. So great is the status of money in the modern world that not even the American empire has been tempted to stand up against it. I have much more to say about this in later chapters (see Chapters Four and Eight).

It is an important question whether power can be tamed. If it can, it must be because there are human needs that operate as a moderating influence on the struggle for status and power. The two needs that come to mind are love and safety. As a moderating influence on the struggle for status and power the human need for love is rather weak. Preaching brotherly love touches the hearts of people, but the practical consequences of two thousand years of preaching the gospel of Christ have been disappointing. It is the human need for safety that supplies most of the drive toward eliminating the clash of empires.

The question therefore arises how it is that those in power can be persuaded to protect those who are their subjects? In the end it is only the possibility of revolt that makes those in power do the wishes of their subjects. The leaders have two means of retaining power irrespective of good works: one, the rack and screw, as it were; the other, brainwashing and propaganda. As the means of control applied by the rulers become more sophisticated, the opposition, to be effective, must increase its vigilance. One important factor here is the size of the group. The leaders nearly always want to extend their sway, ostensively for the safety of the subjects, despite the fact that size makes the group more vulnerable both to external attack and internal disruption. As the Chinese with their doctrine of a mandate from heaven knew long ago, if leadership is concerned only with its own power and wealth it sows the seeds of its own destruction.

4 The higher needs

What is it that matters most to a human being? This depends, as we have already indicated, partly on a person's stage of development and partly on circumstances. Corresponding to the hierarchy of needs there is a hierarchy of ends. (A need, after all, is a species of end.) At the bottom of the hierarchy there is, corresponding to the physiological and social needs, the end of survival. Philosophers tend to deny that survival is the most important end (Cf. Russell [1949]); but if the survival of the species to which the human animal belongs cannot be achieved, all other ends become futile. At the level of survival, all animals including humans exhibit a ferocity and selfishness that is sometimes shocking. Colin Turnbull [1972] found among a group of mountain people who were at the very edge of survival no hint of higher ends. For example, it is clear that at the level of survival, beauty is no object.

Let us turn to consider what happens in society when a class of human beings arises for whom survival is no longer in doubt. At this level beauty becomes a possible motivation. Those who rise above the survival level generally have a habit of saving, and this habit of accumulation is not immediately abandoned once it has served its purpose. In all advanced civilizations, while, on one hand, the poor are concerned only with survival; on the other, the accumulation of goods and the increase in beauty of surroundings remains the major occupation of the rich. The poor live in squalor and ugliness, and the nadir of ugliness is to be found in places where survival is moot such as in Haiti. The rich, on the other hand, collect beautiful paintings and other priceless objects and live in surroundings of taste and elegance. The homes of the rich are apt to be mansions, presided over by a beautiful woman, and set off like a jewel in a natural environment of gardens, woods, and streams. A rich man, in his elegant mansion, finds it easy to persuade an elegant woman to join him and add to its beauty.

The capacity of beauty to bring contentment is not absolute, however; for those who become used to attractive surroundings develop their critical faculties to such an extent that minor blemishes become major faults. Maslow [1971] writes about levels of complaint. It is astonishing to the poor that the rich, who apparently have nothing to complain of, often agree with the preacher in *ECCLESIASTES* that all is vanity. Adapting Maslow's scheme to our own hierarchy of ends, we may say that there are three levels of complaint: instinct, reason, and imagination. The lowest level is the level of instinct where the complaints are about the basic requirements of the body such as food, water, shelter, physical exercise, bodily health, safety, sexual outlets, etc.; the next highest level is the level of reason where the complaints are about the basic needs of the mind such as love, attractive surroundings, mental exercise, self-esteem, power, etc.; finally the highest level is the level of the imagination where the complaints are about the basic needs of the spirit such as art, religion, and morality. In order to have complaints at the level of reason it is required that one has satisfied the complaints at the level of instinct. Moreover, in order to have complaints at the highest level, it is required that one has essentially satisfied all lower order complaints or to have discounted them altogether. Christ, in the *SERMON ON THE MOUNT*, urges his followers to forget the complaints of reason and instinct and live at the level of the spirit. Perhaps what reformers such as Christ do not sufficiently pay heed to is that though power and riches bring only limited satisfaction, only when a certain level of comfort and beauty is reached is it possible for most people to begin to think seriously about the things of the spirit and the welfare of the less fortunate. I do not claim, however, that people who have everything begin to think about how they might improve the lot of everybody: only those who have a capacity for growth reach that stage. The Buddha, who has everything, is saddened by the suffering of the masses. Shelley, the rich landowner's son, bleeds for the downtrodden. Some multimillionaires, such as John D. Rockefeller, when they begin to reflect on their lives, turn to philanthropy. This brings us to the question of what leads people to rise above the level of the rational concern for personal comfort and private beauty and reach to the level of spirit.

It is possible to argue that the basic motivation leading to the level of the spirit is an aesthetic one. Who wouldn't choose to live in a beautiful community rather than an ugly one? Given a sufficient imagination, the artistic urge to create a paradise on earth must overwhelm even those among the rich and powerful who are not paralyzed by the fear that the many will, through envy, rise up and crush them. Shelley writes of the saving grace of poetry: recall his famous last sentence in his *A DEFENSE OF POETRY*, "Poets are the unacknowledged legislators of the world." Poetry, he says, "awakens and enlarges the mind itself by rendering it the receptacle of a thousand unapprehended combinations of thought." The result of such an awakening by poetry is, according to Shelley, a morality defined as follows: "A man, to be greatly good, must imagine intensely and comprehensively; he must put himself in the place of another and of many others; the pains and pleasure of his species must become his own."

5 The Biological Basis for Higher Needs

As indicated in Chapter Two, Goldstein [1939] and Maslow [1954] postulate a drive—which they call "self-actualization"—which impels a person toward higher development. According to this theory, it is a sign of a pyschologically healthy person that he or she is attracted to matters which are higher in the hierarchy of ends. (Being developed is a relative term, however; who among us develops to his or her full potential?) Rogers [1961] calls a fully developed person a "fully functioning human."

But what are the characteristics of a fully functional human? The following three come immediately to mind: cleverness, curiosity, and creativity. Can we not agree that a person who fails to *develop* these traits is not fully functional? I call the property of developing and exploiting these characteristics ***sapience***. Sapience—as the name *homo sapiens* suggests—is what distinguishes human beings from "lower" animals. Other characteristics of a fully developed person such as the ability to transcend the here and now (the trait of "transcendence") were mentioned in the first two chapters. These, which are based on the higher needs, are partly explored in this section, but for an extended treatment the reader is referred to Chapter Six (*What Are We To Do With Our Lives?*).

Higher ends are not overtly utilitarian—they do not, in and of themselves, bring food to the table, nor are they directly useful to society. According to Maslow [ibid.], there is but one higher end which is built into human beings and that is the requirement for "growth." As we have already indicated, other writers, e.g. Murray and Hebb, identify specific higher ends such as the needs for work, play, and understanding. Hebb [1955], in particular, claims that human beings have an inherent requirement for excitement. Expanding on this claim, he writes:

> There is no doubt, either, that problem-solving has some attraction for the rat, more for Harlow's monkeys, and far more for man. When you stop to think about it, it is nothing short of extraordinary what trouble people will go to in order to get into more trouble at the bridge table, or on the golf course and the fascination of the murder story, or the thriller and the newspaper accounts of real-life adventure or tragedy is no less extraordinary. This taste for excitement must not be forgotten when we are dealing with human motivation. It appears that, up to a point threat and puzzle have positive motivating value, beyond that point negative value.

Some philosophers anticipate Hebb. For example, William James [1895] (a psychologist as well as a philosopher) writes: *Need and struggle are what excite and inspire us; our hour of triumph is what brings the void.* Wilhelm von Humboldt [ibid., p. 2] (whom we already had occasion to quote) writes:

> It is in the prosecution of some single object, and in the striving to reach its accomplishment by the combined application of his moral and physical energies, that the true happiness of man, in his full vigor and development, consists. Possession, it is true, crowns exertion with repose; but it is only in the illusions of fancy that it has power to charm our eyes.

Many other philosophers from Aristotle to Bertrand Russell have written in a similar vein. (We return to this point in Chapter Six.)

The biological basis for the higher ends is obscure. Hebb [1955, 1987] himself has attempted to give a biological basis for the need for excitement, however. (As indicated in Chapter Two, excitement is a sign that something in the environment is significant.) Most motives involve some deficiency or lack, such as low blood sugar in the case of hunger or appetite. It is for this reason Maslow suggests the name *deficiency needs* for the lower needs, whether they are physiological, such as a requirement for food, or social such as a need for love and companionship. If Hebb's theory is correct, the need for excitement is no exception. Hebb points out that there is a center for excitement in the brain called *the reticular activating system*. This is a system of nerve ganglia, primarily in the brain stem, which has to be activated if an organism is to remain awake and alert. Hebb theorizes that there is an equilibrium level for stimulation of this system in the same way as there is an equilibrium level for stimulation of, for example, receptors for blood sugar. Like an appetite for food, which is created when the blood sugar level is low, the appetite for excitement is created when the arousal center is understimulated. Moreover, like a loss of appetite due to high blood sugar, a loss of appetite for excitement occurs when the arousal center is overstimulated. In other words, each person tries to maintain excitement at the equilibrium level.

There are, however, some flaws in this theory (at least as formulated above) which are immediately apparent. One is that the appetite for excitement is a function of the state of the organism. An organism does not always seek even a minimal degree of stimulus; for example, when trying to sleep. The requirement of sleep can overcome all appetites, even the craving for food. Moreover, the state of the organism determines what arouses and what does not. The smell of food, for example, is not of interest or importance to an animal with a full belly. It is *prima facie* the case that the behavior of seeking or avoiding anything depends, not on a single physiological parameter, but on the state of the organism *in toto* (Cp. Goldstein[1939]).

Another flaw in the theory is that the excitement that a stimulus arouses is not a direct function of its magnitude, but depends on the *nature* of the stimulus; that is to say, it is not so much the magnitude of a stimulus that causes arousal, excitement, and attention as its information content. The information content of a stimulus gives a cue to an organism of the importance or significance of the stimulus. (It is for this reason that Hebb says a stimulus has two functions: one, the "arousal function"; the other, the "cue function.") Moreover, all organisms, including humans, are highly selective—idiosyncratic even—in their interests. A toad is highly interested in a stimulus if it tells it a fly is near, but other organisms may attribute little importance to information of this type. Because of the human appetite for information and the desire to avoid boredom even bad news may be welcome. For example, people may be thrilled, at least momentarily, at the news of war, not because they are bloodthirsty, but because they have an appetite for—well—surprise.

Apter [1989] seizes on the point that whatever arouses or excites an interest (or is important) is not only idiosyncratic but fluctuates according to the state of the organism and, on this basis, he develops what he calls "reversal theory." According to this theory, a higher organism such as *homo sapiens* functions in one of two modes: one, a stimulus-seeking mode; the other, a stimulus-avoiding mode. Moreover, according to Apter's theory, human beings are constantly switching modes as circumstances change. A switch from one mode to another Apter calls a "reversal." Apter says that it is this tendency toward a reversal of mode that makes the problem of finding happiness or ultimate satisfaction in life practically insoluble.

The waking life of most people, is characterized by Sacks [1992] as "...that eager and anxious tension of anticipation, of intention, that drives us through life." That tension is most noticeable when people, in the midst of carrying out their intentions, are presented with some challenge or opportunity. According to Apter, some people thrive on tension and welcome challenges, while others are characteristically in a stimulus-avoiding mode. There is medical evidence to prove that those who welcome challenge enjoy life more and tend to live longer (Cecil [1985], Cousins [1979], Apter [1989, p.168]). The avoidance of stimulus and challenge is as much a sign of unhealthiness as the avoidance of sex by a neurotic.

Tension and intention are signs that something matters, that something is significant, that something is interesting, that something is *important.* According to the eminent physiologist, Goldstein [ibid.], in a normal individual, "those performances most important for the organism are always fulfilled." For reasons of this character Whitehead [1938] claims, "the sense of importance (or interest) is embedded in the very being of animal existence." So central in Whitehead's philosophy is the notion of importance that he not only says, "The generic aim of process is the attainment of importance...," but "Morality consists in the control of process in such a way as to maximize importance."

As noted in Chapter Two, Whitehead relates importance to interest, significance, perspective, unity, purpose, harmony, intensity, and vividness. There are many other factors that could be used as partial synonyms: nontriviality, utility, weight, relevance, power, beauty, and sacredness. "But [says Whitehead] `importance' is a fundamental notion not to be fully explained by any reference to a finite number of factors." Whitehead's assertion is a reasonable one; for "What is" questions are notoriously unfruitful. Nonetheless, if the notion of importance is to be used, we must characterize it in some way.

Whitehead, himself, makes some attempts to characterize importance. For example, he says, "One characterization of importance is that it is that aspect of feeling whereby a perspective is imposed on the universe of things felt." I regard this characterization as apt because of the role of importance in practical wisdom and since "perspective" is what practical wisdom attempts to develop.

Nonetheless, Whitehead's account of importance is finally inadequate for several reasons. Whitehead not only fails to give an adequate account of how interest and importance relate to purposes and ends, but he fails to note Apter's point, i.e. that what is interesting or important depends on the state of the organism. This brings us to two important questions: one, what are ends?; the other, how do ends, the state of the organism, and importance relate to each other?

Is it really the case that if life did not exist that there would be no end, purpose, or importance in the universe? One does not have to be a "believer" to think otherwise. For example Schopenhauer [1928 ed., p. 93] quotes Spinoza to the effect "that if a stone which has been projected through the air had consciousness it would believe it was moving of its own free will," and adds "the stone would be right." An ***end***, after all, is basically a tendency, and in this sense even inanimate objects have ends. There are two kinds of ends to be considered: one, ordinary ends or requirements which include not only the physiological and social needs, but the higher needs as well; the other, ends like the tendency of living things to adapt to the environment (or to struggle for existence), the tendency for organisms to avoid pain and seek pleasure, and, finally, the tendency (on the part of human beings) to try to make sense out of life. To distinguish the second category of tendencies from ordinary needs I call the latter "*meta-needs*."

John Dewey suggests somewhere that means tend to become ends. A century earlier than Dewey, Wilhelm von Humboldt [ibid., p.47] writes "what has at first been chosen for its utility, in general, becomes ultimately attractive in itself." It is the philosopher Spencer, however, who makes this principle his own. According to Spencer [1879, pp. 185-187], in the course of the evolution of a species each successful strategy (or means) in the service of any end tends to become an end in itself. Spencer does not give an explicit formulation of this principle, but it is clear that this is what he means. On page 487 (op. cit.) he asks: *May we not say, too, that...the use of each more indirect means in due order becomes an end...?* It should be noted that modern psychology has caught up with Spencer and von Humboldt in this regard. For example, Goldstein [1939] attributes to Robert S. Woodworth—the psychologist who coined the word "drive" for the motivating force behind needs—the observation: "The means to an end becomes an object of interest on its own account." Psychologists recognize some needs as innate (inborn) and others as "acquired." Spencer's principle is of use as an explanation for the origin of both kinds of needs or drives.

Consider first an acquired drive, say the drive that many of us have for making money. Making money is an obvious means to satisfying more basic—and most people would say, more "important"—needs. Drives of this character often seem excessive and even pathological when they go beyond what is necessary to satisfy primary needs. Goldstein regards all drives, except the urge toward what he calls "perfection," as the result of a failing in the organism (a dis-ease if you will). For example, hunger is the result of a lack of sustenance or insufficient glucose in the blood. The exact mechanism for the establishment of acquired drives (which range from the drive to make money to the drive toward alcoholism and drug addiction) is obscure, but the principle that acquired drives have their origin as means to more basic ends is clear. I turn next to consider how Spencer's principle applies to innate needs.

Before the use of Spencer's principle can get started, there has to exist at least one general tendency or end; for example, the need organisms have to adapt to their environment. Other ends, even the meta-need for seeking pleasure and avoiding pain, can then be explained as successful strategies (or means) in the service of that end.

As everyone knows, life maintains its existence through physiological mechanisms that provide for metabolism and growth, energy replenishment, defense, and reproduction. As we have already observed, these physiological mechanisms operate primarily by correcting a deficiency (or need) for substances like air, food, and water, and this explains the origin of the so-called "deficiency needs." Maslow notes that deficiency needs account for the physiological and social needs, and we have also noted that they account, at least partially, for those higher needs like beauty and the need for exploration that depend on the need for excitement. The higher needs, however, are remote from the struggle for existence and hence their origin (on this account) still remains to be explained.

Spencer notes that in higher animals, with their well-developed nervous systems, there is one master strategy in the struggle for existence: that of avoiding pain and gaining pleasure. But, as Spencer points out, this strategy tends to backfire; for, as everyone knows, that which gives pleasure can kill you (drugs, for example), and that which gives pain can save you (a rabies shot, for example). As a consequence, in higher animals there has evolved a strategy for avoiding the dangers of yielding to unbridled hedonism. I refer here to the strategy of taking thought. True to form, in some higher animals (especially philosophers) the strategy of taking thought, which begins as a mere means for avoiding danger, becomes an end in itself.

All ends have not been esteemed equally: physiological requirements such as sex, which are shared by nearly all animals, have been despised as vulgar; other ends have been and continue to be more highly regarded, especially those ends that are furthest from a direct struggle for existence (for example, the needs for art, exploration, and religion). It may be noted, furthermore, that a similar remark applies to human occupations; that is to say, the further a human occupation is away from a direct connection to supplying the primary requirements of the human body, such as food and shelter, the more highly it is esteemed. For example, labor such as plowing fields, tending machines, and taking care of children are prized the least, while labor remote from production, such as politics, war, management, and finance are the most highly rewarded. There may be, as Hebb and the philosophers say, a need for work, but examples of this kind tend to show that it is not the utility of the work that causes people to esteem it. No doubt this is what causes Freud [1930] to remark that as a path to happiness, work is not highly prized. Hebb [1953], himself, says "I do not say that men will always yearn to do *useful* work [emphasis in original]." Though prizing may—as the utilitarians say—have something to do with utility, the connection is apparently a tenuous one. I conclude the following: All human ends have something to do with utility, but *the healthiest, most robust, and most successful human beings live in such a way that what is most important to them has only a remote and indirect connection with utility.* Let us call those ends that have only a remote and indirect connection with utility ***nonutilitarian ends***. Some nonutilitarian ends are creative and some are destructive. Those that lead toward politics, war, management, and finance are often destructive rather than creative. As already indicated in the first two chapters, the ends that are important to a person vary not only with a person's momentary state (such as hunger) but in accordance with his or her stage of development. The researches of Maslow and others show that what are called the "*higher ends*" (or needs) are those *nonutilitarian* ends that are most important to the healthiest, most robust, and most *creative* human beings.

One conclusion that emerges from Spencer's principle is that the common conception that ends are more important than means is not true. If ends were truly more important than means, the higher ends would be of minor importance. According to Spencer's principle, all ends, other than the end of survival, have arisen as means in the struggle for existence. Unless we are prepared to say that survival is the most important end, we must conclude that means are sometimes as important, or perhaps more important, than ends. Take the end of the pursuit of knowledge, of which philosophers are fond, as an example. There is no doubt that to our ancestors knowledge was only important as a means toward the end of survival. As human beings have grown and developed, a reversal has taken place and what was once a mere means has now taken over as one of the most important ends, if not *the* most important. It turns out therefore that means are not *mere* means, but under the right conditions can become ends that eclipse in importance most, if not every, other end.

Spencer's principle is of extreme importance, and it can be exploited not only in the problem of the ultimate source of drives or requirements, but in the problem of ordering ends (or at least higher ends) in a hierarchy. I turn next to consider the latter problem.

On what principle or principles can we order ends and what is meant by lower and higher? It is apparent that in Maslow's hierarchy the needs are ordered according to the order of their appearance either in the growth and development of the individual or—what is perhaps the same thing—in the evolutionary development of the human species. On the principle that "ontogeny repeats phylogeny"—i.e. the biological principle that individual development repeats (roughly) the development of the species—one would expect the development of needs in the individual to repeat the history of the development of needs in human ancestry. Moreover, in accordance with another principle, namely that "higher" means "later," this fully justifies the ordering given by Maslow of needs into lower, intermediate, and higher since, among all higher animals, physiological needs appear before social and psychological needs, and higher needs appear, at least in their fully developed form, only in human beings. As we indicated earlier, calling them the "higher needs" reflects a human (and therefore possibly fallible) judgment as to their importance. What remains to be discussed is how to order the various higher needs themselves. It is here that the use of Spencer's principle is critical.

The higher needs, as listed in the previous chapters, are as follows: one, beauty; second, creation and discovery; third, autonomy; fourth, "identity"; and, finally, "transcendence." (The completeness of this list is considered in Chapter Six.) Some have disputed that ends of this kind could have arisen as means in the struggle for existence. Some have even singled out beauty in this regard. As Darwin [1872] noted, however, if a sense of beauty had no other use, it would be of use in selecting a mate whose offspring were likely to be healthy since there is an obvious relation between looks and health. Hence, according to Spencer's principle, a need for beauty arises in many animals because of the obvious effectiveness of the use of a sense of beauty as a means for making it more likely that the next generation has a higher chance of survival. The exact mechanism by which something of use to an organism (like that of a sense of beauty or even the ability to feel pleasure and pain) becomes incorporated into the genetic structure of animals and plants is still a mystery, but this need not concern us here. This is a matter to be settled by researchers in the field.

Instead of discussing the higher ends, as listed above, it is more convenient to discuss the tendencies or propensities that are characteristic of the person who tends to fulfill the corresponding end. These tendencies may be listed as follows: *sensitivity, sapience, self-reliance, dedication,* and *transcendence.* I call these the ***higher tendencies***. I shall not, however, distinguish sharply between the need and the corresponding tendency. *We have already stated, that a need is a species of end and* ***an end is basically a tendency****; therefore, except where there is a reason to distinguish them, we shall treat the names of* *higher needs and the corresponding tendencies as virtual synonyms.* Since in the previous chapter growth is defined as the tendency of a human being to become less occupied with the lower needs and more and more directed by the higher needs, it follows that—insofar as the list of higher ends is complete—the degree to which a person has undergone growth can be measured by the degree to which he or she is characterized by the foregoing tendencies. Accordingly, corresponding to these tendencies are ***norms*** of behavior characteristic of the highly developed person. (See Chapters Six and Eleven.)

Sensitivity (to beauty) is placed lowest in the hierarchy, because the tendency to appreciate beauty—in contrast to the other higher tendencies—is a tendency that is not solely characteristic of human beings (or of other higher animals), but occurs in some of the lower animals. In support of this point, one may quote Darwin [1874]: "When we behold a male bird elaborately displaying his graceful plumes or splendid colors before the female,...,it is impossible to doubt that she admires the beauty of her male partner." Creation and discovery, the item immediately above beauty in the hierarchy of higher needs, is related to *sapience*, which we defined, in the last section as having the tendency to be clever, curious, and creative. As we have noted, this tendency is the defining characteristic of *homo sapiens* and hence is a relatively recent development in the animal kingdom; therefore, on this basis alone, I would tend to place creation and discovery after beauty in the hierarchy of human ends. Another reason for saying beauty, as an end, is prior to the need for creation and discovery is that sapience is probably of more use in the service of beauty than beauty is of use in the service of sapience; in other words, I suggest that creation and discovery is more often a means to beauty than beauty is a means to creation and discovery. (That is not to say, however, that a sense of beauty *may* not be of importance as an aid to the attainment of wisdom—or sapience. For further discussion of this point see Chapters Six and Seven.) On the principle that ends are prior to means, beauty (as the end) is prior to sapience.

Though sapience is of use, especially to an artist, in the creation and pursuit of beauty, there is little doubt that sapience arose in higher primates primarily as a strategy or means in the struggle for existence. According to Darwin [1874, p. 80], "...natural selection...together with the inherited effects of habit, would, under favorable conditions, have sufficed to raise man to his present high position in the organic scale." But according to Spencer's principle, any means, especially one whose cultivation is a the source of not only significance but pleasure, tends to become an end in itself.

Self-reliance, the next tendency on our list, does not require a tendency toward creation and discovery. One can be a rebel without a cause. Moreover, self-reliance is not directly of use in the struggle for existence, though it is an indispensable means to creation and discovery. One cannot imagine a truly creative artist or explorer who is not independent and self-reliant. A creative or explorative person may accept help, but not at the price of giving up autonomy. Through a lack of imagination and understanding, other people are, almost without exception, a drag on innovation and discovery. H.G.Wells [1922], in his novel *MEN LIKE GODS*, describes the scoffing on the part of others that innovators have to face. On returning from a visit to another world the narrator says to himself:

> "If I could tell them what I have seen!...
> "Suppose all men could have this vision of Utopia....
> ....
> "They would not believe it if I told them. No....
> "They would bray like asses at me and bark like a dog!...They will
> "have no other world but their own. It hurts them to think of any
> "world but their own...."

Since self-reliance (or autonomy) is an indispensable means to the fullest development of sapience, it will follow that sapience is prior to self-reliance provided we can argue successfully that sapience, in and of itself, is not a very successful strategy in the achievement of autonomy. Of course, sapience is useful in producing *any* goal, but the question remains whether curiosity, cleverness, and creativity are very successful as means for reaching autonomy. I think the answer must be no unless the creativity is at a level which allows a person to achieve an identity, to BE somebody. Many women are highly developed in the traits of curiosity, creativity, and cleverness, but fail to achieve autonomy. I conclude that autonomy is an indispensable means to sapience, but unless sapience is very highly developed it is often not a powerful means for achieving autonomy. Since the argument is inconclusive, however, we need some other argument before we can conclude that sapience is prior to self-reliance.

Self-reliance is probably a rather late arrival on the scene of human development. Early humans (though already sapient) were not very self-reliant since as social animals they depended on others for direction and protection. It is only when human beings have the leisure and confidence to become solitary explorers, thinkers, and creators that self-reliance becomes important. Self-reliance develops when childhood ends and people grow in sapience. As people grow in sapience and strike out in new paths of their own, elders, priests, and even tradition can no longer be depended on for guidance. I conclude that sapience is prior to self-reliance. We return to the relation of autonomy to creation and discovery in Chapter Six, Subsection 3.3, where both the pitfalls and benefits of self-reliance are explored. I turn now to consider the position in the hierarchy of the next item, that of the need for *identity*.

What is "identity"? As used in this work, the need for identity is the need "to be somebody." This is not as egotistic as it sounds. What is meant here is not a need for self-aggrandizement, but a need to be dedicated to something outside of oneself, a "calling," as it were. Following Maslow one may call people of this kind "problem-centered." These are people who, according to Maslow [ibid.]:

> ...are without exception, involved in a cause outside their own skin, in something outside themselves. They are devoted, working on something which is very precious to them—some calling or vocation in the old sense, the priestly sense.

Persons not dedicated to a cause outside of their own skin, do not, at least in the present sense, have an authentic identity. Now, unless I am very much mistaken, psychological research will back up the assertion that unless for some reason one simply rebels against authority, it is almost impossible to achieve autonomy without a strong identity. That is to say, if one is not to be directed by others, one *must* develop a sense of identity and purpose in life (Cf. May [1953.]). Moreover, while many people, especially the young, yearn for autonomy simply out of rebellion, this motive for autonomy is neither socially acceptable nor likely to be successful. Finally, since self-reliance does not in itself lead to identity, while dedication to a calling as an artist, scientist, or statesman (or whatever) leads almost inevitably to autonomy, I conclude that self-reliance is prior to dedication.

There remains one more need in the hierarchy of higher needs to be considered and that is the human requirement and propensity for *transcendence*. I classify transcendence as the highest of human needs since it lies at the apex of human development, both in evolutionary terms and in terms of individual growth. We need not appeal to Spencer's principle here; rather we appeal to the fact that transcendence, at least in its fully developed form, appears only in highly developed individuals, of which we may take Socrates as an exemplary specimen. Transcendence is what the Stoics, who took Socrates as their patron saint, tried to achieve. Transcendence has two primary uses: one, it promotes sapience; the other, it serves the meta-need for making sense out of life. One might wonder, however, whether identity (or dedication) is prior to transcendence or *vice versa*; for .having an identity like a monk, scientist, or a poet may be a means toward achieving transcendence, but other identities—being a banker, for example—might lead one in the opposite direction. Nonetheless, since having a wide-horizon (in Russell's sense, see Chapter One) is helpful in furthering an impersonal cause, it follows that transcendence can be used as a means of furthering the impersonal ends with which one is identified. Moreover, transcendence leads to religion and religion is thought to be useful in all walks of life. This argument may not be entirely conclusive, but I conclude tentatively that identity (the end) is prior to transcendence (the means). For further clarification the reader is referred to the discussion of dedication in Chapter Six. This completes our examination of the order of the higher needs.

Before we leave the subject of higher needs, there is one topic that needs further consideration. This is the topic of ***meta-need***. The concept of meta-need corresponds to what Maslow [1971] calls "meta-motivation." Maslow's example of meta-motivation is the tendency among human beings to move toward growth or what he calls "self-actualization." I find this idea of such a tendency—though I defend something like it—rather obscure. There are, however, less nebulous examples of tendencies in human beings that I regard as a sign of a meta-need. We have already mentioned several of these including the tendency of living things to adapt to the environment (or to struggle for existence), the tendency of organisms to avoid pain and seek pleasure, and, finally, a tendency of a very different character—the tendency (among human beings) to try to make sense out of life. The tendency to try to make sense out of life is driven by two different motives: one philosophical and the other religious. The religious motive is not easily understood, but the philosophical motive is clear: it is the desire to understand. The religious motive is explored in Chapter Seven: *Beauty, Mysticism, and Religion*. From the point of view of this work, the most important meta-need is the need to make sense out of life.

There is nothing more clearly true than that human beings can never rest content with simply satisfying basic needs such as food, safety, love, and power; for even in a paradise in which human beings can satisfy their every want, there remains one question: what does it all mean? The meta-need to *understand existence*, even more than other meta-needs, including the need not only to exist, but to at least avoid suffering, is, I think, the deepest need of human existence. Nature cruelly implants a desperate desire for existence, but ultimately frustrates that desire. I conclude that humanity's most pressing (i.e. ***the most important***) meta-need, in Maslow's sense, is the need to make sense out of life.

The importance of this need only arises when the lower and intermediate needs are no longer of overweening concern. Those who are engaged in a struggle to live do not worry about the "meaning of life." As Goldstein [ibid.] reminds us, in a normal individual, anything (not of immediate importance) can be neglected if he or she "has something important to do, the neglect of which would bring the whole [person] into danger." I am convinced, however, that the only way for people *who have satisfied their basic needs* to maximize importance and to bring sense into life is through the higher needs. It is for this reason that the focus of Chapter Six (*What Are We To Do With Our Lives?*) is on the satisfaction of the higher needs.

6 Perspectives in practical philosophy

The need for more than one perspective in practical philosophy is becoming more and more realized. For example, Hampshire [1993] writes:

> A complete human being has two faces, one a communicative consensus-seeking, politically active, reasonable face, in the sense of "reasonable" that is opposed to fanatical, the other a private and autonomous, perhaps detached and secretive, uncompromising face of a person pursuing his own distinctive good, perhaps guided in this by a comprehensive morality.

The present chapter may be regarded as a prolegomenon to the construction of two normative systems: one, from the political perspective of society at large; the other, from the personal perspective of the individual. The political perspective is favored by those who argue from the standpoint of "human rights" and "justice"; the personal perspective is favored by those, like Spinoza and Russell, who wish to base ethics on individual development. Each of these perspectives can suffer from partisanship—the political perspective, when it favors one class over another, the personal perspective, when it is narrowly selfish.

Ethics, of course, if it is to be effective in resolving conflicts of interest, must rise above mere partisanship. A question which is relevant to this chapter is whether there are springs of human action which lead one to be hopeful that human beings on this planet are capable of throwing off their narrow interests and living together in peace and harmony. Hobbes, who argues from the political perspective, believes that the only motive of this kind is the need for safety. I, on the other hand, believe there are powerful motives belonging to the personal perspective that can be effective in bringing human beings into unity. In particular, I believe that the most highly developed human beings are not only those in whom such motives are the strongest, but those whose lives are the most enviable. I have already alluded to some of the reasons for this belief in Chapter One, but the matter is pursued further in Chapter Six.

In harmony with a remark of Whitehead, already quoted—"Morality consists in the control of process in such a way as to maximize importance"—we may, roughly speaking, regard ethics as an attempt to decide what is important. (Recall that we stated in Chapter One that an ethical paradigm is appropriate to the extent it promotes what is important to human beings.) As we have seen, what is important, both from the perspective of the individual and from the perspective of society, cannot be separated from the springs of human action and it follows that any attempt, such as that of Kant, to build a practical philosophy on reason alone is doomed to failure.

So ends our prolegomenon to the construction of the two normative systems, one for political organization and the other for individual life. In the next chapter, we turn to the construction of the first of these.

Chapter Four

An Inquiry into the Principles of Social Construction

1 Introduction

When nature is allowed to shape society the design tends, as we have seen in the last chapter, to be that of a hierarchy. In this chapter we consider society as constructed by art. By and large, a conservative favors a society shaped by nature, while a liberal favors a society shaped by art (Cf. Burke [1790], Paine [1791]).

One can question, of course, whether a society as constructed by art is an improvement on society as constructed by nature. After all, it is by interfering with nature that people have polluted the air and sea, overpopulated the earth, destroyed much of the earth's beauty and natural resources, and brought humanity to the brink of nuclear annihilation. Indeed, it seems that every attempt to apply intelligence to human affairs—instead of finally laying human troubles to rest—has added fresh difficulties. One might conclude that intelligence leads to mistakes, and that therefore human beings are better served if they live by instinct, tradition, or religion. It is for reasons similar to this that Kant [1785] thinks that instinct is a better guide to happiness than reason. (I return to these matters in Chapters Seven and Eight.) But it is the ability to make mistakes, and to learn from these mistakes, that is the glory of humankind. Though it is clear that the old models for society are not working, it is up to us to construct (by art) a new one better fitted to solve the problems of the modern age.

At the end of Chapter Two we noted that the art of social construction must be approached with a certain amount of trepidation. Though there may be sound reasons for changing the paradigm by which society is governed, change is not without risk. Whitehead [1927], who calls the paradigm by which a society lives its *symbolic code*, sums up the situation as follows:

> It is the first step in sociological wisdom, to recognize that the major advances in civilization are processes that all but wreck the societies in which they occur:—like unto an arrow in the hand of a child. The art of free society consists first in the maintenance of the symbolic code; and secondly in fearlessness of revision, to secure that the code serves those purposes which satisfy an enlightened reason. Those societies which cannot combine reverence to their symbols with freedom of revision, must ultimately decay either from anarchy, or from the slow atrophy of a life stifled by useless shadows.

The term "society" seems clear enough in the present context; nonetheless, some readers may want a definition. By ***society*** I mean any group of individuals having organized systems of governance, production, etc. I use the terms "state" and "community" as synonyms for "society."

I think nearly everyone agrees that the purpose, if any, of a society or a state is to enable individuals to live better than they could if they lived alone. Socrates, in *THE REPUBLIC* (Book II), says that a *polis* arises out of the needs of its citizens. Aristotle, in *THE POLITICS* (Book III, ch.9), says that the purpose of the state is "to enable all, in their households and their kinships, to live *well*, meaning by that a full and satisfying life [emphasis in original]."

Accordingly, I think we may conclude that the *purpose* of a society is to enable people to fulfill their needs in a way that would not be possible if people lived alone or in smaller groups and that a society is *functional* to the extent it fulfills this purpose. I suggest, as a preliminary definition, that the art of social construction may be defined as the art whose aim is to design a functional society. It will be objected of course that functionality is not the only thing required: one wants, for example, not only a functional society but a "good society." There is no problem, however, if it turns out that a fully functional society is "the best of all societies." I return to this question in Section 3, but I assume, tentatively at least, that the criterion for what is wanted or required in a society is functionality. (This, by the way, is the same criterion that Aristotle applies to the individual.)

It has been said—for example, by Plato—that society can be compared to a human body. To some extent this is true; for just like the human body, if a society is to function well, it needs, in addition to having its subsystems function well, not only to avoid self-destruction but to be stable. But if the idea that a society is like a human organism is taken too far, it supplies a powerful argument—like that given by Plato himself—for an autocratic state. When the subsystems of society carry out the will of a small portion of society it may be called autocratic, but when they carry out what Rousseau called the "general will" of society it may be called democratic. The natural question is: which is better? Since we are assuming tentatively that better means "more functional," the question at this point is: which is more functional? The problem before us therefore is to show that democracy is more functional than autocracy. I regard this as ***the fundamental problem of the art of social construction.***

My plan of attack on this problem is as follows: first, review history with the end of showing that the history of states may be viewed rightly as largely an attempt of people to escape from tyranny and establish democracy (Cp. Marx and Engels [1848, p.9]); then, after outlining the basic principles of democracy, argue that democracy is more functional than autocracy.

2 An Argument for Democracy

Throughout history, the norm of autocratic government has been subject to attack. Thus we find an implicit criticism of autocratic government in the ancient Chinese *BOOK OF HISTORY*, dating from the reign of Shun (B.C.E. 2255-2198), where it says in part: "Heaven hears and sees through our people..." In Europe the first crack in autocracy appears in the fall of Rome, and the consequent friction between the emerging Barbarian rulers in the West with the established Church in Rome. The emerging battle for ascendancy between secular and religious authority weakened both. Hitherto, emperors are all powerful, uniting secular power and religious power into one, and democratic revolt could hardly prevail against them. Eventually religious controversy, combined with the rising power of capitalism, bears fruit in the seventeenth century, when *THE ENGLISH REVOLUTION* establishes the precedent that the people may revolt against an oppressive leadership. After the fact, Locke [1690] announces the doctrine that revolt of this kind is admissible. The largely English immigrants to North America accept this doctrine and, in *THE DECLARATION OF INDEPENDENCE* (1776), criticize the colonial government of King George III on the ground that it neither provided adequately for the "life, liberty, and the pursuit of happiness" of the people nor does it have the consent of the governed.

I think it can be fairly concluded that what the founders of the United States of America wanted was a normative system for their society that satisfies two norms:

(1) *It is required that it shall be the case that the people live in a society that is functional*

(2) *It is required that it shall be the case that the principles on which society is organized have the consent of the people*

Let us define a ***democracy*** as any society in which these two norms are binding.

(Note: The two statements (1) and (2) are **norms** *in the technical sense defined in Chapter Nine, PART TWO.)*

It has to be assumed here that "consent" means genuine consent and not what Chomsky [1987] calls "manufactured consent." The requirement of the consent of the governed was something new in the modern world although it—or something approaching it—was accepted in certain ancient city states. It was hardly practiced, however, in any place except by primitive tribes in places not yet touched by the "blessings of civilization." As I argue later, these two requirements imply that all mature adults, of sound mind, are to be given an equal voice in government. Let us call the first norm, *the principle of functionality* and the second, *the principle of the consent of the governed.* According to the definition of democracy, already given, these two norms are ***the fundamental principles of democracy.***

Because of the rise in recent times of Western-style democracies, history—since the heyday of the Roman Empire and the rise of the nation state in Western Europe—may be viewed as a struggle for democracy. Though democracy has yet to triumph, a definite trend toward it is noticeable. We need only cite examples like the French revolution, the American "rebellion," and recent events in Eastern Europe. This struggle, which Spencer [1892] calls "the man versus the state," is, by and large, the struggle for equality; for when democratic principles are fully implemented, people are blind to status and hence are—in Whitman's words—"no respecter of persons." For example, nowadays, murder or assault is no longer tolerated even if the victim is of low status—(unless of course it is done by institutions that are above the law such as the CIA, KGB, FBI, DEA, ATF, or MI5).

According to some authorities there is evidence to show that before the growth of large states, democracy of one sort or another prevailed. It is claimed, for example, that the chief of a tribe had only the power of personal persuasion and hardly made a move without discussing it beforehand with his advisers (Cf. Fromm [1973].) Furthermore, it is claimed that leaders, living in close contact with their followers, tended to do the popular thing and that leaders were soon challenged if they did not do so. The rank and file individual, so it is said, was not greatly oppressed by the leadership. Nonetheless it is admitted that deviants, with the support of the tribe, were apt to be treated harshly, without due process of law (Cf. Stefansson [1939]). In a traditional tribe, Socrates, for example, would not have lasted as long as he did!

When larger groups form, however, several factors combine to produce an all powerful leadership. In such groups, society can afford specialization and the natural tendency toward hierarchy is exacerbated with the result that society tends to be divided into upper and lower classes. Moreover, the introduction of cultural advantages, open only to the upper class, such as property, education, manners, speech, and clothing, increases the gulf between persons of high rank and those at the bottom. What is more, a leader no longer depends solely on commanding love and respect, but with riches to bestow or deny, surrounded by the trappings of royalty, sanctified by an obsequious priesthood, protected by a powerful guard, and supported by loyal troops, he or she can, if necessary, rule by terror. Rome began as a republic and ended as an autocracy.

In autocratic states, the purpose of the state becomes the purpose of the leader. Recall again the remark attributed to Louis XIV: *L'Etat c'est moi*! History tends to show that the main purpose of the leadership of an autocratic state, no matter how powerful, is further accumulation of power and glory. Typically, Alexander wept when there were no more worlds to conquer. The pursuit of glory in all states proceeds in two ways, however: one, through raw power; the other, through cultural development. I intend to argue that it is the latter that is the Achilles heel of autocracy.

We have already noted some of the historical developments that have contributed to the rise of democratic movements. There is much more to said about the matter, however. In the last chapter we noted that people tend to worship as a god the person at the top of the pecking order. Hume [1977] writes:

> Nothing appears more surprising to those who consider human

> affairs with a philosophical eye, than the easiness with which the many are governed by the few; and the implicit submission, with which men resign their own sentiments and passions to those of their rulers.

Hume concludes that people are governed by *opinion* rather than *force.* It is this that makes philosophy (and cultural development in general) a force to be reckoned with in human affairs. Human beings are governed, not by philosopher kings, but by some dead philosopher who has been dead for ages. Consider, for example, what happened in ancient Rome owing to the teachings of St. Paul. Nietzsche [1886] attributes the rise of democracy to a "transvaluation of values" arising out of Christianity (which he describes as substituting a slave morality for the morality of the master), but this cannot be the whole story since autocratic states have flourished during most of the Christian era. In the latter part of the Christian era, however, there is a development of Christianity which certainly contributed to such a transformation, i.e. the Reformation and its aftermath. Once Roman Catholicism is no longer the only option, Christianity breaks into opposing factions, each certain of the truth, and the sixteenth and seventeenth centuries become a scene of war, disease, and slaughter. Sickening of the slaughter, people begin to consider the safety afforded by law and order more important than religious belief, and this encourages the rise of the sovereign state, the rule of law, and religious toleration. Another important factor is the rise of technology, which makes the state less dependent on slave labor and on mercenaries, and more dependent on the support of the people.

For example, the importation from China of the technology for making gun powder has profound social implications. Arms are called equalizers: all things being equal, an armed woman has a strength equal to that of any man. Another factor is the invention of money and capital, which takes power away from rulers and puts in it in the hands of traders. George Bernard Shaw jokingly suggests in his play *MAJOR BARBARA* that the liberating powers in modern civilization are money and gunpowder!

A technically advanced society depends on education. And as people become more educated and sophisticated, they ultimately become so powerful—and so confident of their power—that they proceed to take over the management of their own affairs. Moreover, given an armed population with some sophistication, the power of the leadership depends either on draconian methods—which have tended to become more and more futile—or on the consent of the governed.

But whatever reasons there be for the rise of democracy, one might still argue that other forms of government are preferable. To complete this phase of the argument, however, I turn now to thesis that, at least as far as modern civilization is concerned, democracy is more functional than the alternative.

John Stuart Mill [1859] in his work *ON LIBERTY* argues persuasively that society functions best when there is a free competition of ideas. The argument is that despotisms are not only not as powerful as democracies, but they do not provide as well for the people. As an example, the Third Reich failed to be functional, in part, because it failed to provide an environment in which culture, particularly the arts and sciences, flourish and, in particular, drove most of its best scientists out of Germany. According to Tocqueville [1862], when state power is concentrated in few hands this tends to impoverish the people, not so much because of excessive taxation, but by tending to stifle the people's potential for development. Tocqueville writes:

> It is certain that despotism ruins individuals by preventing them from producing wealth much more than by depriving them of what they have already produced; it dries up the source of riches, while it usually respects acquired property. Freedom, on the contrary, produces far more goods than it destroys; and the nations which are favored by free institutions invariably find their resources increase even more rapidly than their taxes.

An autocratic state fears the innovators of society, and therefore state power is inimical to cultural development and creativity. Neither ancient Rome nor traditional China (in modern times) were very innovative, at least when compared to ancient Greece. Galois, one of the greatest geniuses of all time, was killed at the age of twenty probably because he was a revolutionary in politics as well as in mathematics. Nowadays, Chomsky, the great linguist, is hardly popular with the establishment and cannot even get a letter published in *THE NEW YORK TIMES*. A despotism is in a double bind: first, since the glory and perhaps the very existence of a state depends on the achievements of its people, it must, through education, develop the potential of the people; but then, when they develop themselves, and increase in learning and sophistication, the people begin to ask questions that make them a threat to the leadership. In the old days, dissidents who were not killed were either banished or they themselves escaped into exile. As the world fills up, there are fewer places for dissidents to move and begin anew, and therefore the state has more dissidents to contend with. As a consequence of these factors, the pressure for individual freedom and "justice" begins to build. We see this happening in China today where the leadership is attempting to foster economic development without democracy.

Though a society cannot solve all the problems of people, it can facilitate their solution by providing an environment in which people can not only deal with their own problems successfully but lend a hand in helping others. Such an environment, if I am not mistaken, is one in which there is no underclass, slavery or serfdom is unknown, and people of all kinds—whether male or female, gay or straight, black or white—are encouraged to develop themselves in an atmosphere of freedom. In a thoroughly functional society, the intrusiveness of government into the lives of private citizens is reduced to the minimum necessary for carrying out its proper activities.

One condition, already mentioned, for a functional society is that it be stable. No state can be formed and be stable unless there is some consensus on its constitution, and hence it follows that no society can be functional unless somehow it obtains the consent of the governed. But what makes people consent to laws and institutions? There are basically two ways to make people consent to their lot: one, enforced consent as in an autocracy; the other, consent of the kind obtainable by improving the lot of the people. I think we may say that history shows that in the former case the people tend to revolt; hence the only viable option is the latter. As Chomsky argues, it is possible for a ruling elite to manufacture consent, but I do not think this can last in the long run.

It is obvious that collective enterprises cannot be organized and maintained (or be stable) except through unity and cooperation. Moreover, our argument tends to show that, under modern conditions, the requisite unity and cooperation can be obtained only if rulers are able to advance the interests of the common people and obtain the unmanipulated consent of the governed. I conclude that for a society to be functional in this day and age it must not violate the principle of the consent of the governed. It follows that the two principles of democracy are not independent. More importantly, I conclude that autocratic societies—that is to say, societies that violate the principle of the consent of the governed—are, properly speaking, dysfunctional and therefore the principle of functionality favors democracy. But can a society be designed solely on the basis of functionality? This is the subject of the next section.

3 Completeness

Utilitarians claim that no normative principle other than *the principle of utility* is needed in establishing the norms of society (Cf. Mill [1871]). In my view, the principle of utility is merely a form of the principle of functionality since any society that suffers unhappiness unnecessarily is *ipso facto* dysfunctional. In this section I intend to show that we can rely solely on the principle of functionality only if the principle of functionality is conceived rather broadly. Moreover, as I intend to show, there are cases in which even under a broadened conception the principle does not give a clear answer. We must decide what to do in cases of this kind.

It is clear that people cannot develop what Russell [1916b] calls "the excellence that is within them" without society; for people do not develop in a vacuum and it is difficult or impossible for human beings to develop their maximum potential unless society itself nurtures them. As Hillary Clinton [1996] says, "It takes a village," but she might also have added that advanced individuals appear only in an advanced village. Plato developed in ancient Athens only because Athens provided him with the education and surrounding culture that made his accomplishments possible. Without the scientific culture of Europe, an Einstein could not have been developed. To be fully functional, therefore, a society cannot simply cater to the basic needs of its citizens; on the contrary, to be fully functional a society must promote the growth of its citizens. Moreover, it is not clear that crass utilitarian considerations totally justify some of the activities that are imperative for the growth and development of society and its members. When new things, like space exploration, which require, for their success, a large expenditure of energy and resources, are proposed they are often opposed on utilitarian grounds. But the things that are most significant in human life sometimes appear to be totally useless. As we have already had occasion to note in the last three chapters, human beings have in their hearts the need to go beyond the mundane. According to Wells [1913],

> All this world is heavy with the promise of greater things, and a day will come, one day in the unending succession of days, when beings, beings who are now latent in our thoughts and hidden in our loins, shall stand upon this earth as one stands on a footstool, and shall laugh and reach their hands amid the stars.

Even the most primitive society must tend to satisfy the basic needs of human beings—the physiological and social needs. Moreover, a primitive society may develop powerful individuals such as the great conquerors and great prophets—Genghis Khan, Alexander, Buddha, Christ—but greatness in an advanced society is not measured by power. On the contrary, greatness in an advanced society is measured by the extent to which the society not only develops democratic institutions but develops thinkers and artists such as the great geniuses—Plato, Shakespeare, Mozart, Einstein. The criterion I suggest as a measure of the greatness of an advanced society is the extent to which the society tends to foster human development and to satisfy the higher needs of human beings. Recall that the higher needs are the needs for beauty, creation and discovery, autonomy, identity, and transcendence. Let us agree that the most advanced societies are those that tend to fulfill these needs to the fullest extent.

The principle that the proper function of society is to foster human development is clearly enunciated by other authors, the earliest of whom is perhaps Wilhelm von Humboldt in his work *THE SPHERE AND DUTIES OF THE STATE.* According to Humboldt [1854, p. 87], the state is not itself an end, "but is only a means toward human development." I conclude that in order that functionality be considered an adequate design principle it must be interpreted in such a way that it commits society to the development of higher things including but not limited to art and science. Accordingly, I propose that we measure the degree of functionality of a society by the degree to which it tends to satisfy the higher needs of its members.

As indicated in previous chapters, especially Chapter One, among the higher needs of human beings is the need for morality. According to Kant [1785], morality requires us to aim at a society in which all the members of society are treated as ends in themselves and nobody is treated as a mere means. If people are to be treated as ends in themselves and not merely as means it would seem that all people, regardless of sex, race, status, or national origin, must be given equal treatment. Consequently, all things being equal, morality upholds the aim of an egalitarian society. Yet some questions remain. In particular, one must ask to what extent egalitarianism is compatible with functionality?

There is another way of looking at the moral perspective that is also due to Kant: according to Kant a rule is moral only if it can be applied universally. There may be a certain amount of justice in this view—people do object to a lack of universality in the application of rules on the ground that "justice is blind"—but unless a rule is very complex, certain exceptions are inevitable. To take a simple example, one does not treat the old and infirm the same as the young and healthy. Nonetheless, all things being equal, one may agree with Kant that there is something irrational in favoring one class of people—or an individual—over another. But extreme egalitarianism may be neither functional nor feasible. For example, for the Soviet Communists, the goal of universality did not turn out to be feasible and their society was not functional. For reasons such as these, I believe that a lack of arbitrariness, rather than universality, is a better test of rationality. Let us agree therefore that a society is designed on "rational" principles provided the guidelines (or design principles) under which it is constructed not only include the criteria of feasibility and functionality, but rule out any trace of arbitrariness and special pleading.

How does one test for arbitrariness and special pleading? I believe there is no better test of a law than having a referendum on the socio-political results that flow from it; for however imperfect the common people may be as rational beings, they are more able than politicians—or even philosophers, who often engage in special pleading—to determine arbitrariness or irrationality in any scheme. An arbitrary system is likely to lose, in the useful Chinese phrase, its "mandate from heaven." I conclude that a socio-political system that depends on the consent of the people is the least likely, in the long run, to suffer from irrationality.

From this discussion it is clear that not only does morality favor egalitarianism but, insofar as egalitarianism is compatible with feasibility and functionality, so does rationality. But the extent to which a functional society can go toward egalitarianism is a question that at this point remains unsettled. We return to this question when we consider in further detail the character of subsystems of society.

Besides functionality, there are many other characteristics of a social system that may be proposed as the aim of social construction. For example, some authors want a "just society," others a "moral society," and still others a "great society." All these one may suppose are worthy aims. We have already discussed morality and greatness. As for justice—for reasons already given in Chapter Two—I do not believe "justice" is a proper criterion for developing a society; nonetheless, I return to questions of justice in the next section.

We have concluded that if functionality is to work as a criterion for the excellence of society it must be broadly conceived. But it is not necessarily the case that functionality, however broadly conceived, gives a clear direction in every area. One may ask, what then? That is to say, what do we do when functionality does not give us a clear answer on the wisdom of imposing a law? For the purpose of discussion assume M is a norm and suppose it cannot be proved to the satisfaction of all (or nearly all) that functionality requires either M is the law or some norm contrary to M is the law; then I claim it is not reasonable to impose M as a law. Many possible examples of M come to mind, two of which are: one, "Under certain circumstances (as specified by law) it is admissible that it shall be the case that a person convicted of a crime is put to death"; the other, "It is required that it shall be the case that there are no abortions." To impose a law when it is not clear that functionality requires a law is, in my opinion, an irrational intrusion of state power over the lives of individuals. Functionality actually requires that the number of laws be kept to an absolute minimum; for an excess in the number of laws is not only inimical to happiness, but places a great restraint on initiative, and as a result the economy suffers, people become frustrated, angry at the government, and outbreaks of violence may occur. There is a principle in metaphysics, called Occam's razor, which, when paraphrased, applies to legislation. Instead of the metaphysical rule, "Entities are not to be multiplied beyond necessity," one may say, "Laws are not to be multiplied beyond necessity." The unnecessary laws are those that do not make a clear substantial contribution to life, liberty, and the pursuit of happiness. One way to cut down on the number of laws is to use the rule, best expressed in the vernacular: "If it ain't broke don't fix it." Too much regulation not only results in bureaucratic bungling, but very often has disastrous consequences. It is for this reason that the United States Constitution contains the provision that the legislature shall make no law regarding either religion or the press. I conclude that for the sake of functionality there has to be constraint in the making of laws. (Cf. Humboldt [ibid., p. 87], Spencer [op. cit.])

One may suspect that no system adequate for a theory of socio-political principles is going to be complete. New situations continually arise and laws have to be amended to make sure a system continues to be functional. It is continually the case that new problems arise that require the application of intelligence for their solution. Old laws, no longer functional, need to be abandoned and new ones imposed to cope with new situations. Moreover, normative (or constitutional) principles are by their very nature vague and in need of interpretation. If this were not true we would have little need for the Supreme Court Justices, constitutional lawyers, and constitutional scholars. Unfortunately, this fact dooms any attempt to derive, in a totally scientific manner, a complete set of norms starting from a simple set of axioms.

To sum up: We agree that a society is functional to the extent it tends to facilitate individual human development and the satisfaction of the higher needs. I have argued that a normative system for the principles of social construction cannot be complete because of the inherent vagueness of constitutional law and because a constitution must be continuously adapted to new circumstances. Furthermore, I have contended that a complete normative system for society would tend to have a chilling effect on freedom and, therefore, for the sake of functionality there has to be constraint in the making of laws. Very often constraints of this kind are built into the *constitution* of the state in the form of *laws of impotence.* Further discussion of these matters is to be found below in Chapter Eleven (***Part Three***).

4 The Structure of Society

As we have noted, many authors, including Plato, have compared human society to the human body. The human body—as well as a society—is a self-regulating system which has a tendency to maintain itself in a constant state; for example, the body maintains its temperature at about 98.6 degrees Fahrenheit. Biologists call this process "homeostasis." The equilibrium state for a society is the state in which it fulfills its proper function which, we have claimed, is to provide *an environment of freedom for the fullest development of its members.*

A society functions through the use of subsystems, each of which has a particular function. These subsystems are by no means independent of each other. This is shown, for example, by the relation of the system of communication and control, which is set up to coordinate the activities of society, and the justice system, which has the function of maintaining order and settling disputes between citizens. Besides the two already mentioned, there are at least five other important subsystems in society. (In large states or empires with many subdivisions—provinces, counties, cities, etc.—each subdivision forms a subsystem.) One of these is a system—called the economy—for procuring and distributing goods and services (including medicine and entertainment); another is a system for education and learning; a third is a system for defense against outside forces; a fourth is an ecclesiastical system for the purpose of assembly for worship and ritual; a fifth is a job-allocation system for the selection of people to work in the various subsystems—(including the job-allocation system itself); etc.

A society functions well only if its subsystems fulfill to perfection their proper function or intended purpose; subsystems seldom, if ever, live up to this standard, however. For example, one can say that the proper function or intended purpose of an economic system is to provide for the material welfare of the people, but it is common to find an economic system perverted into an instrument for private enrichment. Similarly, the proper function of the system for communication and control is to coordinate the activities of all citizens, but it usually is perverted into an instrument for one class to have power over the others. The proper function of the justice system is to ensure domestic tranquillity and to see to it that the laws of the land are faithfully executed, but the system is often corrupted into a system for profit and oppression. The proper function of the system for education and learning is first, to see to it that the new generation is able to function well, second, to promote learning and research, and last, but not least, to serve as a resource for improving the function of all branches of society, but the subsystem is often perverted into an instrument for the propagation of received systems of thought and suppression of opposing views. The proper function of the system for defense is to meet the challenges to territorial integrity posed by other societies, but it is often converted into a machine for the domination of people both home and abroad. The proper function of the ecclesiastical system is to help bring "meaning," beauty, and fellowship into the lives of people, but too often it is perverted into an instrument for class oppression and intellectual stultification. The proper function of the job-allocation system is three-fold: one, ensure that each person has a position in society commensurate with his or her talents; two, see to it that the teams of workers in each subsystem are selected for their ability to work together effectively; three, see to it that society makes the best use of the available human resources; but one suspects that no system of employment comes close to fulfilling all these goals.

The function of the subsystems of society may be analyzed in terms of social needs: if the economic system works well it satisfies the need of society for goods and services (including medicine and entertainment); if the system of communication and control works well it satisfies the need of society not only for coordination but for a system of laws and regulation that promotes freedom; if the justice system works well it satisfies the need for, well, "justice"; if the system for education and learning works well it satisfies the need for learning; if the system for defense works well it satisfies the need of society for protection from enemies; if the ecclesiastical system works well it satisfies the need for religion; if the job-allocation system works well it satisfies the need of society for division of labor, etc.

Of all the social needs, the one that is most controversial is that of justice. The difficulty with the word "justice" is that, like the word "good," it appears to be indefinable. There are almost as many definitions of justice as there are philosophers. On one hand, Rawls [1971] applies principles like those of Kant—universality, lack of arbitrariness, and rationality—to arrive at a view of justice as "fairness"; on the other, Plato defines justice as "having and doing what is a person's own and belonging to him or her." If one accepts Plato's view it is not surprising that Nozick [1974] defends as "just" entitlement principles that lead to great inequality of wealth, the very pillar of the established dominance hierarchy. As mentioned in Chapter Two, Plato arrived at his definition of justice by first constructing a model for society (*THE REPUBLIC*), and then subjecting the model to analysis. What he accomplishes, however—as already indicated in Chapter Two—is a demonstration that one can design a society very well without mentioning the term "justice" (or its synonyms). Thus instead of saying that the justice system satisfies the need for "justice," one can simply say that the justice system satisfies the need to settle disputes and uphold the laws of the society or state. Hence before one can speak of justice, one must settle what is "having and doing what is a person's own and belonging to him or her"; that is to say one must have in hand a constitution and a body of laws.

But who determines laws? If the people are strong enough to rid themselves of tyrants, bosses, and kings, they determine the laws; if not, it is a select few who determine the laws. Therefore, in a sense, the sophist, Thrasymachus, is correct when he says "justice is the interest of the stronger"; for whatever faction is the stronger in society determines the laws of the society or state and therefore settles what "having and doing what is a person's own and belonging to him and her" consists in.

The next six subsections are devoted to individual subsystems. (No treatment of the ecclesiastical subsystem is given because it is treated in Chapter Seven (*Beauty, Mysticism, and, Religion*).

4.1 Communication and control

We have already claimed that a society may be compared to a self-regulating system that maintains itself within certain parameters. Any self-regulating system operates according to certain rules; e.g., one can say that the heating system of a modern house obeys the rule: keep the house neither too warm nor too cold. A society is *not* like an ordinary self-regulating system, however; for to a large extent it determines its own rules and learns from its mistakes. I do not say that a society, even one constructed by art, determines *all* its own rules since a society, no matter how sophisticated, operates by two kinds of rules: one, the rules imposed by nature; the other, the constitution and laws which the society chooses for itself. The latter rules determine the structure of society including the kind of government and the kind of economic system by which the society operates.

In order to determine the system of government by which a society is to operate, the initial system of communication and control must set up a *constitution*; that is to say, the list of the *fixed* commands by which the society is to be governed. The fixed commands are like the wired-in requirements of a nervous system. Examples of fixed commands of a society are found in the constitution of *The United States of America.* In the initial stage the fixed commands are set up by means of a constitutional convention or equivalent; in the later stages the constitution may be amended either by revolution or by constitutional means. The make-up of the initial body that is to invent the initial constitution must be rather inchoate, at least until a proper job-allocation system can be set up or organized. In the case of the United States the initial body was determined by twelve of the original thirteen States. Some societies allow emendation by a direct vote of the people in a plebiscite. In order to maintain stability, it is usually made very difficult to change a constitution. In the case of the *United States Constitution,* an amendment to the constitution may be proposed by two-thirds of both houses of congress, or by a constitutional convention called by two-thirds of the States, but in any case must be ratified by three-fourths of the States.

I assume here that the state set up by the society's constitution is an electoral democracy, but postpone a discussion of electoral systems until we come to the job-allocation system. Before any electoral system is to work, however, one must determine *who* is a citizen. The United States Constitution gives certain privileges to "persons," and the courts have interpreted the notion of person rather broadly. The fourteenth amendment says, in part, that no State shall "deprive any person of life, liberty, or property, without due process of law..." This, unless I am mistaken, is interpreted by the Supreme Court of the United States to apply to corporations as well as people.

The interpretation of corporations as persons makes a mockery of the principle of equality of persons in a democracy. According to Greider (Cf. Mander & Goldsmith [1996]):

> Above all, a corporation by its nature possesses political resources that very few individual citizens can ever hope to accumulate. Thus, if corporations are to be regarded as citizens, they are equipped to hold the first rank in U.S. politics, and nearly everyone else will inevitably become citizens of the second class.

The status of corporations as persons tends to convert a society into a plutocracy—government by the rich. Who, but corporations—in a capitalist system like that of the United States—controls the media and other means of communication? If corporations have free speech, the rich tend to drown out the speech of ordinary individuals, and the system of communication and control tends to become, to a great extent, part of the economic system. We return to this matter in two places: once, when we discuss ways to tame bureaucracy; the other, when we come to consider the economic subsystem.

Like the nervous system of the human body, which consists of a central nervous system together with a peripheral autonomic system, the system of communication and control in a society has several subsystems: on one hand, the central government; on the other, peripheral systems of control (and communication) ranging from church, school, press, local organizations and family to the unorganized and spontaneous firestorms of gossip and opinion that rage through society. The central government has the following three subsystems: an intelligence-gathering system, a deliberating system, and an executive system. The information-gathering system gathers information on how far society departs from equilibrium; the deliberating system assesses what must be done to bring society into equilibrium and issues commands to do just that; the executive system carries out the commands. I consider each in turn.

The phrase "intelligence gathering" may make one think of spies and bureaux such as the CIA and the FBI; however, most intelligence, to which the government has access, does not come from undercover surveillance. On the contrary, most intelligence belongs to the public record and comes from sources, such as the press, to which the public has access. Undercover surveillance is to a large extent inimical to democracy. One can cite such instances as Nixon's covert investigation of his enemies and J. Edgar Hoover's files on such figures as Martin

Luther King. On the other hand, a government needs to be informed of plots to subvert its legitimate functions. If the United States had been informed in advance of Japan's plans to bomb Pearl Harbor, it could have taken defensive action. Hence intelligence gathering of a covert nature by such agencies as the FBI and the CIA are sometimes helpful in preserving democracy, but the abuses of these agencies are so enormous that one would hope that ways will be found to curb them. One way to curb them is through the power of the purse, a topic which is pursued below and in Chapter Eight (*Several Suggestions for the Salvation of the Planet).*

The deliberating system has two subsystems: a law-making body and a decision-making body which may be called the ***tribune***. The tribune is mostly concerned with day-to-day matters while the law-making body (or ***legislative system***) is the subsystem which (by and large) makes the standing rules of the society. The tribune must make its decisions relatively quickly, while the legislative system is usually free to proceed at a deliberate pace. Both systems are ordinarily constrained by a constitution. I consider each of these systems separately, beginning with the legislative system.

Once a law making body is formed, it needs to confirm or deny rules for forming the government found in the constitution. Usually, of course, the constitution sets limits to how far it can do this, especially when there is a federal system as in the *United States*. Once the constitutional rules are established and confirmed, the legislature can go about its main business which is to complete setting forth the laws of the land, in accordance with the directives of the constitution.

The laws with which a legislature is concerned are of two kinds: one, those that provide guidelines for the subsystems of society; the other, the norms of behavior that regulate the activities of individuals. The latter (which are not completely subject to legislation) are treated in the next chapter (Chapter Five: *Social Ethics*), and hence I concentrate on the former here. The subsystem which is of greatest concern to legislators is that of the economy, particularly in so far as the economy hinges on the disposition of property.

In industrial democracies the notion of "*property*" is taken for granted and seldom if ever debated, and yet it is fundamental to the very idea of justice and law. Even Plato, who defines "justice" as having and doing what is a person's own and belonging to him or her spends little or no time explaining what can be owned and what ownership entails. Generally speaking, people own anything that they can appropriate without someone objecting; for it is said that "possession is nine points of the law." For example, if I pick up a shell on a beach, I own it (if I want it) as long as no one else claims it. The function of the law is to settle disputes (over claims and other matters), and the function of the legislature to determine laws which allow judges to settle disputes. The law tends to uphold tradition, but if legislators are radical they may propose to alter tradition; for example, Plato in the *REPUBLIC* [449c] opposes the tradition that a baby belongs to its mother. In any case, the law is ineffective unless the state is prepared to back it up by force. This brings us back to what Thrasymachus said: justice is the interest of the stronger (or "might makes right").

Primitive people have little conception of "property rights." North American Indians, for example, could not (in days of yore) imagine owning the land since the land—with little exception—belonged to the community. Primitive people are territorial, however, and will dispute with anyone trying to take away their territory. In this case it is extremely clear that ownership (or economic power) depends on military power. According to Russell [1938]

> ...the economic power of private persons depends on the decision of their government to employ its armed forces, if necessary, in accordance with a set of rules as who shall be allowed access to land [or other capital]; while the economic power of governments depends in part on their armed forces, and in part upon the respect of other governments for treaties and international law.

It is well to remember that at one time the Constitution of the United States allowed (what is unthinkable today) the ownership of people. Considerations of this kind make it clear that ownership (and justice) is a nebulous concept and means whatever the laws of a state make it mean. On the other hand, I claim that in a democracy *it is required that it shall be the case that the legislature make no law that does not tend to promote the free development of all people in* (as Humboldt [ibid., p. 65] says) *their "richest diversity."*

The tribune, the other deliberating system to be considered, is, like all subsystems of control, difficult to keep democratic. Unless people in society take to the streets—or either the legislative or the judicial system overrides it—the tribune is the final arbiter and, as Harry Truman put it, "the buck stops here." The tribune is usually headed by one dominant individual (called the *head of state*) and therefore takes the form of a dominance hierarchy. Under the head is a board or cabinet, which usually consists in the heads of subsystems. In the structure of the tribunal system a subsystem is usually called a "bureau" and hence a government characterized by excessive regulation is often called a *bureaucracy.*

One of the functions of a bureau is to advise and inform the head of state. Each bureau mimics the structure of the central government and hence has three subsystems: information-gathering, deliberating, and action-making. (The bureau head may or may not employ a committee—or set of committees—to take over some of the deliberating functions.) The bureaux have access to an enormous amount of information, and hence, through them, the people have an opportunity to feed back to the command-making system knowledge of the effects of governmental commands. In addition to obtaining information and advice from heads of bureaux, the head of state has his or her own private advisors, with their own expertise.

Fundamentally, bureaux are set up in order to execute the commands of the command-making system. This brings us to the final system of the three subsystems of government to be considered: the *executive system.* The executive system translates the various laws of the legislature together with the commands of the head of state into directives and makes plans—which ideally adhere to the guidelines set down by the laws of the land—to implement the directives.

A government is nonfunctional to the extent it does not serve the people, but as a consequence of the ways bureaux are administered, the people and the bureaucrats (i.e. the members of the bureaucracy) are natural enemies. A bureau tends to have the administrative structure of a dominance hierarchy. (For a discussion of dominance hierarchies see Chapter Three.) It is for this reason it is easy to turn people against the bureaucrats; e.g., politicians in the United States love to run against "Washington" and even in Mao's China, Mao's Cultural Revolution was at least in part a revolt against the bureaucrats.

Because of experience with bureaucrats, it is tempting to move toward anarchism—the ultimate revolt against bureaucracy. Many people regard the existence of a dominant group in society as simply a result of a conspiracy. A milder form of conspiracy theory is advocated by Chomsky who points out the ways the established elite, in a kind of preestablished harmony, act in concert to manufacture consent. Be that as it may, one need not believe in conspiracy to see that in some form or other a dominance hierarchy is almost inevitable in any organization. My reasons for this conclusion are as follows.

According to the theory set forth in the last chapter, equality is, in a sense, unnatural since the principle of organization in nature is the dominance hierarchy. It is apparent therefore that if we look to biology and nature as a basis for judgment, there is a *prima facie* case for inequality. Another argument for inequality is one based on history. The existence of inequality of wealth and power has in the past led to both cultural and economic development. There is therefore an empirical basis for the argument that the accumulation of wealth and power in a few hands—if not carried too far—can redound to the benefit of society as a whole. Inequalities of wealth and power (within reason) seem to be the price that modern societies pay for the economic base on which to build a culture which promotes life, liberty, and the pursuit of happiness. A third argument for inequality stems from the failure of autocratic attempts to impose equality from above. At least in large states such as China or the former Soviet Union, the attempt to ensure equality of property and privilege is self-defeating since it leads to an enormous bureaucracy and fixes a great gulf between the leadership and the people. Finally, there is an argument for inequality and the dominance hierarchy based on size. According to economists, specialization and mass production are a function of the size of the market. Besides providing a large market, large states have many advantages among which are the following two: a general impunity from conquest as well as a general ability to enforce the peace; the other, an ability to attract both able people and investment from other countries. Equality and avoidance of bureaucracy is difficult to obtain in a large state. It may be concluded therefore that, as long as states have to be large to thrive and as long as higher and higher levels of economic productivity are so much in demand, there is no way to eliminate, in a substantial way, the dominance hierarchy and thereby approach absolute equality. Perhaps the only reasonable question therefore is how to eliminate not the dominance hierarchy itself, but the excesses of bureaucracy. The question is: What ways can be found to humanize administrative structures—whether small scale or large scale, global or local, government or corporate—and promote equality, i.e. have people treated as ends in themselves rather than merely as means?

I believe there are at least five ways to tame power, not necessarily independent. These are: one, separate the ceremonial function of government from its judgmental function; two, diffuse power; three, make unequal treatment unconstitutional; four, educate; five, revolt. I need only consider four of these here since education is discussed under the heading of the educational subsystem.

The advantage of separating the ceremonial function of government from its judgmental function is well known. Where this is not done the tendency of people toward hero worship makes it difficult to keep a popular leader or a demagogue, such as Hitler, from assuming dictatorial powers. The separation may be done either by having a limited monarchy as in Britain or Sweden or by electing a head of state for a fixed term. In those countries that do not choose to have a limited monarchy, it is therefore admissible, if not required, that it shall be the case that the people at large elect a purely ceremonial head of state. Of course, there also has to be an administrative head whose duties are not ceremonial. Moreover, in order not to have a hamstrung administration, I suggest the principle that the administrative head (or prime minister) be also the leader of the law making body. This may be done (as in Great Britain) by having the ceremonial head of state nominate for prime minister one of the leaders (not himself or herself) of parties in the law making body who then has to be confirmed by the law making body by a majority vote. In order to maintain stability it is necessary that the prime minister not be ousted easily. For example it could be required that the prime minister not be removable for at least two years after appointment unless he or she is voted out by (say) two thirds of the legislature. The ceremonial head of state could either be elected for life or for (say) a period of five years.

I turn now to the second of our ways to tame bureaucracy and the dominance hierarchy: diffuse power. The rise of the middle classes in Europe and the consequent rise of democracy has had much to do with the diffusion of power in the West. Democracy, which ranges from the town meeting form of New England to the more prevalent form of a representative body, has always been in conflict with the power of the rich and well born. On the other hand, as we have already said, those who have been able to accumulate wealth have had a traditional role in creating new industry, subsidizing culture, and encouraging the dissemination of new ideas. This can often be done more effectively through private initiative than by the state since an individual can, more easily than government, avoid bureaucratic inefficiency and corruption. Moreover, it has already been noted that the method of Communism, in which theoretically the means of production are in the hands of the people, has not led to "the original and free development of individuals" for which Marx [1977 ed., p. 191] hoped. But other methods are still being tried.

One method, popular in the West, is that of using ombudspersons and countervailing organizations. Governments themselves are notorious for evading the law, and hence there needs to be some way of gathering evidence of unlawful behavior on the part of government officials. The judiciary cannot itself investigate cases of unlawful behavior—that would prejudice its capacity as an impartial jury. Moreover an administration rarely can be trusted to investigate itself. Hence the agency has to be independent not only of the judiciary, but of the rest of the government. The agency has to have the power to subpoena documents and to require officials to answer questions under oath. Freedom of the press is vital, in this regard, in order that there be no barrier to alerting both this agency and the public that there is something unlawful going on. In a democracy, however, *it is required that it shall be the case that the charge of such an agency be so circumscribed as to prevent it from prying into purely private affairs that have little or no bearing on the conduct of government.* Moreover, this agency cannot function well unless it is required that it shall be the case that it be independent like the judiciary and therefore I suggest that the head of this agency be appointed in much the same way as judges.

It is ironic that we speak of Western democracy and yet more and more adults are spending most of life in a place where they have little or nothing to say about what goes on. This is the workplace. In the workplace the entitlement of the managers to exercise control and the entitlements of the investors to reap profits is in conflict with the workers' entitlements to have a say in their work and enjoy the fruits of their labor. In an attempt to forge a new contract in which these entitlements are altered, workers have reached for help of various kinds. One is the formation of a union; another, which is becoming increasingly popular, is for workers to become owners. Unions, in a global economy, are becoming less and less effective; moreover, the formation of a countervailing organization, such as a union, to rectify the lack of democracy is unfortunately like fighting fire with fire. In joining a union, a person soon finds he or she, instead of having one master, has two. Nonetheless, a union can be a device not only for demanding more individual control, but to ensure the existence of ombudspersons whose job is to investigate complaints against the leaders and to try to bridge the gap between the leaders and followers. Unions defend the worker from being abandoned by an employer, and help workers to achieve not only equity, but a voice in management. We return to these matters in Chapter Eight.

Two other methods of diffusing power are discussed in Chapter Eight. One, which has become more and more urgent as we go toward a global economy, is decentralization and breaking up of large units into smaller ones (including federalization, trust busting, and devolution). The other is a reform of tax policy. The latter policy has had little success in the United States, largely because of corporate power and flaws in the electoral system. Perhaps such a policy can only have success in small nations such as Sweden where the propaganda of the rich has not reached its most virulent form and people can elect members of a law making body that represent their own interests. Only such a body will be able to pass laws aimed at a redistribution of wealth and power that will result in equal treatment of everybody. At any rate, since according to the United States Supreme Court corporate power is built into the Constitution of the United States, we are brought to the third method of promoting equality: constitutional reform.

The Constitution of the United States yields certain protections to "persons." In some respects corporations are treated as persons and in others not. Individuals may be executed, but not dismembered; corporations, according to anti-trust laws, are subject to dismemberment, but not to execution—except perhaps by excessive fines. According to Greider [ibid.]

> When corporations are convicted of crimes, they lose none of their abilities to act in politics. Corporations are "citizens" who regularly offend the law—both in the criminal sense and in the civil terms of flouting regulatory statutes. Yet their formidable influence on political decisions goes forward undiminished, along with the substantial financial reward they harvest from government.

Corporate abuse of power could be substantially reduced if there were a constitutional amendment that denies corporations first amendment "rights." This might be thought to have a chilling effect on newspapers, which—after all—are run by corporations. There are probably few people who are not aware that Jefferson [1787] wrote: *were it left to me to decide whether we should decide to have a government without newspapers, or newspapers without government, I should not hesitate to prefer the latter.* Perhaps if Jefferson were confronted by today's spineless press he would not have made such a remark. It should be noted that Jefferson prefaced his remark with a statement urging that people be given "full information of the [affairs of state] through the channel of the public papers, and ***to contrive that those papers should penetrate the whole mass of people*** [emphasis added]." This is hardly possible in today's press owing to corporate monopoly and the tendency of the establishment reporters to echo each other. Establishment reporters and commentators engage in a frog and mouse battle over relatively trivial matters and are oblivious to almost everything else. If one really wants to be informed of what is going on, one has to consult the alternative press, a press almost hidden from public view. But the reader misunderstands me if he thinks that I fear only the power of corporations and do not likewise fear the power of government.

Except when they achieve monopoly status, the power of newspapers to express opinions must be protected, but the opinions—if they are to receive first amendment protection—have to be viewed for what they are: the opinions of individuals (either anonymous or signed) and not of corporations. Media monopoly is most egregious; for there is, in the United States, a restraint of trade in discussion. Democracy cannot long flourish in a society in which the amount of media time allotted is determined by the amount of money corporations—or powerful individuals—are willing to pay for access. The United States is now dominated by a punditocracy: the Safires, the George Willses, the David Broders, the Tom Friedmans, the Cokie Robertses, the Maureen Dowds, the Ted Koppels, the Bob Novaks, the Tim Russerts, and the rest of the usual suspects. (See Alterman [1992].) Worse still, all pundits talk (or write) about the same thing (even Charlie Rose, who holds forth when all responsible people are in bed!). The moguls, who control the media, find room for the opinions of the far right, but seldom the far left, though the media pretends to have—as an entertainment!—a battle between the left and the right. (As far as the media is concerned, Clinton is a leftist!) The only place on television that one occasionally finds an intelligent presentation by a knowledgable *person* on a single *topical* subject is C-Span. (That is not to say that *all* other television is trivial: for example, there are *academic* programs on educational channels; moreover, cable regularly presents marginal subjects like biographies, and on public television one finds, from time to time, compelling investigative reports.) The public says it is sick and tired of one issue television, but the media rolls on its majestic way, oblivious that it is in the sewer. There must be some way to destroy the media monopoly which has—despite the constant proliferation of more and more channels (devoted to more and more of the same)—reduced public discussion to the lowest level it has been since the introduction of television. Does the public really need more pundits pontificating on the Monica Lewinsky affair? Why not require the media, not only to promote the freest expression of opinion, but to open up the channels to a wide range of people and to a wide range of topics? Why not force them to free the channels of news and comment from endless discussions (by the very same people) on a single topic. If this were to happen, government by newspaper (or media) would cease.

The method of constitutional restraint, which we have been discussing, is generally applied to governments and not to individuals and corporations. The first amendment to the United States Constitution, for example, starts out with the phrase "Congress shall make no law..." This method has begun to be extended beyond government, however., and applied to other subsystems of society. Great strides have been made, for example, by denying to subsystems of the United States the power to discriminate against minorities because of race, color, or national origin. What I am suggesting here is that the method of constitutional restraint be broadened to deny corporations and powerful individuals overweening power over the lives of ordinary citizens.

The monopoly that the rich and powerful exercise over communication sells the people's political birthright: elections go to the highest bidder. In order to prevent this I propose that it be required that it shall be the case that the media be opened up to all comers. I think there are ways to do this and still avoid a cacophony of voices; for example, by freedom of assembly and petitions signed by a sufficient number (say one percent?) of voters. Why not, therefore, pass a constitutional amendment to require equal access? Democracy requires no less.

Finally, we come to the fifth method of taming power, the method of last resort: revolt. This method is at once the least appealing and the most exciting, a method that spawns the most admirable *and* the most despicable of human beings. Revolt has the sanction of the American revolutionary, Thomas Jefferson, who (as we noted in Chapter Two), wrote "...a little rebellion, now and then, is a good thing..." Reasons for such rebellion are ever present; for a government is only as democratic as its agencies and, moreover, unless bureaucratic agencies are brought up short from time to time, they tend to assume dictatorial powers. Consider national agencies such as the CIA, FBI, FDA, DEA, ATF, INS, the Federal Reserve, and even international agencies such as the IMF. The CIA topples governments; the FBI and ATF kill defenseless people without trial (as at Ruby Ridge and Waco); the FDA locks up the unorthodox (e.g. Wilhelm Reich); the DEA threatens with imprisonment perhaps half the African-American population (together with a large percentage of courageous physicians); the INS deports people (even to certain death) without due process of law; finally, the IMF, in an effort to collect debts, destroys the ability of the common people in debtor countries to make a living. The history of the effort of people to free themselves from bureaucratic control is a long one. In ancient times, reasons for revolt increased as state power grew and people from other tribes or states were enslaved, but the classic revolt of the slave against the master began only with Christianity. Christianity, in the early days, was effective as a countervailing power, but when it became coopted by the Empire, it became a tool in the hands of the powerful. To this day, however, some of the devout, such as the Quakers, exhibit the initial revolutionary zeal. Revolts of the day vary from the methods of passive resistance favored by pacifists, such as demonstrations, sit-ins, boycotts, picketing, and strikes, to more violent confrontations such as riots and armed rebellion. In pacific countries, people resort to violence as a method of solving grievances only out of extreme anger or despair. But violence begets violence; e.g. the murder of Civil Rights leaders and the violent destruction of the *Branch Davidians* encourages the growth of organizations like the *Black Panthers* and the militias. It is to be hoped, however, that despite regional conflicts and the revolt against the central authority, local groups all over the world will be able to come to agreements that enable them to build a structure that fulfills the motto of the United States: *e pluribus unum*. For my views on how this may be done, I refer the reader to Chapter Eight (*Several Suggestions for the Salvation of the Planet*).

4.2 The Economy

Marx and Engels (Cf. Engels [1878]) state that the economy forms the basis of all social structure and is the final cause of all social changes. While this may be an exaggeration, what is true is that economic power tends to be the deciding force in industrial democracies. As we have already noted, the simple reason for this is that in most industrial democracies, corporations dominate the means of communication and therefore exercise so much control over the affairs of society that there is not a clear cut separation between the economic system and the system of communication and control. (Coolidge, who is perhaps misunderstood, is reported to have said that the business of America is business.) We have already noted that economic power (or ownership) depends on the law and the *stronger* faction (or factions) in society determines the laws. It is natural to suppose therefore that economic power rests ultimately on the force of arms. As an illustration of this thesis, Russell [ibid.] writes:

> Why must a tenant farmer pay rent for his farm, and why can he sell his crop? He must pay rent because the land "belongs" to the landowner. The landowner owns the land because he has acquired it by purchase or inheritance from someone else. Pursuing the history of his title backwards, we come ultimately to some man [or group of men] who acquired the land by force—either the arbitrary power of a king exercised in favour of some courtier, or a large-scale conquest such as the Saxons and Normans.

(The reader who wishes to pursue the matter can consult the work on power by Russell (already cited) which he calls "*POWER: A NEW SOCIAL ANALYSIS.*")

The economy divides into two divisions: domestic and foreign. The two cannot be entirely separated; nonetheless, I begin with a discussion of the domestic economy.

Economics, the dismal science—sometimes called "the science of wealth"—is primarily the science of debt. Debt is a form of obligation and hence depends on an ethical notion: a promise to pay. Traditionally, economics confines itself to the consideration of debt of the kind which can be quantified. The assignment of numbers to debt is often somewhat arbitrary, however; and in cases of dispute it is up to the courts—or to the law—to assign a number to a debt. In the *TORAH* [Exod.22.1], for example, it says: *If a man steal an ox, or a sheep, and kill it, or sell it; he shall restore five oxen for an ox, and four sheep for a sheep.* The quantification of debts seems to give debt a concreteness which it does not deserve. Debt is an abstraction: it cannot be seen, smelt, tasted, heard, or felt. Capital, on the other hand, is concrete. *Capital* may be defined as anything that may be used to pay a debt. Capital can be either something that is prized on its own account (such as land or goods) or can be in the form of tokens (such as coins, dollar bills, or credit in a bank account). *Money* may be defined as capital in the form of tokens.

The (monetary) *value* of a token (such as a dollar bill) is measured by its power to retire a debt (or pay for a purchase). This value is subject to fluctuations; when the value decreases it takes more tokens to retire a debt and this is called "inflation" and when its value increases it is called "deflation." Money is based on faith—the faith that certain tokens will retain their monetary value.

Business wants inflation low for many reasons including the fact that higher prices are bad for business and cause workers to demand higher wages. The measures that control inflation are extremely onerous, however, and the cure may be worse than the disease. The value of money is kept high (and inflation low) by high interest rates. A high level of interest may delight the rich since it brings in a great return on their investments; however, the cost is bankruptcy for government, business, and the general population. When the burden of debt rises, businesses fail, productivity decreases, and unemployment rises. Moreover as the population ages, the high cost of finance bankrupts the modern welfare state (cf. Thurow [1996, p.100]). Education, infra-structure, research, and the arts go underfunded. Among European nations, only oil-rich Norway is not in difficulty. The irony is that in the modern global economy fears of inflation are misplaced. Lester Thurow [ibid., p.129] calls inflation "an extinct volcano."

Many economists believe that a nation can thrive in an environment of a moderate amount of inflation (cf. Baumol & Blinder [1985], p. 106)). According to Thurow [ibid., p.188], the optimum rate of inflation is around 2 percent per year. The rich, who, as we already noted, are in control of the government and the media, are able to persuade the public that inflation is to be avoided at all costs. On the other hand, according to Wells [1931, p.391]:

> Inflated money favours the worker and the active producers against the creditor, because one need not do so much or produce so much to earn a particular sum of money as one would have to do were the circulation not inflated. It is easier to pay off debts, to pay fixed rents, debentures, interests on loans, in a phase of inflation. Also the holders of national securities, while receiving the monetary interest they had bargained for, get in effect nothing like the original purchasing power of their money. The community as a whole therefore is able to carry its debts more easily.

[Recall the *Lord's Prayer*: "forgive us our debts"!]

> It is true that higher money wages will be eaten into or even eaten up by higher prices, but these in their turn are made up for by increased employment. It is the family wage and not the individual wage that determines the level of working-class prosperity.

It is said that "money is the root of all evil." Perhaps it should be said that "debt is the root of all evil." The world is being crucified on a cross, not of gold, but of debt. According to Wells [ibid., p.434]:

> The economic mechanism of mankind groans under a vast burden of debt now, by habit and custom, not by necessity. Debt prevents plenty, it is a restraint and a subtraction, but debt is no more essential to economic life than was human sacrifice to the building of a bridge or the raising of wheat. Yet for ages men were unable to disentangle one from the other.

In the past, two kinds of economic doctrine have struggled for dominance: the free market in the West and the managed economy in the East. In the case of a managed economy, the economic system is just one more government bureau. But not even the free market allows the economic system to fend for itself since the government prints the money and charters banks.

The idea of a market economy has triumphed over the idea of a managed economy mostly because the so-called West (which now includes Japan!) is more productive than the erstwhile East. There are at least two reasons for this: one, in the market economy the existence of many independent units allows for salutary competition which both improves the quality of products and makes them more accessible to more people; the other, the free market, unlike the managed economy, appeals to the human passion of greed and therefore motivates people to produce goods not only in large quantities, but in high quality. In addition to increased wealth, there are at least two other benefits that are traditionally associated with the market economy: one, lower population growth; the other, democracy and freedom. All three benefits—wealth, lower birth rate, and freedom—are connected to each other: increased wealth together with freedom causes a lower birth rate, and a lower birth rate results in more goods for fewer people and hence more wealth per capita; moreover, on one hand, freedom in the market encourages people to make decisions and hence—in theory at least—become more open to democracy and free speech, and on the other, democratic governments tend to encourage freedom not only in the market place but in the bedroom.

Birth control is the key here. In developed countries, where people have plenty to eat, the fight against birth control by the Vatican and others is losing. In countries like China desperate measures such as abortion are being tried, but unless civilization retrogrades toward callousness, strong opposition to abortion will long remain. If the world standard of living rises and methods of birth control are perfected the need for more desperate measures will recede.

Though history has shown that there is no better way to supply material wants than private enterprise, the benefits of private enterprise are not spread equally. The *NEW TESTAMENT*, in an oft quoted passage [Matt:25.29], describes the situation aptly: *For unto every one that hath shall be given, and he shall have abundance; but from him that hath not shall be taken away even that which he hath.* Though the situation remains today largely unchanged, there has been a long struggle in the nineteenth and twentieth centuries to ameliorate the situation. In the nineteenth century, after industrialization exacerbated the situation, Marx and Mill took up the cause. The tradesmen had won their struggle against the lords and princes, but now farmers, small merchants, and the poor began a struggle against central banks and those whom Roosevelt later called the "economic royalists." Though thinkers like Wilhelm von Humboldt, Herbert Spencer and Friedrich Nietzsche teach that government has no responsibility for the positive welfare of the citizen, conservatives feel pressure to do something to avert a revolution. By the 1880s, Bismarck introduces national social security and health insurance in Germany, and before World War I, Teddy Roosevelt implements anti-trust legislation in the United States and Winston Churchill instigates the Unemployment Insurance Act in Britain. The ***welfare state*** is born.

As Adam Smith taught, the market economy does not work well unless there is competition between many independent units. For this reason, if for no other, democratic institutions have a need to regulate business. Monopolies and cartels and even large states become fat and lazy and use their power to benefit the few at the expense of the many. The degree of democracy in a capitalist (free market) country can be measured by how well government succeeds in ameliorating the worst excesses of capitalism. As measured by this yardstick, the United States of America, the world's oldest democracy, is the least democratic of all advanced countries. For example, the United States stands alone among advanced countries in not having universal health care. There are at least three reasons for this: first—as has been already stated several times—in the United States corporations control the media (cf. Miller [1996], [1997]) (and therefore politics); secondly, the United States voting system, which is almost unique among democracies, substantially cuts the level of public participation in elections; finally, not even legislators understand money and how it is manipulated for the benefit of money lenders by the central bankers (see Greider [1987]).

Financial activity has a pivotal role in the government of society. In modern society, funds are supplied by the government according to political demand. For example, in wartime there is no question that the funds are available to do what is necessary. Both taxation and the creation of money tends to shift money (and therefore power) from one sector of the population to another sector or from one sector of the economy to another. The refusal to create money or raise taxes to satisfy a political demand signifies that that demand is given a low priority.

The banking system, which in part controls financial activity, is, in the United States, appointed by the head of state and confirmed by the legislature. There is a small measure of political control here because the appointments are not for life.

There are two main roles of government in financial activity: one, achieving distributive justice; the other, regulating the level of economic activity. These two roles are somewhat incompatible and political parties tend to emphasize one but not both. One party emphasizes distributive justice and the other emphasizes economic activity at the expense of distributive justice. When there are two dominant parties in a democracy it is inevitable that one of them favors the haves and the other at least pretends to favor the have-nots. Each of these parties, when in power, sows the seeds of its own destruction; for on the one hand, if the have party is in power the very rich get richer and the less rich get poorer, swelling the ranks of have-nots and ensuring the election of the have-not party; on the other hand, if the have-not power is in power the poor get richer, swelling the ranks of the haves and ensuring the election of the have party. This process may be explained more fully as follows.

The have party, when it achieves power, lowers taxes on the wealthy, neglects research, deregulates commerce, increases policing power, does everything it can to destroy the unions, encourages the exporting of jobs to other countries where there is cheap labor, neglects the poor and homeless, raises the cost of health out of the reach of the workers, and neglects the infra-structure. As long as the economy does fine, people are not unhappy that some people have more money than others; for it is thought that from wealth comes the industry which generates the people's jobs, and, moreover, as long as those at the top are able to spend money, a little of it trickles down to the poor. It is thought that when people are well off, having money frees them from dependence on others, encourages them to think large thoughts from the economic point of view and to invest their energies in productive enterprises. But when enterprise becomes very expansive (said to be "overheated") and workers are in a position to demand more money things change. Those who have money become afraid of inflation and they call for tight money. Banks like to lend money to those who have money, and without the ability to borrow, business comes to a halt. When business comes to a halt, money stops trickling down, the demand for distributive justice revives and the have-not party (called liberals, socialists, or laborites) has a chance to become ascendant. In order to fight unemployment and deflation, the new administration eases credit, and with the help of the legislative branch, engages in public works projects, or otherwise puts money into the hands of the poor; consequently, business is revived, a wave of purchases again stimulates the economy, and some people again become sufficiently rich to start the cycle all over again. Theoretically this process is not completely operative in the United States, however, since the banking system is to some extent insulated from politics.

By and large, we have been discussing the domestic economy. I turn now to consider the economy in relationship to other states. Throughout human history, it has been the desire of each state to increase its wealth above that of other states. The old-fashioned way to obtain wealth was by war. Plunder was made into an art form by some tribes such as the Vikings, but by the end of the nineteenth century Europeans accepted trade as the most feasible means of becoming wealthy. Trade is less risky than war, and at least has the possibility of increasing the wealth of all countries. But economic rivalry, like war, has winners and losers. Great Britain was the pioneer in industrialization and hence was the first to build an empire on free trade. In this century and the last, merchants have enriched themselves at the expense of those whom Mark Twain [1962 ed.] laughingly called "the people sitting in darkness."

An argument for free trade, which goes back to Adam Smith [1776], is that prosperity depends on specialization and that specialization, which is a function of the size of the market, increases under free trade. It is supposed that, given two trading partners, increasing specialization increases what Adam Smith calls the "opulence" of both trading partners and does not impoverish one at the expense of another. History shows that this statement is not always true. Industrialized countries have always had the advantage in trade over nonindustrialized countries.

The disadvantage of free trade to an undeveloped country is that it makes it difficult for an undeveloped country to industrialize since infant industries are generally no match for foreign competition. Moreover it may be foolish for a locality to sacrifice its autonomy for the sake of a higher standard of living. What's more, it is not only foolish, but dangerous, to allow one locality or combination of localities to dominate over others. Shelley [1811], arguing for self-sufficiency, writes:

> How would England, for example, depend on the caprices of foreign rulers if she contained within herself all the necessaries, and despised whatever they possessed of the luxuries of life? How could they starve her into compliance with their views?

For the sake of freedom, and in particular, in order not to be clients of an industrial giant or oligarchy, all regions need to undergo development. Historically, the United States was only able to build up its industrial base by raising tariffs against manufactured goods. Japan has resisted buying abroad, by one ruse or another, and now has become an industrial giant with a virtual monopoly on many products.

Money, especially in modern times, is the life blood of the economy, and there is—despite Adam Smith—a need for measures, other than the use of trade, to free up its circulation and prevent excessive capital accumulation in some parts of the globe and capital deficit in others. Within a single country, a government uses two measures to increase the circulation of capital: one, deficit spending; the other, tax those who are doing well and invest in projects like the TVA, the space program, the internet, or even the military. Similar things cannot be done globally unless there is a world government. A poor country that engages in deficit spending and prints more money does not enrich itself; moreover, unless it can join a cabal like OPEC, it hardly has the power to extract money from rich countries. What I think is needed is for a world government to levy a tax against capital (especially that stashed by moguls in Swiss bank accounts) and invest the money in local projects. For this to work for the benefit of the people two things are necessary: one, the projects must be run for the benefit of the local people *by the local people*; the other, the local government must be so free from corruption that the money does not end up benefiting only the local elite.

The financial power of the government enables it to set priorities. Traditionally, the highest priority of government is the glory of the state. Nowadays the highest priority seems to be the support of multinational corporations. We have said that the glory of a state can be achieved in two ways: one, through raw power; the other, through cultural development. With the development of nuclear energy the limits of raw power in a democracy have been reached, and one would hope that at long last leaders are ready for cultural development.

If so, it is none too soon. Civilization itself is on trial. We are faced with possibly overwhelming problems of poverty, crime, pollution, global warming, and overpopulation. We may be like lemmings on the road to disaster, some bloated with overconsumption, others severely deprived, and all too many brutalized to the point where Markham's question in *THE MAN WITH THE HOE* (already quoted in Chapter One) is appropriate: "What to him are Plato and the swing of the Pleiades?" The situation is not, however, completely hopeless. What can be done about it is the subject of Chapter Eight (*Several Suggestions for the Salvation of the Planet*).

4.3 The defense system

Technically, a military (or defense) establishment is part of the action-making system of the government. Nonetheless, because of its importance, I consider it separately here. Conservatives admire the military and yet the military is the acme of bureaucracy. The military is the ultimate primitive pecking order organized in accordance with human nature at its most primitive with little or no concession to what makes people truly human. It is inevitable therefore that life in the military tends to clash strongly with the expectations of people brought up to respect equality and freedom. Those whose deepest feelings are for autonomy, freedom and the higher things of life cannot but have a revulsion for the military life-style. That is not to suggest that the military can be completely eliminated from modern life. What is needed is to keep the military in perspective and not to exaggerate its need or importance.

Perhaps the most important reason to keep the military in perspective is that the military mind-set poses a threat not only to democracy but to world peace and distributive justice. All over the world we have seen the military rise to power, inevitably coupled with an eclipse of the elementary principles of freedom. Even when this does not happen the military mind-set can lead to the institution of laws such as the *Alien and Sedition Laws* (of the John Adams Administration) and acts of folly such as the *Bay of Pigs*, etc. What is perhaps more important, the military mind-set leads to an exaggerated sense of the needs of defense. Such a mind-set, mostly fueled by conservatives who foolishly wish to block out any threat to the excesses of capitalism, results in an imbalance in the allocation of a nation's resources and escalates the search for weapons of ever greater destructive power. In recent history, mutual fear of each other led the United States and the former Soviet Union to an arms race of enormous proportions. The toll of the arms race between the United States and the former Soviet Union was so great that the leaders of the latter eventually threw in the towel. As a result both nations are poorer and war has become more destructive. The only ray of hope is that war is now so destructive that all out war has become unthinkable.

It cannot be said that nothing can be learnt from war. War, at the very least, shows what can be done when people feel common interests are at stake. On the other hand, except when there is a military threat, it is hard to mobilize people to defend society. Not all lessons from war are easily understood, however. World War II should have taught us how we can mobilize ourselves to counter a threat to our well-being *even when there is no well-defined enemy.* The answer, in the case of World War II, was deficit spending. Because of deficit spending, the economy of the United States blossomed despite the enormous waste of the war in both human and material resources. It apparently dawns on few economists that public investment in education, infrastructure, the environment, and public institutions for the arts, science, health, recreation, and entertainment, not only gives profitable and productive employment to people, saving them from the ravages of idleness, drugs, and other mischief, but produces great wealth. This wealth is not only in the form of capital, but what is more important, in the form of culture and civility. It is the flowering of civilization that I consider to be society's real defense. This matter is further discussed in Chapter Eight (*Several Suggestions for the Salvation of the planet*).

It might be hoped that if all states were democratic, the world would be peaceful and a department of defense would not be needed. This hypothesis needs further testing. But hardly any states are completely democratic in any meaningful sense. Democracy in the United States has, at least until recently, been tepid, excluding, on one hand, the poor and, on the other, women, American Indians, and African-Americans. Democracy did not prevent the *Indian Wars* and *The War Between the States.* Perhaps the most intractable wars occur within states rather than between states. In Chapter Eight I discuss the need for a world government to regulate commerce, monitor pollution, and broker disputes between different states or localities.

4.4 Education

Since the human animal does not live by instinct, it is necessary to teach the young how to take their place in the world and become effective citizens. To be successful, education not only has to impress on the young a culture, but encourage them to transcend it. For example, young mathematicians not only learn the tools of the trade, but learn to think critically. In this way people, once they have thoroughly absorbed what the past has to offer, are encouraged to be critical of that tradition.

Whitehead [1929] talks of "rhythmic cycles of education" and describes a process of learning in which there are three stages. McKenney [1955] basically accepts Whitehead's idea but emphasizes the role of skepticism. What is needed is to accept Whitehead's first two stages, but modify the third to allow for critical evaluation. The three stages may then be described as follows: first there is the motivational stage of learning in which the student is introduced to some area of study in such a way as to pique the student's interest; second we have the stage that Whitehead calls the "stage of precision" in which the current paradigm is thoroughly absorbed; the third is the critical stage in which what has been learnt is reevaluated. In the first stage one is told of the heroes and heroines of the past, how they arrived at the current paradigm, and how the paradigm itself is glorious with vast, perhaps infinite, possibilities; in the second stage hard reality sets in and one toils, as it were, with blood, sweat, and tears to master the techniques of the current paradigm. But the third stage is the real shocker where one learns that not everything is known and perhaps the current paradigm is not adequate to meet the challenges of the future. In the end the student learns to break free of the shackles of the past and to think expansively once again.

It may be taken for granted that the primary purpose of education is to increase the sapience of the young. The trait of sapience is not something entirely genetic, but something that can be cultivated. It is often the case, however, that schooling rather than encouraging sapience tends to discourage it. That is to say, education tends too often to decrease native curiosity, creativity, and cleverness. But cleverness for its own sake need not be encouraged; for it is only in connection with discovery and creation that cleverness becomes truly important. I assume therefore that the primary purpose of education is to foster creativity and curiosity—together with the fruits of the latter such as discovery and the pursuit of knowledge—rather than mere cleverness. The three aspects of sapience are not completely distinct, however; for there would be no creation without curiosity and cleverness. If we accept the idea that curiosity (embodied by Eve) is female and cleverness (symbolized by the serpent) is male, then creation is the marriage of the two.

The drive toward creativity and discovery generates a very broad class of activities ranging from the ordinary (as exemplified by a cat sniffing out its environment) to the extraordinary (as exemplified by the various explorations of Leif Ericsson, Einstein, and St. Teresa). This drive is, at least partly, the drive toward beauty. Keats [*ODE ON A GRECIAN URN*] writes:

"Beauty is truth, truth beauty,"—that is all
Ye know on earth, and all ye need to know.

There are many obstacles to learning that must be overcome if education is to succeed. Perhaps the greatest obstacle is human nature itself. According to traditional Christian beliefs, owing to original sin, human nature is flawed and therefore human beings are prone to seven deadly sins: pride, covetousness, lust, anger, gluttony, envy, and sloth. For the pupil the relevant sin is sloth. Threatening the student with hell fire used to work to some extent, but one searches for more effective and kinder means. In the modern age, where we pride ourselves on the knowledge of human psychology, it is certain that ways can be found for motivating students short of putting them on the rack.

We need not consider each of these seven sins, but before we leave this topic let us consider one other sin: that of lust. There are a few short years before lust sets in with a vengeance and these years may be used to prepare youths to handle the problems of love and lust. There are two main problems here: one, problems of prevention of disease and pregnancy; the other, problems of insuring that, when they are more mature, the students are each able to establish and maintain a caring relationship. The first problem, which is merely one of absorbing the available information about the process of reproduction and its pitfalls, should not cause great difficulty; for owing to the overwhelming curiosity of youth concerning these matters teachers who are comfortable with the subject should find little difficulty in teaching it. This has been the experience of the Netherlanders (Cf. Perutz [1992]). Moreover the second problem—that of encouraging students to develop, in youth as well later in life, deep, loving, and steadfast relationships—hardly merits separate attention since it is inseparable from the general problem of ensuring that the young grow into individuals who are sensitive, trustworthy, and gentle.

This brings us to the problem of character. One of the traditional aims of education is to improve the character of the young. According to Russell [1926] the ablest educators have in the past been generally successful in achieving their aims. It is critically important therefore that the aims of education be thoroughly examined. For the sake of a (foolish?) consistency (that hobgloblin of small minds), I repeat once more what was said to be the true purpose of the educational subsystem: to promote learning and research, to see to it that the new generation is able to function well and to serve as a resource for improving the function of all branches of society. This is not the place to argue the case, but I believe that it is not hard to show that these aims are best promoted by producing individuals who govern their lives in accordance with the growth paradigm (or paradigm of self-development) which is discussed briefly in the previous chapters and which is discussed in greater detail in Chapter Six. Such people have (among others) the following three characteristics: first, they are knowledgable; second, they are creative; finally, they are judicious, i.e. they are able to critically evaluate paradigms and, when necessary, come to the conclusion that "the emperor has no clothes." I think it is not hard to show that lack of a cultivation in the young of personal growth not only stands in the way of personal happiness but of the development of a functional society.

The desire to improve the character of the next generation has two motives: one, the desire to improve society and help the young to get on in life; the other, to enrich the lives of the next generation. The young adapt better to life if they control their emotions and are respectful of others and, moreover, society is improved if people are brought up to be honest, trustworthy, and industrious. But these are to a large extent matters that are not well handled in formal education and usually left to parents (and society in general) or to the ecclesiastical establishment. As I indicated above, I believe that the way not only to enrich one's life, but to improve one's character is to become what Maslow calls "growth-motivated." For this reason, I believe that for the sake of personal human development and the development of a functional society, it is critically necessary for growth to become the primary end of education. (Cf., e.g., Russell [1916b] and Dewey [1902, 1920].) There are many difficulties that stand in the way of bringing about personal growth in a substantial segment of the population. Of these I consider three: one, the difficulty of disadvantaged heredity or environment; two, the difficulty of a lack of role models; three, the difficulty of finding the technical know-how of producing growth-motivated people. I consider these in order.

What makes a Hitler? Is sociopathy inherited like a predisposition to cancer? Or does it appear at random in the population? Certainly a lack of love in infancy and childhood can have devastating effect on the prospects for social adjustment later on. Among the mountain people studied by Colin Turnbull [1972], a lack of the necessities of life caused a complete breakdown of normal social relations. Though there are some people who develop into decent human beings despite deprivation in youth, the prospects for a growth-motivated population arising out of extremities of poverty are nil. Hence we need to concentrate on providing the circumstances that are most encouraging for growth. Before learning can properly begin, the environment in which people grow up must be made propitious for growth and development. Those whose upbringing or environment has been less than propitious need special treatment. This brings us to the second difficulty: that of the lack of role models.

The further development of a people obviously depends on previous developments. Newton stood, as he said, on the shoulders of giants. Children absorb the prevailing paradigm, and can only transcend its limitation by questioning it. Fortunately—even though this has its pitfalls—it is not hard to get children to question the ways of their elders. Nonetheless, where there is no tradition of critically examining the current paradigm, it is difficult to make changes, let alone cultivate a judicious attitude, and in such a case criticism must come from an outside source. In the modern world such questioning comes in the form of books and other information, often from surrounding cultures, and hence it is important that sources of information not be cut off. Outside influences are not all innocuous, however; and the hapless student has—so to speak—to separate the wheat from the chaff and other contaminants. This brings us to the final difficulty to be discussed: the difficulty of finding the technical know-how for the development of growth motivated people.

It can be shown (see Chapters Six and Seven) that there are techniques such as meditation that, if properly used, not only confer benefits to health but help remove some of the impediments to the development of character. Though such training has undoubted benefits on the spiritual side, the example of India shows that something in addition is needed for material progress. A high civilization which produces works of beauty and intelligence needs a material base, and therefore people must be inspired to produce the abundance necessary if all are to live well. Let us first look at the techniques that are available for inspiring people to develop the technical skills that are necessary for a high civilization.

In the modern age there are many techniques for doing this. The use of the cinema and computers in education has yet to reach an acme of development. The improper use, in the past, of the cinema for propaganda purposes need not deter us from the use of cinema in all phases of education. Myth and story can inspire an interest in subjects of all kinds; master teachers, captured on film, can explain technical details in a way that is understandable to students who are prepared for it; but most important of all, the story of those who have struggled against all odds against a prevailing paradigm can inspire critical thinking. Films, however, have the disadvantage that the student is passive when he or she views them. Computers, on the other hand, are hands-on and can be used to encourage critical thinking. In order to use a computer one has to know what one is doing and one cannot get away with slip-shod work, and yet the young seem to enjoy the challenge that computers afford. With a computer students are competing against themselves or with the facts as they are and not with other students, and each student receives the encouragement that progress in skill naturally confers.

To sum up: The most effective teacher is one who has the traits that enable him or her to be a role model. The teacher who acts as though every thing is known, and students have only to absorb it, does the student a great disservice. Whitehead remarks somewhere that the philosopher is an ignorant person thinking and adds (Cf. Whitehead [1951, p. 699]): *The self-confidence of learned people is the comic tragedy of civilization.* Students, on one hand, need to be taught that little is known and that there are no boundaries to knowledge; on the other, they have to be taught to honor the past and their teachers. Growth (or what Maslow calls "self-actualization") involves learning to live with such dichotomies. Other dichotomies that must be taught if human beings are to grow include the following: to be ambitious and yet mistrust ambition; to be aware that the world is dangerous and full of murderers and deceitful persons, and yet to trust that democracy can work and the meek shall inherit the earth; to learn that the world is a place where sorrow abounds and everyone has to face death, and yet be aware that there is ecstasy and meaningfulness for those who seek it. Above all, if the earth is to be a haven for human beings in the future, students have to be taught that the greatest happiness is to be found in two things: one, pursuing those things that make life significant; two, working for a world in which one's children (or the children of those whom we love) can live in health and peace and harmony with others of their kind.

4.5 The job-allocation system

There are two sorts of positions to be allocated in any society and these are of a radically different sort. One consists in positions of power; the other in the remainder. I begin with the latter.

If job-allocation were scientific—such as in Huxley's *BRAVE NEW WORLD*—job-allocation would be done by a government bureau which, after routine testing, would assign to each young person his or her role in life. In a democracy, however, job-allocation is fortuitous and not everyone finds his or her place in life. Finding a job in a democracy—at least in its lower reaches—is like the game of musical chairs: those who do not jump at the nearest empty seat (if there is one) may be lost. People unable to find a job in a democracy can only hope like Mr. McCawber that something will turn up. In truth, for most positions, the job-allocation system in a democracy is no system at all—although somehow the jobs get taken and the work gets done.

I turn now to consider positions of power. Positions of power in society are of two kinds: one in which holding the position depends on having some technical skill; the other, in which the position of power depends solely on possessing certain social skills. (Here I include among "social skills" having a certain social position—or in having enough money or the right friends and relatives.) Perhaps the prime example of power depending on technical skill is the power of a physician. In anti-utopias such as Swift's *Laputa*, power depends solely on technical skill. (Recall that the Laputians were ruled by mathematicians who mastered flight and ruled from the clouds.) The power of a bureaucrat usually depends on social skill, but in some cases may depend in part on technical skill. Having the job of a legislator or of a tribune generally is owing entirely to having social "skills" such as the right looks, a glib tongue, belonging to the right group of people, or being rich or famous.

For either personal or nonpersonal reasons, people or groups of people seek positions of power in any society. There are at least three sources of satisfaction in personal power: one, it is more pleasant to give orders than to receive them; two, wealth and possessions accrue to those in power; three, the holders of power tend to be worshipped even in a democracy. In a democracy, however, the seeking of power for personal rewards is discouraged. In this subsection I make some remarks on democratic procedures for gaining power.

The art practiced by those who seek positions of power in government is called *politics*. Those who hold these positions may be said to have *political power*. In a democracy there are rules of politics that are meant to ensure that the candidates for positions of political power are qualified to hold the office they serve and that the public has a fair chance of evaluating the candidates on the basis of their character and their stands on the issues. Ideally, the rules for seeking power allow the public to choose between competing programs and goals and minimize the chances of electing someone who abuses the public trust.

A society is democratic to the extent that persons in positions of power in government carry out the will of the people. It is unlikely that democratic norms would be implemented by any law-making body that does not represent a wide spectrum of the people. Moreover, this can be achieved only by direct participation or by representative government in which the only qualification for voting is reaching the age (say eighteen) in which a person is able to make an independent judgment. In a sufficiently small society, the legislature can comprise the adult society as a whole, but in larger societies this is unwieldy and other ways have been tried to get the will of the people, the most popular of which is a legislature consisting of elected representatives.

Washington [1796] in his *Farewell Address* warns of the "baneful effects of the spirit of party." Nonetheless, no device has been found which is the equal of a political party when it comes to giving the people in a large country a chance to have a voice in public policy. Consequently, I think that a rule disallowing political parties is inadmissible.

Electoral systems which allow for a third party of any electoral significance suffer from at least two possible drawbacks: one, the election may lead to an unstable government owing to the fact that no party has a governing majority; the other, the government that is elected to power has only minority support. The first difficulty is found in Italy where the government collapses regularly and the second difficulty is found in Great Britain where the ruling party is usually supported by only a plurality (c. 40%) of the voters. There are voting systems that have been proposed which enable one to overcome these difficulties. One such system is for the voter to order candidates for an office by order of preference and to eliminate candidates who do not receive enough votes to qualify. Of course, if one candidate receives a majority of first place votes, all the rest of the candidates are eliminated. If no candidate receives a majority of first place votes, there are various systems that may be used to eliminate candidates. One such system is as follows: eliminate the names of those candidates who do not appear on a majority of ballots in either first or second place, and renumber the choices on the ballots accordingly. (For example, on a given ballot, if x is the first place choice and y the second and x is eliminated, treat y as the first place choice.) If not all candidates are eliminated, proceed as at the beginning—i.e. if one candidate now has a majority of first place votes, that candidate wins. Such a system is not foolproof, but if it should turn out that all candidates are eliminated by this procedure, one could either have a run-off election or award the race to the candidate that has a plurality of the original first place votes.

It is difficult to make elections fair when some of the constituents are poor and others are rich. Incumbents find it easier to raise money than challengers, and elections in places like the United States are generally won by the candidate who spends the most money. Another difficulty in making elections fair is that there is a tendency for accusations of misconduct—so-called mudslinging—to take the place of a discussion of issues. Perhaps even more unfair is the practice of distorting the opponents' positions. Any program that a candidate puts forward, unless it is extremely simple to explain, is liable to be distorted by the opposition, e.g. Governor Brown's "flat tax" proposal in the Presidential primary campaign in the United States in 1992. In order to be fair, it is necessary that qualified candidates be given adequate media coverage, and it is necessary that distortions of positions and accusations of misconduct be critically examined by the press. Legislation needs to be enacted to ensure that campaign statements are subject to critical scrutiny by independent investigators.

Candidates face the difficulty that instead of being judged on their effectiveness as representatives of the people, they are judged on their "character." One of the character flaws often cited is that of "inconsistency." There are two kinds of inconsistency: one, the inconsistency of those who say what they need to say in order to be elected; the other, the inconsistency of those who after mature reflection have a change of mind. An example of the first kind is the infamous "Read my lips" pledge of President Bush. Opinions may differ, but I think that an example of an inconsistency of the second kind is that made by Governor Brown who, having immersed himself in seeking big campaign contributions and seeing at first hand the corrupting influence of such contributions, decided to limit his war chest to contributions of only one hundred dollars or less. The well known remark of Emerson [1841] is worth quoting in this connection: "A foolish consistency is the hobgoblin of little minds,..."

The electoral process tends to distort the legislative process. Politicians, in the struggle to be elected, often take the short view of things and, for example, are satisfied to cut any program as long as it is not one that directly affects his or her constituents. Moreover, because of electoral pressure, or because of a wish to keep their rich contributors happy, no limit is imposed when it comes to spending vast sums of money on harsh measures—measures that play to the fears of the electorate—e.g. building up armaments, supporting covert action, and expanding the criminal justice system, etc. How different it is when it comes to spending money on gentle projects such as those that benefit education, culture, the environment, health, and the poor! In the latter case, the politician tends to remember fiscal responsibility.

The electoral process, as it now exists in most places, does not work to produce a representative legislature. Women, minorities, and the poor—those who do not belong to the elite—tend to be left out. Powerful incumbents are elected year after year to serve the interests to which they are beholden, which is rarely that of society as a whole. These interests are not entirely the monied interests but may be the interests of a particular region of the country, say Houston or Peoria, or they may be (say) the interests of powerful union bosses. In order to remedy this one might experiment with the selection of the membership on a random basis from a pool of qualified candidates. In this way one might select occasionally an Abraham Lincoln, a Hypatia, or a Benjamin Franklin. The terms of the representatives would have to be of sufficient length so that they have time to learn their jobs, and staggered so that (say) two-thirds of them are not neophytes. Another branch of the legislature could be an elective body chosen to serve regional interests and to function as a check on the other branch. This function would be served if the elective body could veto, by a vote of (say) two-thirds of its membership, the decisions of the other body.

Perhaps the greatest difficulty in politics, and one which gives the term politics a "bad name," is that, in many cases, the election goes to the least able or least honest candidate. Politics is a game in which too often sleaze pays off more than ability or moral character. A person of great ability whose personal growth is out of the normal range is not understood and may even appear to be—in Ibsen's phrase—"an enemy of the people." For this reason, a person of this kind is successful in seeking power only in extraordinary situations—for example, Thomas Jefferson and Benjamin Franklin in the early days of the North American Republic, Lincoln at the time of the threatened breakup of that republic, and Vaclav Havel in the struggle for democracy in Czechoslovakia. In ordinary times, however, ordinary people prefer leaders more like themselves.

Nonetheless the real inspiration behind political change comes from those exceptional persons called "poets" by Shelley [1821]. According to Shelley, "poets are the unacknowledged legislators of the world." Shelley's "poets" are creative geniuses who have a way with words and whose names reverberate down through the ages: not only Homer, Sappho, Dante, and Milton, but prophets and mystics such as Confucius, Laotse, Buddha, Plato, Christ, Mohammed, and Teresa of Avila. It is by no means easy to give a complete list of those who are "poets" in Shelley's sense, however; for the list must include not only the figures given above, but writers such as Adam Smith, Thomas Jefferson, and Karl Marx, and many novelists, playwrights, and songwriters. The inspirational message of a "poet" is at first spread by songs, plays, lectures, and discussion groups, and then as a movement forms and sweeps society, through rallies and marches. For example, the sexual revolution was begun by poets such as Shelley and scientists such as Darwin, Freud, Marie Stopes, and Havelock Ellis, then explicitly formulated by writers such as Mary Wollstonecraft, Shaw, Wells, and Russell, and finally brought to completion by popular figures such as Margaret Sanger, Hugh Hefner, Elvis Presley, and the Beatles. The welfare state in England has been attributed to Shaw, Wells, and Russell.

In the last analysis, it is the character of a people that is on trial in every important election. It is tempting to think that the people get the kind of government that they "deserve." People cannot be swayed by selfishness, racism, or narrow-mindedness unless they share those characteristics. But unless *homo sapiens* can be persuaded to *outgrow* some primitive instincts that were perhaps useful in previous ages but *no longer are so,* it is difficult to see how human life can long remain on this planet.

Let us not despair, however. It is time to remind ourselves how far we have come. It was not long ago that slavery was accepted everywhere. The threat of nuclear annihilation has been lifted at least temporarily, and perhaps this is a harbinger of greater things to come. In geological terms human beings have been on this planet for an infinitesimal time and have only begun to develop their powers. There is still hope that men and women can develop their moral fiber to such an extent that we can overcome the social and environmental problems that threaten to overwhelm us.

4.6 The justice system

The justice system may be divided into two parts; for, on one hand, there is the subsystem of government, the judicial branch of government—which I call the *judiciary—whose function is to determine the meaning and application of the laws;* on the other hand, there is the subsystem of government, called *the criminal justice system*, whose function is to deal with those who break the law. I consider each in order.

The judiciary has to be independent of political pressure for obvious reasons. A great measure of independence is often assured by making the term of the members of the judiciary, who are ordinarily called *justices,* be for life so that they cannot be removed except for malfeasance. Traditionally, justices are selected by the head of state, and confirmed by the legislature. This process has not worked very well in recent years in the United States since the appointments tend to be political. There is need for an independent nonpolitical body to make a list of the most qualified people (say 10) out of which the head of state is allowed to choose one. In order to ensure the consent of the governed, confirmation by the legislature remains necessary.

The judiciary, not being an administrative structure, has an opportunity to be more egalitarian in its structure than other systems since the members tend to be insulated from politics. As in most systems, however, the structure of the judiciary is often a hierarchy with a chief justice at the head. But the authority of the chief justice, at least in the United States, depends on moral authority and intellectual persuasion rather than outright dictation. Moreover, the matter with which the judiciary deals, such as the interpretation of a constitution, is, ideally, subject to quiet deliberation and intellectual analysis. This ideal becomes more difficult to obtain when the principles on which the decisions are based are not firmly rooted in the constitution, and members are chosen on the basis of their views, rather than on the depth of their scholarship and on their intellectual acumen.

A criminal justice system—to which we now turn—would be unnecessary in a perfect society, hence the present discussion might be omitted if we only sought a model for an ideal society. For this reason I first thought to give the subject a chapter of its own, but since I do not intend to treat it extensively, I consider it here.

There are two classic motivations for crime: one, the desire for gain; the other, the desire for excitement. In a perfect society the desire for gain might be minimized, but the desire for challenge, struggle, and excitement in young and healthy individuals is infinite. (This seems to be part of the reason that Plato regarded the infinite [*apeiron*] as evil.) Unlawful behavior is not only extremely exciting in real life, but the excitement it generates supplies the main source of entertainment in books, operas, plays and motion pictures. Violent crime is mostly confined to youth because it is among youth that the desire for excitement is the strongest. Even the old and infirm may resort to crime to maintain or promote their status, however. Love of excitement, particularly of the violent sort, is exacerbated, as psychological studies have shown, by the example of violence in the culture, including that found in media such as cinema and television. As pointed out in Chapter Three, the desire for challenge, struggle, and excitement can only be eradicated at the price of producing lives characterized at best by boredom and at worst by depression.

Perhaps the main reason for breaking the law in a democracy, however, is what is called "sociopathy." The sociopath may be defined as a person who has no conscience and therefore never has feelings of guilt. In objective terms sociopaths are those who make plans without scruples about damaging others. Normal people sometimes treat others merely as means (and not as ends in themselves), but differ from sociopaths in that they have pangs of conscience about it. From the point of view of the sociopath it must seem that normal people are either faking or are unaccountably blinded by sentiment. The lack of a conscience, characteristic of sociopaths, is caused—or so say many psychologists—by a lack of love in early infancy. There are degrees of sociopathy which may be measured by the extensiveness of a person's social bonds: some people are concerned only for themselves, others only for their family, still others only for their own race or social group.

There remains one more reason for breaking the law that we have to consider: a protest against "injustice." A happy citizenry has less reason to break the law than an unhappy one. If citizens are deprived they will break the law, e.g. those who are hungry will steal. Moreover, when the laws are "unjust" is it not reasonable to disobey them? If society were perfect a reason of this character would be absurd, but no society is perfect. The breaking of the law as a protest against injustice is advocated, for example, by Thoreau [1849], and practiced successfully by Ghandi and Martin Luther King. It follows that the law may be broken not only for the lowest of motives but for the highest.

A criminal justice system has at least two functions: one, to prevent crime; the other, to apprehend and bring to trial those who break the law. The first is, I think, the most important and I consider it first.

In a democratic society one might expect that opposition to the law would be at a minimum. This may not be the case, however; for where there is hope of overturning the law there may be more agitation against some of its features than when opposition is hopeless. For confirmation of this one has only to look at examples such as the former Soviet Union. Nonetheless, I believe that the most effective measure that may be taken to prevent breaking of the law is to have a society which the citizens feel is "just"; that is to say, a society in which the number of grievances citizens harbor in their bosoms is reduced to a minimum. A democratic model for a society should fulfill this bill; for to the extent a society is egalitarian and democratic the grievances that encourage crime such as poverty and social alienation are removed. Moreover in a democratic society most avenues of protest are lawful—even destruction of symbols of the state, e.g. the flag.

Another means of cutting down law breaking is to reduce the number of laws. There are many laws on the books in nominally democratic states, presumably made for the protection of citizens, which not only create a large class of criminals, but fail in their object. This includes laws against breathing or ingesting certain substances, laws against abortion, and laws against gambling, pornography and prostitution. Alcohol used to be banned throughout the United States, but the Federal ban was lifted when it was found to be counterproductive. According to Humboldt [ibid., p. 112]: *If it were possible to make an accurate calculation of the evils that police regulations occasion, and of those they prevent, the number of the former would, in all cases, exceed that of the latter.*

One may advance the principle that when a certain item becomes contraband, and a large profit can be made from selling it, imposing the contraband is a mistake. (That is to say, if x is the item in question, "Ban x!" is inadmissible.) The reason I say this is that a ban on traffic in illegal items has the effect of creating a new class of criminals whose large profits enable them to buy off the police and to hire hoodlums who terrorize poorer classes of neighborhoods and take the rap when there are arrests. The spread of substance abuse may take one of two forms, in neither of which is a contraband required: one, a sizable and politically potent part of the adult population abuses the substance; the other, it hits only a fringe of the adult population. In the former case substance abuse cannot be stopped, and in the latter it can be handled as an illness without criminal sanctions. If substance abuse is decriminalized, the money saved in fewer police, judges and jails can be used to educate the public on the dangers of substance abuse and to treat those who are the victims. I return to this topic in Chapter Eight (*Several Suggestions for the Salvation of the Planet*).

Two means of cutting down crime were found in ancient Rome: bread and circuses. A democratic society will deal adequately with the first, so let us turn to the second. We have already noted the need of people for excitement as a motive for crime, particularly among the young. William James [1911] wrote of "the moral equivalent of war." It is a mistake for modern society to try to remove all sources of danger and all outlets for violent activity; for there are those whose appetite for violent and dangerous activities is such that if it is not given a relatively harmless outlet, it will erupt in ways that are extremely destructive.

It still remains to be considered what to do when law breaking does occur. The sanctions that are used against criminals have three main aims: one, revenge; two, deterrence; three, rehabilitation. I discuss these in order.

Revenge has been condemned by all the great teachers of the world. According to Humboldt [ibid., p. 52], "revenge is but a new wrong." Nonetheless it is perhaps utopian to believe that society will ever come to the point that it heeds the words of the Buddha and Christ and endorses the idea of forgive and forget. Even utilitarians can approve of a system where "justice" is achieved through punishing a criminal since it can be argued that in such a system more happiness is gained than is lost; for when a criminal is punished the multitude receive the satisfaction of revenge while—except perhaps for family and friends—only one person suffers. Nonetheless, I believe the damage that revenge does to society far outweighs the satisfaction that punishing an enemy might bring; for it encourages a spirit of meanness and leads to witch hunts and a search for scapegoats.

The emotion of anger sometimes has a positive side: it may supply the energy needed if the situation that brought about the anger is to be resolved. Anger is particularly effective when parents need the energy to protect small children, but often it is less appropriate when it supplies the motive for public action. Nonetheless, the fear of wrath can serve as a deterrent to unsocial activities, and this brings us to the second aim of sanctions in criminal cases: deterrence.

There is no doubt that punishment acts as a deterrent to misdeeds, especially in smaller communities where it is difficult to hide. When punishment becomes uncertain, as is often the case in larger communities, punishment becomes a less effective deterrent to crime. There is also some question of how far one can go in using the extremes of punishment before a revulsion against cruelty and a sense of carrying things too far sets in. In modern countries torture has been ruled out as a deterrent to crime, although the spanking of children still has a large even if diminishing constituency. In some Muslim countries the cutting off of hands is thought to be an effective and appropriate deterrent to stealing, but in most of the world, the spirit of revenge has to be satisfied by depriving the criminal of liberty or, as in the United States, by an occasional execution.

This brings us to the question: is the death penalty ever "justified"? (That is to say, is the order "Execute that man (or woman)!" ever admissible?) The arguments against the death penalty are strong. Since revenge is to be rejected as one of the aims of criminal sanctions, and since rehabilitation of a dead criminal lies in the hands of God, we are left with deterrence as the only relevant aim of the death penalty. But the threat of life imprisonment, which is less brutal, is probably an equally effective deterrent. Moreover, it has been suggested that in the United States, where the courts are tied up in appeals against the death penalty, that life imprisonment is cost effective as well. Life imprisonment need only be imposed on the class of convicted criminals for whom nothing less acts as a sufficient deterrent.

Probably the most telling argument against the death penalty, however, is that it too late to say you're sorry when the wrong person has been executed. This does not seem to sway a majority of voters in the United States whose sense is that "justice" has not been done unless a life is taken for a life. (Is it not then "just" to execute women who have deliberately aborted a foetus?)

I conclude that except for a spirit of revenge, there is little to be said for a system of punishment that goes beyond the imposition of fines, the requirement of community service, and the incarceration of those who are liable to commit further violent deeds. There are three types of offenders that need to be discussed further: one, the rich and the well off; two, the poor; three, the sociopaths. The rich and poor cannot be discussed in isolation since one class can only be defined in relation to the other. I leave the discussion of sociopaths to the last, though there are sociopaths among all economic groups.

The rich and well off live in neighborhoods that are relatively free from crime since the affluent have no wish to destroy their neighborhoods for petty gain. Before they are tempted to engage in activities that tend to jeopardize their social status, the affluent need an extra incentive such as winning an election, landing a lucrative contract, or embezzling a million or two. Those who have much, have much to lose, and it is therefore not hard to think of an appropriate punishment for what is called "white collar crime." It is difficult to punish the rich and powerful, however, since they can hire expensive lawyers and can even buy off legislators and stack the laws in their own favor. Moreover, the influence of the rich and powerful on the election and appointment of judges tends to ensure that they are given more than "fair" consideration in the courts, and even if convicted, the rich have the means to make it likely that the conviction will not stand in the higher courts. A recent example in the United States is the Iran-Contra case in which the special prosecutor Walsh was unable to obtain the conviction of people who diverted funds and lied to the people's representatives. As long as the rich and powerful exist, "justice"—as Thrasymachus said—is (or at least tends to be) "the interest of the stronger." Hence without a truly democratic government it is not possible to deter the crimes of the rich and powerful.

We may conclude that the rich cannot be deterred from crime unless ordinary people are not so poor that they cannot run for elective office without being bought out by the rich. On the other hand, it stands to reason that a poor person, if desperate or ambitious, cannot easily be deterred from engaging in illegal activities. Hence, as long as people are separated into rich and poor, the prospect of deterring criminality remains dim. I conclude that the problem of criminality cannot be solved in isolation from reform of the whole social system.

Sociopathy is a disease that can only be cured, if at all, by love rather than hate. There is, of course, some danger in treating sociopathy as a disease. For example, there may be a temptation, as in the former Soviet Union, to incarcerate those whom the government wishes to silence on the ground they are insane. This, however, is not a realistic option in a genuinely democratic state. Perhaps a more serious question is, in treating sociopathy as a disease, are we not in danger of "coddling criminals"? This concern needs an answer.

In my opinion, criminality, except in those cases in which it takes the form of social protest, IS a disease and hence, if we are honest, we have no alternative to treating it as one. What has to change is the idea that the insane cannot be guilty of crimes. Genuine criminal behavior results from one of two conditions: one, a lack of the normal capacity to form social bonds; the other, what philosophers call "weakness of the will." Let us consider these two conditions separately, beginning with the latter.

I consider weakness of the will under two headings: one, yielding to the temptation to do something that is hurtful in the long run to oneself; the other, yielding to the temptation to do something that is hurtful to others. Again I consider these two alternatives separately, but begin with the former.

As an example of the former consider the so-called sin of gluttony. When appetite tends to overcome the scruple against overeating, the resulting disease is obesity. There are many ways of treating this disease. Physicians prescribe appetite control medicine of various kinds. Another way is to control the diet in such a way that one eats less appetite-inducing foods such as sugar. Others include forming a "gluttony anonymous" club in which one turns to help from friends or to the grace of God. The point is that weakness of the will in cases of this kind is a disease of the appetite control center of the brain and can be treated by such means as psychotherapy, diet, or drugs. I think similar things could be said in all cases in which "weakness of the will" leads to personally harmful actions. This concludes my remarks on cases of this kind; it remains to consider the other cases.

Is lust for power, like that of Napoleon, a disease or is it merely a normal condition of human beings? In lower animals, such as chimpanzees, lust for power is considered normal, but is it not a disease when it drives a human being to lie or even to commit murder? In order for there to be a democracy, lust for power must be curbed, and I think that those who yield to the temptation to lie in order to gain power suffer from an illness. When blind ambition led President Bush in the Presidential election in 1992 to spread a vicious rumor that Clinton had met during the Vietnam war with KGB agents in Moscow, I think Clinton was correct when he described Bush's innuendos as "sick." An inordinate lust for power, in a very virulent form, is found in some young males, who are driven by their hormonal system, but the disease is also found in old men and in women. Males who have a history of rape have been given hormonal treatment, but so far as I know this has not been successful. The only known treatment for an inordinate lust for power is involuntary restraint in cases like rapists (or Napoleon) or, in more ordinary cases, defeat—such as suffered by Bush.

Rapists and perhaps dictators bring us back to the consideration of the first class of criminals mentioned above—sociopaths—i.e. people who have difficulty forming social bonds and whose only use for other people is to take advantage of them. The disease of sociopathy is perhaps the most important problem faced by society and one for which, as yet, there is very little in the way of treatment. What I have to say about the matter will be found in Subsection 4.4 in which I discuss the education subsystem.

It is time to consider what is *required* in a criminal justice system.

Certainly "Rehabilitate," is a command, which if carried out, offers the greatest possibility of the return of society to equilibrium. It may be that the damage done to the psyche in infancy which results in sociopathy cannot often be entirely undone, but it is probable that much more can be done than has been done in the past. Greater research on this problem is needed. (The possibility of moral reformation is discussed in Chapter One.)

Institutions for housing criminals, as they now exist, can only be compared to old-fashioned zoos or insane asylums. One problem is the size and crowding of the institutions. The society of desperate men or women cannot be other than explosive. We put in one small place people who hate human society and are surprised when they come out hating society even more. It is no doubt brutalizing to be kept in a place in which guns are trained on you, there is no provision for privacy, and there is no acceptable outlet for strong sexual instincts. If human beings are to be humanized they must be treated as human beings and not as caged animals. Each must have a private space and the necessities of life such as food, water, the exercise and development of skills, and access to social life including freedom to accept such sexual partners as are willing to accept them. Prison life can probably be made even less hazardous than the life which most criminals lead outside. If prisoners can be made to enjoy life, without drugs or alcohol, they may not wish to take up their old destructive habits when they are released. A prison must be made into a village so constructed that while in prison the inmates are productive citizens who are self-supporting. Forced labor is to be rejected, however. It is a safe bet that most people if given a chance will work in productive occupations if they can earn either money or other credits. Evangelical religion should perhaps be encouraged in prisons since it often seems to work wonders in rehabilitating a life. If a prison village is sufficiently productive it could employ psychotherapists whose job is to prepare the inhabitants to join ordinary life outside the village.

Security is admittedly of great concern in a prison, and more research needs to be done on this problem. How can one make a prison secure without brutalizing measures? Even now, prison riots and the taking of hostages is an enormous problem. If prisoners were able to live a more normal life perhaps they would not take such desperate measures so often. If privileges are greater, the prisoners are perhaps less likely to support policies that may lead to the privileges being withdrawn. On the other hand, the demands of those who take hostages usually include a guarantee of no retaliation. But a promise made under duress cannot be taken seriously and the prisoners must come to expect that such demands are useless.

To sum up: There are two basic motives for complying with the law: one is love of neighbor and the other is respect for duty. Kant found only the latter virtuous while most people feel that true virtue comes out of Christian charity. There are times when love of neighbor causes one to oppose the law, but in a well-designed society one would not expect this to happen. Much can be done in decreasing crime through reinforcing the motives for respect for the law, but no matter what is done to reduce crime in society there will probably always be a residue of "Jack the Ripper" types that must be put away to protect others.

The danger in any criminal justice system is that both society at large and the custodians of prisoners become brutalized in the process. It is for the sake of the keepers as much as for the prisoners that the prisoners must receive as much loving care as is compatible with their confinement. Moreover, it is important to remember that opposition to the law may arise from other than criminal motives and, in any case, the certainty of guilt is not absolute; therefore those who hold others in their care are well advised to remember that it could happen that they "entertain angels unawares" [Heb: 13.2]. The bottom line is that the emphasis in a democratic system is not on fitting human beings into a system, but on adjusting the system to fit the needs of the individual.

5 Concluding remarks

After a constitution, spelling out the liberties, responsibilities, and entitlements of both people and governments, is designed, one may say, if one likes, that the resulting society exhibits a certain virtue, say "conservative," "liberal," or "just." This observation is similar to Plato's dictum that "justice" is written large in his *REPUBLIC*; however, I reject the idea that the word "justice" (or "liberty" or "responsibility") can receive a final definition. Aside from serving as a tool for propaganda, such labels serve very little purpose. Nonetheless a system that proposes as a criterion for the "sphere and duties of government" that of the "*human development in its richest diversity*" may aptly be called "liberal." We explore liberalism further in Chapter Eight (*Several Suggestions for the Salvation of the Planet*).

In the next chapter we take up the mores that regulate individual human conduct. Some of these are subject to legislation, others lie outside the jurisdiction of law-making bodies. It is to these we now turn.

Chapter Five

Social Ethics

1 The New Decalogue

The present chapter and the last form a unit, held together by a common theme. This theme is the principle of functionality. In the last chapter the principle of functionality is used to argue for democracy; in this chapter it is used to argue for norms of human conduct.

I take ten principles that seem to me to be most important as a basis for discussion: **gentleness, awareness, responsibility, integrity, health, altruism, trustworthiness, cooperativeness, piety**, and **industry.** Since there are ten principles I call it the "New Decalogue." The number ten is rather arbitrary since—as anyone who has studied the original decalogue knows—the breaking down of the principles of human conduct into separate maxims may be done in various ways. Unlike the original decalogue, the principles are not absolute but may conflict; when this happens, it is to be understood that the earlier ones are to take precedence over the later ones. For example, the principles of integrity and trustworthiness come before the principle of cooperativeness which means that it is only imperative to be cooperative when it is possible to be cooperative without sacrificing integrity and trustworthiness. The logical form of these principles is discussed in ***Part Three,*** Chapter Eleven. Since the validity of one of the principles depends on not violating one of the preceding principles in the list, it is obvious that the principles are not independent. The completeness of the list is discussed in Section 2 below.

Each of the principles is based on a characteristic (such as gentleness) which may be called a *virtue*. A detailed description of the virtues that correspond to each of the principles is given below, but in order to forestall a possible misunderstanding, I say a word here about what I mean by "piety." I apologize for the use of this word, but there does not seem to be a word that is more suitable. The word is used not in the sense common in religious circles, but in a sense somewhat like that of Einstein [1930] who uses the phrase "cosmic piety." As understood here, piety means being governed by the following maxims: first, "Be respectful"; second, "Develop empathy with others"; and third, "Develop toward the cosmos a relationship of piety." Piety as implied by the third maxim includes respect but not fear. What is meant here by piety resonates with the oft quoted remark of Kant [1788] (in the conclusion to his *CRITIQUE OF PRACTICAL REASON*) in which he says that two things fill the mind with ever increasing wonder and awe: the starry heavens above and the moral law within. It also resonates with what Maslow [1970b] calls "resacralization." As understood here, piety means not only the capacity for awe and wonder, but a firm tendency to tread lightly on the beliefs and aspirations of others, to respect parents, avoid cynicism, feel a sense of belonging, and to respect the earth, the biosphere, and the environment. It means a passionate regard for things outside of oneself and the realization that what happens to oneself is not of cosmic importance. At its highest reaches it means a mystic transcendence of oneself and the ability to identify with the whole. Piety in this sense is found in people, like some Quakers I have known, who sacrifice themselves for great causes.

We examine each principle in turn. (In the following the terms "principle," "norm," and "requirement" are treated as virtually equivalent.)

Gentleness has been taught by all the great teachers of the world. Perhaps the first expression in literature of the imperative to be gentle is the famous passage in the *PENTATEUCH* [Lev. 19.18]: "Thou shalt love thy neighbor as thyself." Soon after we find in *THE BOOK OF TAO* (cf. Yutang [1942, p. 618]): *I have Three Treasures; Guard them and keep them safe: The first is Love...* Lao-tse taught nonstriving and nonviolence. He wrote [op. cit., p. 600]:

Even in victory, there is no beauty,
And who calls it beautiful
Is one who delights in slaughter.

Meanwhile, a contemporary of Lao-tse, Confucius [op. cit. p. 831] taught the golden rule in the form "Do not do unto others what you do not want others to do unto you."

A deeper expression of the norm of gentleness is found in another contemporary of Lao-tse, Gotama Buddha [op. cit., p.363], who is reported to have said:

> If a man foolishly does me wrong, I will return to him the protection of my ungrudging love; the more evil comes from him, the more good shall go from me; the fragrance of goodness always comes to me, and harmful air of evil goes to him.

Five hundred years later we find Jesus of Nazareth teaching the same doctrine. One must admit that though the founders of these great religions are honored throughout the world, there are few adherents of these religions who follow in the spirit of their masters.

Though arguments from the standpoint of the individual are irrelevant in this chapter—since we are arguing from the standpoint of society—I cannot forbear quoting some remarks of Lao-tse recommending gentleness on the basis of consequences to the individual. There is the well known maxim, quoted by Laotse [op. cit., p. 607], "The violent man shall die a violent death." (Cp. Matt. 26.52) Moreover, Lao-tse [op. cit., p. 622] notes that resistance can be self-destructive, remarking, "When a tree is hard, it will be cut down." Finally, Lao-tse [op. cit., p. 615] notes that "The Female overcomes the Male by quietude."

It is clear, I think, that to be gentle is to exhibit thoughtful consideration, good humor, and patience, and consequently a gentle person has the following three traits: first, a willingness to accommodate; second, a slowness to anger; and last but not least, a considerateness of the kind that encourages others to be in charge of their own lives. It is therefore clear that gentle people obey Kant's principle; that is to say, they treat others as ends in themselves and not merely as means. Kant, of course, regarded his principle as the essence of morality.

The virtue of gentleness cannot come to full flowering in an autocracy. A ruling class that denies people dignity, freedom and equality is not gentle and therefore true gentleness is incompatible with a rigid hierarchy. Moreover, I think most philosophers would agree that a lack of gentleness upsets the equilibrium of society. For example, according to Hume [1751, pp. 11-14], part of the merit of benevolence and sympathy (and hence gentleness) is *...its tendency to promote the* *interests of our species and bestow happiness on human society.* Russell [1954] is even stronger in backing up this position:

> ...we have reached a moment in history in which, for the first time, the mere continued existence of the human race has come to depend upon the extent to which human beings can learn to be swayed by ethical considerations. (Cp. Matt. 5:5)

Now by "ethical considerations" there is no doubt that Russell means benevolence and sympathy, and therefore gentleness. In a nutshell, the argument for gentleness is roughly this: even supposing baseball manager, Leo Durocher, is correct when he says "nice guys finish last", nevertheless, from the viewpoint of what is in the interests of society, Lao-tse (See Yutang [1942, p. 603]) is correct, when he says "Gentleness overcomes strength." More precisely, the argument is that the main requirement for a functional society is that citizens obey the command to be gentle. Arguments of this type are made more precise in Chapter Eleven, ***Part Three***, where the tools of imperative logic are employed. Empirical support for this argument can be given by a comparative study of societies.

Awareness, the next norm (or virtue) on our list, is—roughly speaking—curiosity guided by intelligence. The need for both curiosity and intelligence is obvious in a democracy; for without curiosity a citizen is uninformed, and without intelligence, being informed is useless. But curiosity needs to be tempered by something more than intelligence. Thomas Gray wrote: "When ignorance is bliss, 'tis folly to be wise," and there are indeed cases in which it is necessary, for the sake of gentleness and democracy, that curiosity be kept within bounds. For example, few would wish to see the news media in everybody's bedroom.

The need to curb invasions of privacy, whether by busybodies or governmental agencies, is widely recognized; for example, for governmental agencies, by the United States constitution. Without limitations on governmental action, it often happens that a minority of the population is singled out for harsh and unequal treatment or even to be subjects of unspeakable experiments. In cases of these kinds, there is a denial of life, liberty, and the pursuit of happiness, in direct violation of the principle of functionality. It follows that curiosity has to be tempered by sensitivity and gentleness. Paraphrasing Russell [1925], who wrote: "The good life is inspired by love and guided by knowledge," we may say: *awareness is curiosity, guided by rationality, and tempered by sensitivity and gentleness.*

In a classless society there is a lesser need for awareness to counter abuses. But even in a free society, there are, as Tocqueville [1848] points out, three battling classes: the rich, the poor, and the middle class. In this battle the rich have the advantage. Unless the poor are aware they are easily misled by a power elite—the government, politicians, the economic royalists, together with their hired hands, the newspaper writers and broadcasters. When the power elite is not engaged in an actual cover up, it engages in what is called "management of the news." By controlling the press and other media, the power elite can get the ignorant to give up their liberties either by misdirection, appeals to patriotism, or promising a mess of pottage. Therefore, the *cliché* "Eternal vigilance is the price of liberty."

The principle or requirement of *responsibility*, the next norm on our list, is a companion principle to the principle of awareness. The responsibility of the citizen is not only to maintain political awareness but to be politically active. I interpret the principle as implying two maxims: "Do all you can to keep others from abridging the freedom of yourself and others;" the other, "Exercise your freedom."

Of course, gentleness, awareness, and political responsibility do not exhaust the requirements of a civilized society. Civilization depends on, among other things, the observance of the principle of *integrity*, the next norm on our list. To see why this is so one need only recall the implications of integrity and examine what a society in which people lacked integrity would be like. Without integrity people would not be honest nor keep their commitments. If people do not have honesty, they do not pay their debts or taxes and they become delinquents if not criminals. If physicians, lawyers, bankers, tradespeople, and farmers do not honor their commitments, they are not patronized and the economy collapses. If officials are not honest, the republic becomes corrupt and neither government nor the economy can be carried on efficiently. As a result, the state is not able to defend itself against either nature or other states and ceases to exist.

Kant [1785, 422] argues in a similar vein. He points out that the institution of promise-making could not exist unless people, as a general rule, had integrity. What Kant ignores, however, is that this result would not be germane if the moral law had no relation to consequences.

Now a society in which integrity is not the norm is in disequilibrium and cannot exist as a society, and therefore we may be tempted to conclude that the principle of integrity is absolute in the sense that there is an irrevocable commitment to not lying. There are reasons to doubt this, however. If we accepted such an absolute requirement we would have to accept as an absolute commitment the *prima facie* obligation not to lie. To what extent is such a commitment valid?

Integrity implies, as we have indicated, honesty, and honesty, in turn, implies admitting mistakes freely, telling the truth, and keeping promises. Kant regarded integrity as an absolute principle, and even to save an innocent life, would not sanction a lie. I do not share this view. On the contrary, I think civilization depends even more on politeness than on candor, and a rigid integrity is to be rejected if and when it is used as a license for harshness. But even if we reject integrity as an absolute principle, and admit "Bend to the wind" as sometimes admissible in the conduct of life, that does not invalidate the principle of integrity as a general rule since, as we have noted, without the expectation that people will live up to their commitments, civilization becomes impossible.

It is easy to see that Kant [ibid.] is right when he points out the socially pernicious consequences of lying. Let us suppose, on the other hand, that *pace* Kant, there are conditions in which it may be necessary to lie. Such exceptions to the rule against lying are widely accepted. A society may well function more smoothly if some of the rules are not followed too rigidly, and in particular, it may be that in certain cases the truth (at least as told by individuals) will be more disruptive to society than a lie. (For some lies, of course, this is not true—Plato's "royal lie" for example.)

The argument for the requirement of integrity is continued in Chapter Eleven, ***Part Three***. (In that chapter *the formal apparatus of imperative logic* is available.)

The *health* requirement, the next principle to be considered, is not absolute; for it is sometimes incumbent on a citizen to sacrifice, for the sake of others—or for the sake of integrity—not only physical but mental health. In certain cases the sacrifice may be total. Socrates, for example, yielded up his life rather than make an ignominious escape to a dubious freedom. As David Hume points out, people who make such sacrifices are generally commended; for without a modicum of self-sacrifice society would either collapse because of internal dissension or be subject to destruction by outside forces.

On the other hand, it can be shown that whenever people in society are not, as a rule, healthy, society has a tendency to be out of equilibrium. Indeed, it is easy to see that if a society is to be able to provide for the common defense and promote the general welfare it must have citizens who are developed in body and mind. A loss of bodily vigor and zest is debilitating, but a loss of rationality is fatal to the social order. A further explanation of the health requirement is given in the Chapter Eleven, ***Part Three***. We turn now to consider the principle of *altruism*, the next norm on our list.

In an ant culture altruism is total, but this degree of altruism is not to be recommended; for a program of total sacrifice cannot completely fulfill the higher needs of human beings. There are three maxims that I believe the principle of altruism entails to which I subscribe. The first is the maxim of temperance: *Be temperate toward oneself, but generous toward others.* Temperance is a kind of austerity or niggardliness toward oneself that is necessary for the sake of health. There is no need, however, to couple austerity toward oneself with niggardliness toward others. The second maxim is a corollary of the awareness principle and reads as follows: *Do not presume to know what others want, but allow them to choose for themselves.* This maxim is necessary to avoid generosity of the kind which, though it may please the giver, does nothing for the receiver. The third maxim is: *Give joyfully or not at all.* The last maxim is not intended to preclude generosity, but to preclude the kind of self-sacrifice that results in hidden resentments and insincerity. (Nietzsche calls it *ressentiment.*) Hidden resentments are not in the interests of society, since they result in hypocrisy, lies, and a loss of integrity and may eventually result in personality disorders like that which psychologists call *passive aggressiveness.*

The next principle—the norm of *trustworthiness*—may be argued for on a basis similar to that of the principle of integrity. Indeed trustworthiness overlaps with integrity since no one can be trustworthy who is not honest and honorable. Honor is, I think, one of the deepest principles ingrained in the human psyche. The function of a code of honor is to facilitate the growth of a society in which people can trust one another. A sense of honor in the human psyche has two possible sources: one, it is innate and only needs the right environment to make it manifest; the other is that it is instilled by training. It probably matters very little which is true since we can at least say that children are born with an innate capacity to learn a code of honor. People tend to be unhappy when they are unable to live up to the code of honor to which they have become accustomed. Some may even commit suicide.

Nonetheless, though most people want honor, they also want other things, some of which they can't get—or think they can't get—without breaking the code of honor to which they have become accustomed.

Much unhappiness is produced by having a code of honor that is too rigid, both in those who abide by it and in those who do not. Those who abide by a rigid code tend to be cold, ungentle, and unhappy; moreover, they tend to destroy the lives of those who—so to speak—make mistakes. Christians (and apparently many Muslims) in general—despite the strictures of Christ against casting the first stone—err on the side of rigidity. A rational code of honor is one that is neither so rigid as to be ungentle, nor so lax as to be detrimental to the welfare of society.

Without *trust* a society is destroyed. But to a *sociopath*, of course—who is a genuine member of no society and is concerned only with getting his or her own way—the destruction of society would mean nothing. A true member of society is one who obeys the principle of trustworthiness; for a *genuine member of a society* is a person whose sense of self is so bound up with the fortunes of his or her society that to betray the trust of society would be unthinkable. (See Johanson & Puligandla [1972].) It might seem therefore that for such a person there is little difference between being "selfish" and being "altruistic" since the operative self (for him or her) is an expanded self of the kind that Clifford [1886] calls a "tribal self." Nonetheless, I claim that—except in the very trivial sense in which everyone is selfish—those whose fortune is bound up with the fortunes of their community are not properly called "selfish." On the contrary, one might even define an "unselfish person" to be a *genuine* (or sincere) member of a society.

The difference between integrity and trustworthiness is that one can be an outsider or an outlaw and retain one's integrity, but to be trustworthy one must be loyal and obey the law. The principle of trustworthiness implies that it is necessary to be sincere, to be loyal, and to obey the law. But in Hitler's Germany, loyalty to the state meant committing acts that are completely at variance with the principle of gentleness. Relative to such a state, we may conclude that the principle of trustworthiness is invalid. Moreover, the matter of unquestioning loyalty to a state again raises problems of morality of the kind that I address in the next chapter and I say no more about it here.

Cooperativeness—the next principle on the list—aims at producing harmony in society. Though harmony is important for society, it is not the end-all and be-all; for the strength of a society depends partly on the ability of individuals to remain steadfast in their convictions and to defend aggressively not only their own freedoms but that of others. This sometimes requires bringing "not peace, but a sword." For this reason cooperativeness is eighth on the list of principles, following integrity and citizenship. The road toward harmony is eased by following some subsidiary maxims such as "Be cheerful," "Be not contentious," "Do not complain overmuch," "Do not make unnecessary trouble for others," "Avoid excessive self-reference," "Be not boastful," "Avoid self-conceit," and "Be ready to help when asked."

Piety, the next principle to be considered, is one of the strongest forces that bind the social order. Society has always demanded a simple piety: a piety that led to the observance of religious rites and honoring parents, country, and the gods. Such a faith had the function of producing a national identity and forging a nation out of otherwise isolated individuals. But nowadays, as Catholics fight Protestants and Sunnis fight Shiites, simple piety is an obstacle to peace. Myths, which used to unite, splinter the people into many factions. If we are to have a democratic society in which people are capable of happily uniting into a whole, there is a need for a new piety that unites rather than splinters.

The first stage in forming the new piety is the destruction of the old myths. The work of destroying the old myths on which the old faiths are based is effectively done through textual scholarship and science. Holy books, however inspired, are found to be written, not by the gods, but by men and women as prone to error as oneself. But when the destruction of the old myths is accomplished, something else has to be put in their place if we are not to be left with isolated individuals, cynically looking for their own advantage, unable to identify with other members of society. The serpent of skepticism, having destroyed the common faith, throws each person onto his or her own resources. I suggest that a new piety be put in its place.

The norm of *industry*, the final principle in the new decalogue, is listed last because it takes the lowest priority of all the principles. Industry, when it is raised to high priority, can be used to justify tyranny, and hence the necessity to temper industry by democratic procedures. It seems probable, however, that in the long run, those work-places in which workers share in making decisions about how the work is to be done are also the most productive. As corollaries to the principle, I would include the two maxims: one, "Be frugal;" the other, "Develop your talents and use them to capacity." The first corollary is easily seen to follow since productive capacity is dependent on savings. The second corollary is obvious and needs no comment.

This completes our preliminary discussion and defense of the principles of human conduct. As already noted, the reader will find further discussion on a more formal basis in ***Part Three***. We turn next to a discussion of the completeness of the set of principles.

2 Is the New Decalogue complete?

The *New Decalogue* may not be complete, but it carries us a good way. In many important cases, a characteristic that seems to be omitted from the list may be subsumed under one or more of the items in the list. For example, consider competitiveness. In modern societies a certain amount of competiveness contributes to its success, but this, in the main, is true only if the competiveness is of a certain kind. One must distinguish competiveness of the creative kind—found in sports, the arts and legitimate business—from competiveness of the destructive kind found among gangsters and certain entrepreneurs. The first is an expression of vitality, good spirits, and a desire to do one's best; the second, an expression of greed, aggression, and double dealing. Healthy persons, full of vim and vigor, exhibit the first kind, but not, I should say, the latter. Thus, it seems feasible to subsume competiveness, of the creative kind, under the principles of health and industry.

As already indicated, however, I make no claim that the *new* decalogue is complete. What is mainly missing are principles that belong to what may be called personal ethics. As stated in the second chapter, according to Spencer, ethics may be divided into "two great divisions, personal and social." Recall that in social ethics, courses of action are judged according to the viewpoint of society, while in personal ethics courses of action are judged according to the viewpoint of the individual. The principles in the *New Decalogue* obviously belong to social ethics. This does not mean, however, that one may not derive some general rules of personal conduct from the *New Decalogue*. For example, consider the principle of gentleness. It is perhaps true that functionality is very little affected if a single individual in society is not gentle, but boorishness is an infection that tends to spread and there is little doubt that for *maximal* functionality it is required that it shall be that case that *all* people are gentle. Nonetheless, it is not hard to show that the *New Decalogue* does not provide a complete guide to ethics.

In order to show this is the case, we need only consider the problem of abortion. If the sole end of a human being were to reproduce, there would be no problem. When the ends of a mother conflict with that of maintaining a fetus to term, there is a problem. In the case of a conflict, how does one evaluate the relative importance of promoting: one, the possibilities inherent in a mother's fetus; the other, the ends for which the mother herself strives? In case the unborn child can live outside its mother's womb there is no problem, but do the potentialities in an embryo always outweigh the ends of the mother?

In an imperfect world, there is a temptation to have society intervene in behalf of the unborn human being. But a woman is not just a producer of other human beings: *she has her own agenda.* Can one treat the unborn foetus as an end in itself and not interfere with the ends of the mother? How much does society owe to the foetus and how much does it owe to the mother? In the case of two human beings, in which one does not live on the life blood of another, the ends of each can be treated equally, but can they be treated equally if one is an appendage of another? I do not doubt that a climate of opinion in which mothers are encouraged to take their potential offspring to term contributes to the equilibrium of society. But as indicated in the last chapter, I believe it is probable that the intervention by society in the decision of a mother to take a pregnancy to term causes more disequilibrium in society than a *laissez faire* policy and for this reason I conclude that a *constitutional law of impotence* applies to abortion. It follows that the New Decalogue is incomplete and, moreover, is essentially incomplete.

As examples of incompleteness one could take any of a number of other cases in which a constitutional law of impotence applies. For example, drugs and alcohol. I think that in cases such as this, in which the public feels easy neither with an all out license for the individual nor with intervention by the state, it is possible for society to come into and maintain its equilibrium only if there is a general change of consciousness. It may be, of course, that this never happens, in which case one or the other of the alternatives, such as prohibition or legalization becomes a perennial plank in the platform of a political party. We return to issues of this kind in Chapters Six and Eight.

A constitutional law of impotence leaves the conduct of life in certain matters of importance up to the private decisions of individuals. I conclude that the normative system (or model) based on what I call the *paradigm of (societal) functionality*, which has been the subject of the last two chapters, serves very well as a guide to social organization, but does not provide a complete basis for ethics. *Pace* Plato, one can only conclude that there is a need for a normative system for individual conduct as well as a normative system for the governance of society. A normative system of this kind (based on the *growth paradigm*) is the subject of the next chapter.

Chapter Six

What Are We To Do With Our Lives?

1 Introduction

The subject of this chapter is an inquiry that I call "ethics." According to the rough definition given in Chapter One, ethics, as an inquiry, is a search for a pattern by which to live one's life. Since I believe we can only rest easy in our inquiry if we can find a pattern of life that satisfies the desire, common to most of us, to have a reason for living, I conclude that our definition virtually identifies ethics with personal ethics, or philosophy of life, and not with social ethics—the subject of the previous chapter. As Karl Popper emphasizes, however, definitions don't matter; accordingly, let us begin the inquiry.

We concluded in Chapter One that a pattern by which to live one's life (or an ethical paradigm) is appropriate to the extent it promotes what is important to human beings. One difficulty is that what is important depends to a large extent on circumstances. To a starving person only food is important. (I expand on this point in Subsection 3.4.) In this chapter we inquire into what is important to people who no longer have to worry about food, shelter, and the basic necessities of life. In Chapter Three, we concluded that for people in those circumstances, the most important thing is to make sense out of life; moreover, we concluded that the only way this can be done is through "growth." Recall that we defined ***growth*** as the tendency to become less and less occupied by the lower needs and, correlatively, to be more and more directed by the higher needs. In this chapter I attempt to spell out in detail what this means.

2 Precursors to growth

Who does not remember the story of Siddhartha, the founder of Buddhism, who after undergoing extensive mortification of the body, decides that unless one first attends to one's fundamental needs, higher development of the mind and spirit cannot occur? According to Russell [1916b],

> Men, like trees, require for their growth the right soil and a sufficient freedom from oppression. These can be helped or hindered by political institutions. But the soil and freedom required for a man's growth are immeasurably more difficult to discover and to obtain than the soil and the freedom required for the growth of a tree.

There is no reason to doubt Russell on this score. We have already made the point (perhaps once too often!) that the higher needs only have meaning for those who had the opportunity to satisfy the basic requirements such as having plenty to eat, a warm comfortable bed, and love of their fellow men and women. These requirements, which I call the *precursors to growth*, are the needs that ordinarily must be satisfied before higher needs attain real importance in the lives of people. The classic expression for precursors of this kind is the maxim: *mens sana in corpore sano*. In satisfying this maxim (and others like it) one may seek guidance from works on health (e.g. Andrew Weil [1995]), but in addition one needs to develop social skills and what Aristotle calls "virtue," e.g. "cardinal virtues" such as the requirements of fortitude, prudence, and temperance. According to Emerson, Russell, and others, excessive prudence or sophrosyne is to be avoided, however—but this is to anticipate some of the developmental principles that go beyond the precursors to growth.

One of the precursors to growth which serious minded people, myself included, might tend to overlook is the need to have fun. To develop properly a person needs joy in their (his or her?) life. Unless boys, and probably girls, have the opportunity to get into harmless mischief they are probably, in their adult lives, going to suffer from a lack of zest. The kind of mischief I have in mind is well depicted in the works of Mark Twain, works like *TOM SAWYER* and *HUCKLEBERRY FINN*. There is no doubt that fun is also important for adults, but probably not as crucial as it is for children. Insofar as possible in a world full of pitfalls and disasters, it is healthy to avoid taking oneself too seriously and, moreover, to avoid the feeling "life is real and life is earnest."

On the other hand, desire for fun and excitement, taken too far, can be one of the most certain ways of preventing growth. Many people find fun and excitement in what may be described as trivial or destructive ways such as gambling, violence, and drugs; these, if spiritual growth is to be given a chance, have to be brought under control. The question of the need for the control of violence and egocentric activity is something we need to discuss further, however, and we return to this topic in connection with the norms of higher development, to which we now turn.

3 The growth paradigm

In previous chapters, I propose, as a hierarchy of higher ends, the following list: *beauty, creation and discovery, autonomy, identity*, and *transcendence*. My intention is to list them in their "natural order," i.e. the order in which they appear in human development, whether in the individual or the species. It may be, however, that some of these ends arise simultaneously in human development; for example, the need for discovery and creativity might arise simultaneously with the need for beauty. On the other hand, in Chapter Three I argue that beauty comes before creation and discovery on the ground that ends come before means and creation and discovery are more often a means to beauty than beauty is to creation and discovery. Similar arguments are given in Chapter Three for the order of the other ends.

Since each item in the hierarchy is an *indirect* means to the fulfillment of the items lower in the list, it is to be expected that an item in the list normally becomes an end-in-itself only when the preceding items tend to be fulfilled. Just as the higher ends themselves only become important once the precursors to growth such as the needs for health, food, safety, and companionship are provided for, I suggest that each item in the list of higher ends tends to become more important as the ends that precede it (in the list of higher ends) are fulfilled; conversely, a lack of satisfaction of an item lower in the list tends to make it of greater concern than those items that are higher in the list. For example, if the need for autonomy is frustrated, I suggest that the need for identity becomes more important as a *means* to autonomy than as an end in itself.

Note: I do not claim that the proposed list of higher needs is complete. It seems likely, however, that it is complete enough to characterize what Maslow calls "self-actualization." See Section 4.

As stated in Chapter Three, corresponding to the hierarchy of higher ends there is a list of *higher tendencies* that characterize the person who tends to fulfill the corresponding end. These are: ***sensitivity, sapience, self-reliance, dedication,*** and ***transcendence***. The norms I call "the norms of higher growth" are based on these tendencies.

Note: According to Chapter Eleven (***Part Three***), a ***norm*** *of higher growth* is defined as a statement of the form "*It is binding that one acts in a manner characterized by* X" where X is one of the higher tendencies. For a precise explanation of the meaning of the terms "norm" and "binding," the reader is referred to Chapter Nine in PART TWO. The normative system based on the five norms of higher growth is called *the growth paradigm*. A formal study of the growth paradigm may be found in Chapter Eleven in ***Part Three***.

All the higher tendencies are to be explained in detail, in Section 4 below, in the course of defending the choice of norms. Nevertheless it will not be amiss if I make a few preliminary remarks here.

By sensitivity I mean "being mindful of aesthetic quality." Now by being mindful of aesthetic quality I do not mean what may be called "an excessive concern for appearance," i.e. a concern for appearance of a kind that results in behavior such as trying to outdo others on decor and the possession of beautiful objects or choosing a companion on the basis of physical attractiveness. Moreover, instead of enjoining one to seek the possession of beautiful and expensive objects, the norm of sensitivity, as defined in this work, instructs one to be *aware* of and to *appreciate* aesthetic quality, both (like Whitman) in everyday objects and (like Einstein) in the unusual and mysterious.

Sapience, defined in Chapter Three as being *clever, curious,* and *creative*, connotes an explorative and thoughtful mentality, a willingness to experiment and try new things. Sapience is what makes people teachable, playful, and explorative in adult life. As we indicate in Chapter Three, sapience is the essential cognitive and manipulative trait of *homo sapiens*, and, as such, is the most important single indicator of a fully developed human.

Self-reliance and dedication, the next two characteristics in our list, are also important ingredients in the development of a self. In the piquant phrase of Ernie Mott in the film play *NONE BUT THE*

LONELY HEART, a self-reliant person is a "human animal who don't look for a master." As an example of such a person, let us consider Socrates. The life of this philosopher exhibits both the triumphs and the set-backs of self-reliance. On one hand, Socrates, a man who follows his own inner daimon, is celebrated as the father of Western philosophy; on the other, he is regarded as a public nuisance who asks too many questions and as a consequence is convicted of impiety and corrupting youth, and is condemned to death.

As the discussion in Chapter Three makes clear, contest, even more than variety, is the spice of life. Dedication to a calling is a way of diverting the impulse toward contest away from socially destructive and trivial uses. Those familiar with the work of Rawls [1971] will note that the norm of dedication is related to the so-called "Aristotelian principle," which Rawls states as follows:

> ...other things being equal, human beings enjoy the exercise of their realized capacities (their innate or trained abilities), and this enjoyment increases the more the capacity is realized, or the greater its complexity.

The four traits of sensitivity, sapience, self-reliance, and dedication are—or so I contend—not only important but necessary for the full development of a self.

After a self is developed, the next stage of growth is to transcend it. The idea of transcendence, in the present context, is not easy to explain and I am afraid that it will be dismissed as nonsense by many. *Transcendence* is defined in Chapter One as that which frees us not only from the here and now but from narrow self-interest. People who have the characteristic of transcendence develop feelings very much like those which Whitehead [1933] attributes to "peace," which he calls "the barrier against narrowness." Whitehead describes "peace" as follows:

> It is a positive feeling...hard to define and difficult to speak of. It is not a hope for the future, nor is it an interest in present details. It is a broadening of feeling due to the emergence of some deep metaphysical insight, unverbalized and yet momentous in its coordination of values. *Its first effect is the removal of the stress of acquisitive feeling arising from the soul's preoccupation with itself.* Thus peace carries with it a surpassing of personality...*There is* [in peace] *a grasp of infinitude, an appeal beyond boundaries. Its emotional effect is the subsidence of turbulence which it inhibits...*

> [In peace] *the 'self' has been lost, and interest has been transferred to coordinations wider than personality.* [emphasis mine]

The state I call "transcendence" and the feelings Whitehead calls "peace" occur when people enter a realm of human behavior that Goldstein [1963] calls "the sphere of immediacy." According to Goldstein:

> When we are in this sphere, subject-object experience remains more or less in the background and the feeling of unity comprising ourselves and the world in all respects and particularly in relation to other human beings is dominant...[This state] may be achieved only by surrendering ourselves to the world with which we come in contact without fearing to lose our relation to the ordered world...[T]he sphere of immediacy creates a deeper existence that affords not only the possibility of living in the static condition of the abstract-concrete sphere but also tolerating uncertainty without losing our existence...The feeling of unity in the sphere of immediacy is the deepest foundation for the experience of well-being and for self-realization.

Some arguments were given in Chapter One for the principles of higher development, especially for transcending one's own narrow boundaries. In the next five subsections additional arguments are presented. I try to show that those who not only fulfill the precursors to growth, but enjoy in their lives the maximum attainment of importance, govern their actions by the norm of transcendence as well as the other four norms of higher development.

3.1 Aesthetic sensitivity

Perhaps *aesthetic sensitivity* needs no defense, for it is said: *Beauty is its own excuse for being* (Emerson [*THE RHODORA*]). Nonetheless, let us see what can be said on its behalf.

The principles of higher development are, in a sense, rather *elitist* since they are principles for those and only those who have fulfilled the precursors to growth. Most of the people to whom these principles apply live in rather comfortable circumstances and for people of this sort, especially those who do not grub for a living, it is hard to exaggerate the importance of beauty. Indeed, for some of them the very "meaning" of life is to be found in its aesthetic quality. For example, there is a character (who happens to be an Arab) in Somerset Maugham's [1915] *OF HUMAN BONDAGE* who, when asked the meaning of life, produces as an answer a Persian rug. Be that as it may, no psychologist can deny that once most material wants are satisfied, the need for beauty in a person's life tends to increase. According to Freud [1930], the "aesthetic attitude to the goal of life offers little protection against the threat of suffering, but it can compensate for a great deal." The importance of beauty lies in its capacity to generate excitement, light up the life of those who experience it, and thereby bring a sense of significance to a life that otherwise would be empty. I conclude that for people who have fulfilled the precursors to growth, maximal attainment of importance is not possible without what Maslow [1954] calls a "continued freshness of appreciation." It follows that sensitivity is a requirement of higher development. I return to the nature and importance of aesthetic appreciation in the next chapter (Mysticism, Beauty, and Religion).

3.2 Sapience

The method to be used in arguing for the validity of the norm of *sapience*—as well as rest of the higher norms—is similar to the above. Thus to show the validity of the norm of sapience we need only argue that—like the absence of aesthetic appreciation in an otherwise fulfilled life—dullness (i.e. having neither cleverness, curiosity, nor creativity) makes a dull life.

In *GENESIS* it is written: *And God said, "Let us make man in our image,..."... So God created man in his* ***own*** *image, in the image of God created he him; male and female created he them* [emphasis in original]. It is also written [Shakespeare's *HAMLET*]:

> What a piece of work is man! How noble in reason! how infinite in faculty! in form, in moving, how express and admirable! in action how like an angel! in apprehension how like a god! the beauty of the world! the paragon of animals!

One has only to look at one's friends and relatives to see that this is true! Joking aside, let us agree that the *great constuctor,* evolution, has endowed humans with sapience and to that extent people are like gods. Some might think that, accordingly, people would be preening themselves like peacocks, but this does not happen as often as one might think. This denouement occurs, I think, because people have a horror of *hubris* and in their heart of hearts would like to be carefree like the beasts. People tend to think that, in many ways, the human way of life is against nature and that the human intrusion into the domain of the gods is an occasion for an apology. Consider, for example, the Greek myth of Prometheus. Zeus—already somewhat miffed at Prometheus for introducing art and science to people and, moreover, for the sin of trapping all human troubles and putting them in a box—becomes angry when Prometheus offers to mere mortals the god-like ability to make fire. Prometheus is condemned to eternal damnation; moreover, the box of troubles is opened, and humanity is condemned to toil and trouble. Woman, of course!, is partly to blame. Pandora, like a curious (sapient?) woman, opens the box as Zeus knew she would. *GENESIS* tells a similar story with the serpent in the role of Prometheus and Eve in the role of Pandora. The name of Prometheus stands for foresight and, similarly, the serpent is a traditional symbol for wisdom, not only in the Near East but among Hindus and the original native Americans. Like Prometheus, the serpent helps the ancestral human—a weak and helpless beast—to become *homo* ***sapiens*** and thereby destroys forever the ability of people to live in accordance with nature. Already made in the *image* of God, the first man and woman yield to the temptation to *become* like gods. Myths of this kind tend to show that the achievements that distinguish human beings from other animals generated in ancient people not only an understandable pride but a tremendous feeling of guilt. This sense of guilt, exacerbated by Christianity, remains with us to this day.

I remarked above on the human desire to be "bestial," i.e. to be like the beasts. And who does not resonate to the famous passage from Walt Whitman's *SONG OF MYSELF?:*

> I think I could turn and live like the animals,
>   they're so placid and self-contained,
> I stand and look at them long and long.
>
> They do not sweat and whine about their condition,
> They do not lie awake in the dark and weep for their sins.
> They do not make me sick discussing their duty to God,
> Not one is dissatisfied,
>   not one is demented with the mania of owning things.
> Not one kneels to another,
>   nor to his kind that lived thousands of years ago,
> Not one is respectable or unhappy over the whole earth.

Whitman's poem, though not entirely accurate, is without doubt poignant and moving. Similar expressions against sapience and for living in accordance with nature are found throughout literature and are part of the common wisdom. Goethe's Faust, tiring of the search for knowledge, looks for something that will make him say "Stay: thou art so fair!" Koheleth, in *ECCLESIASTES*, finding in wisdom "vexation of spirit" and "much grief," turns next to—among other things—aesthetic delights. In addition there are adages like "Curiosity killed the cat" and "When ignorance is bliss 'tis folly to be wise." Perhaps more to the point, much that is distressing about the world today can be attributed to advances in knowledge. Nonetheless, there are of course compensations for being like a god, and it is to these we now turn.

Russell [1972] (who should know) writes about the godlike powers of the eminent mathematician as follows:

> In the sublimity of abstract thought, he finds a joy which outweighs the shortcomings of the world of will and emotion; in its chiselled perfection of form, his thirst for worship finds an outlet. And so long as any perfection, however rare and however difficult, can enter into human life, he feels that the world does not exist in vain.

Moreover, Russell [1918] contrasts the aim of disinterested creation with the aim of personal gain. He writes:

> A life lived in this spirit—the spirit that aims at creating rather than possessing—has a fundamental happiness, of which it cannot be wholly robbed by adverse circumstances. This is the way of life recommended in the Gospels, and by all the great teachers of the world.

Russell, in this passage, echoes his godfather Mill [1871] who writes: *Next to selfishness, the principal cause which makes life unsatisfactory, is want of mental cultivation.* In another work, Russell [1928] features a quotation from Voltaire which says: *Aimer et penser: c'est la veritable vie des esprits.* Freud [ibid.], curiously enough, also associates thinking with sex. Freud (as is well known) makes the claim that the pursuit of art and science is a "sublimation of the instincts." Be that as it may, what he has to say about creative activity is worth quoting in full. He says:

> One gains the most if one can sufficiently heighten the yield of pleasure from the sources of psychical and intellectual work. When this is so, fate can do little against one. A satisfaction of this kind, such as the artist's joy in creating, in giving his phantasies body, or a scientist's in solving problems or discovering truths, has a special quality which we shall certainly one day be able to characterize in metapsychological terms. At present we can only say figuratively that such satisfactions seem "finer and higher."

Spinoza—as well as many a modern psychologist—argues for sapience on the practical ground that knowledge can help one overcome passions and obtain peace of mind. Spinoza's thought is a rather ancient one, e.g. in ancient Greece, the classic maxim, "Know thyself," was reputedly written on the walls of the temple of Delphi.

Schopenhauer [1901 ed.] remarks that the two foes of human happiness are pain and boredom. It is easy to see that sapience helps people to overcome both; for it is sapience that enables people to be teachable, playful, and explorative in adult life. Being teachable, people learn practical wisdom, and being playful and explorative, they avoid boredom. Though dedication to creation and discovery may be frustrating as shown by the lives of artists like van Gogh and Michaelangelo, a life without what Rogers [1961] calls the Eureka experience—"This is it!," "I have found it!," or "I have expressed it!"—is a comparatively dull one. Scientific studies have proved the obvious: an exploratory person, whether child or adult, has a more interesting time than a fearful one (cf. Rogers [ibid.]).

Freud comments that the satisfaction afforded by creative pursuits is tepid at best and open to only a few (cp. Russell [1972]). He is wrong on both counts. Perhaps he never had his hair stand on end when in the throes of creation or on making a discovery. I had a friend who jumped up and down in joy when he made (or thought he made) a mathematical discovery. And recall Archimedes rushing naked out of his bath shouting "Eureka!" Moreover, I think it should be understood that one does not have to be a van Gogh, an Einstein, or a Saint Teresa to be creative. Maslow [1968, p. 136] writes:

> ...one woman, uneducated, poor, a full-time housewife and mother, ...was...original, novel, ingenious, unexpected, inventive. I just *had* to call her creative [emphasis in original]. I learned from her that a first-rate soup is more creative than a second-rate painting, and that, generally, cooking or parenthood or making a home could be creative while poetry need not be; it could be uncreative.

> Another of my subjects devoted herself to what had best be called social service in the broadest sense, bandaging up wounds, helping the downtrodden, not only in a personal way, but in an organization which helps many more people than she could individually.
>
> Another was a psychiatrist, a "pure" clinician who never wrote anything or created any theories or researches but who delighted in his everyday job of helping people to create themselves...
>
> From another man I learned that constructing a business organization could be a creative activity. From a young athlete, I learned that a perfect tackle could be as esthetic a product as a sonnet and could be approached in the same creative spirit.

A complete defense of sapience cannot be given at this point, however; for it must be admitted that a life of sapience has all the drawbacks noted by Whitman and can lead one into a life of trouble like that of Russell. A more nearly complete defense must await our discussion of the principles of self-reliance, dedication, and transcendence. Nonetheless, we may conclude, pending the results of those considerations, that for those who have achieved a certain degree of development, a life without sapience cannot possibly exhibit the maximal attainment of importance. This conclusion may be backed up by scientific research such as that of Hebb [1953, 1954, 1967] and Maslow [1954], and by the clinical observations of Rogers [ibid.]. Therefore I conclude (tentatively) that the norm of sapience is valid.

3.3 Self-reliance

The struggle of the *self-reliant* person against society is the theme of Clifford Odets' [1945] film play, *NONE BUT THE LONELY HEART*. In the play, the character Ernie Mott (played by Cary Grant), claiming to be an inventor, is asked by his skeptical girl friend, "What you invent?" Momentarily nonplussed, he draws himself up and says, "I happen to be working on my greatest invention at the present time. A human animal who don't look for a master." He grins and adds, "Ain't easy!" Later on Ernie's friend (played by Barry Fitzgerald), wishing to help Ernie, suggests that they meet again. When Ernie wants to know "What for?," his friend replies, "Maybe help you get more adjusted to your environment." Ernie bristles and says, "Now wait! Why not help my environment get more adjusted to *me*!" Remarks of this kind are characteristic of self-reliant people, who typically do not seek to adapt to their environment at all costs.

According to psychologists the adult who has been well-loved as a child tends to be not only more exploratory but more self-reliant (Rogers [ibid.], Hebb [1987], Maslow, [1954, p. 347]). In most societies, however, lack of love and a fear of rebellion in teachers and parents teach people to be timid and immature. The immature adult, like a neurotic child who stays close to its mother, fears the challenges resulting from the freedom that self-reliance entails (Kierkegaard [1843], Fromm [1941], Sartre [1943]). Emerson [1841] in his essay on self-reliance claims that a self-reliant person has to fight society. He writes: *Society is in conspiracy against the manhood of every one of its members.*

But the reluctance of people to become self-reliant and thereby become what Rogers calls "a fully functioning human being" is only partly owing to the opposition of society. The sort of life one can obtain by becoming "adjusted to society" and giving up directing one's own life can be made to sound very attractive. One might even question self-reliance as a requirement and say, as I think Mark Twain said somewhere about martyrdom: except for the honor of it, I'd just as soon not have it. This raises the following question: Is it admissible to cede ultimate control over one's life in exchange for a life of ease and pleasure? John Stuart Mill asks a similar question when he asks, why not settle for the sort of happiness that a pig might have?

Let us suppose that there is a way you can relinquish control of your life and yet obtain everything you ever wanted. Such was the situation of Faust who—you will recall—for such a prize, sells his soul to the Devil. Owing to the advances of science, we may soon all be in the position of Faust, as may be seen by reading Huxley's [1932] *BRAVE NEW WORLD*. In such a case the state will take the place of the Devil and supply without effort on our part everything we want on the condition we are "good" citizens.

As an example of a person to whom all is given, consider Koheleth in *ECCLESIATES.* He writes:

> I made me gardens and orchards, and I planted trees in them of all kinds of fruits:
> I made me pools of water, to water therewith the wood that bringeth forth trees:
> I got me servants and maidens, and servants born in my house; also I had great possessions of great and small cattle above all that were in Jerusalem before me:
> I gathered me also silver and gold, and the peculiar treasures of kings and the provinces: I gat me men singers and women singers, and the delights of the sons of men, as musical instruments, and that of all sorts.

Still Koheleth found his life wanting. Clearly, having everything one needs in life including all imaginable aesthetic delights is not enough to justify life. According to Bertrand Russell [1930, p. 28]:

> The human animal, like others, is adapted to a certain amount of struggle for life, and when by means of great wealth homo sapiens can satisfy his whims without effort, the mere absence of effort from his life removes an essential ingredient of happiness.

Even Koheleth finds some solace in effort. He says: *There is nothing better for a man, than that he should eat and drink, and that he should make his soul enjoy good in his labour.* The need for work and struggle is akin to the power drive, and accordingly it might be thought that the power drive is one of the sources of the need for self-reliance. But the power drive, which causes people to seek domination over their associates, has the result that a person becomes less self-reliant and more dependent on others; for a person with power over others comes to rely on others to fulfill his or her wants, especially when his or her wants are of either a material or social nature. The need for self-reliance, properly understood, is a need to be largely independent of others. A driving force behind self-reliance is the drive toward sapience. The drive toward sapience is composed partly of curiosity (the exploration drive) and partly of a drive toward creative self-expression. Accordingly, let us discuss the drive toward exploratory, creative, and self-expressive activity as a motive for self-reliance. There are two questions to consider: one, do exploratory, creative, and self-expressive activities really call for self-reliance?; the other, because of the obvious dangers of self-reliance, why bother to explore, create, and express oneself?

That exploratory and creative activities call for self-reliance seems to be so obvious as to be hardly worth stating. We need only repeat what is said in Chapter Three, where it is pointed out that autonomy is an indispensable means for creative expression. Let us therefore turn to the other question: why bother?

To hold one's own against all the forces of society takes courage—or stubbornness and foolhardiness. According to Rollo May [1969]:

> The hallmark of courage in our age of conformity is the capacity to stand on one's own convictions—not obstinately or defiantly (these are expressions of defensiveness, not courage) nor as a gesture of retaliation, but simply because they are what one believes. It is as though one were saying through one's own actions, "This is my self, my being."

The *development* of an identity, strong enough to produce a set of beliefs that are truly one's own, involves, it is true, a certain amount of risk. But if safety were the end that outweighs all others, people would not voluntarily become detectives, explorers, fire fighters, physicians, nurses, lion tamers, mountain climbers, missionaries, stunt pilots, soldiers, bomb defusers, prostitutes, police, criminals, etc. I, of course, have no wish to deny the importance of health, but when material and social needs are already well in hand, becoming dedicated to some task becomes more important than safety. This, however, is to anticipate, in part, the defense of the next norm on our list, the dedication principle.

We may conclude, pending our investigation of the dedication principle, that the principle of self-reliance is valid. That is to say, for those like Koheleth who do not have to struggle for a living, a necessary condition for not living in vain is to be a fully functioning human being who makes his or her own decisions.

3.4 Dedication

As already mentioned, the *dedication* principle lends powerful support to the preceding principle (of self-reliance); for it is impossible to have an identity if one is directed by others. Those who wish to accomplish something in any given field (or just to *be* somebody) have to persist despite the discouragement of others. As Maslow [1970a] says of the "growth-motivated person": *...people* [of this kind] *may actually be* ***hampered*** *by others. The determinants of satisfaction and the good life are for them inner-individual and not social* [emphasis in original]. I would add to the last sentence that honors, status, etc. become less important than the task at hand. It therefore turns out that having a strong identity is a powerful (and almost indispensable) means toward self-reliance (or autonomy), and it is for this reason that in the discussion in Chapter Three it was decided to put identity after autonomy in the hierarchy of higher ends.

The need for dedication—a calling as it were—is a curious need, however. A calling tends to supply for each person a "personal identity," and therefore satisfies a need of people to *be* somebody rather than to just *do so*mething. The need for dedication is a need characteristic of the human species, occurring nowhere else in the animal kingdom. The search for "personal identity," previously the province of aristocratic gentlemen, has become widespread because so many people have fulfilled all their fundamental needs including family and children and consequently are looking for something more in life.

They are looking for something other than sex or drugs, that will—so to speak—turn them on. The search for "personal identity"—whether it be as an artist, a politician, an inventor, an entrepreneur, an explorer, a journalist (like I.F.Stone), a mathematician, a magician, a detective, a kite-maker, a printer, a scientist, a gardener, a sailor, an educator, a religious recluse, a missionary, a born parent, an actor, a shaman, a singer, a curator of a museum, etc.—is, I think, the search for what really matters (or is of real significance or importance) in a person's life. Various writers use different phrases to describe what is searched for. Russell [1916b] speaks of "individual excellence." If I am not mistaken—pace Brendan Gill [1989]—it is being dedicated to what really matters that Joseph Campbell [1988] has in mind when he says, "Follow your bliss." I believe that personal "bliss" in Campbells sense may be identified with ultimate importance. According to William James [1900], life is significant: *...only when...inner joy, courage, and endurance are joined with an ideal.* But whether described as individual excellence, "bliss," ultimate importance, or an inner ideal, it differs from person to person and, like a key that opens a lock, opens up or energizes a given individual, but may not open up or "turn on" anyone else. (According to James [ibid., p.292], "...there is nothing absolutely ideal: ideals are relative to the lives that entertain them.") For example, if your thing is football, that would seem deplorable to me, but no matter: what matters is that it be important to you. There is little doubt that life becomes most purposeful—and incidentally offers the greatest joy—when you are doing that for which you are—or seem to be—idiosyncratically made.

That is not to say that, without dedication, life is always insignificant. Either life can be so beautiful or it can be so profoundly precarious that to search for a mission in life is superfluous. No one in the throes of romance, in the first blush of parenthood, in the heat of battle, or in a peak experience of any kind needs to justify life. Moreover, some people can, like a cat in the sunlight, be content to just BE. (My father seemed to be of this character.) In particular, people for whom life is precarious (generally speaking) do not need to be given a reason for living. This is illustrated by Russell's change of attitude when he was recovering from double pneumonia during his stay in China in 1921 (cf. Russell [1968]). He writes: "Lying in my bed feeling I was not going to die was surprisingly delightful. I had always imagined until then that I was fundamentally pessimistic and did not greatly value being alive. I discovered in this I was completely mistaken, and that life was infinitely sweet to me." The point is that dedication to a calling or an ideal—as William James well knew—is not all that makes life significant. There is a transcendent beauty in life that goes beyond dedication and striving. But this is to anticipate the next requirement on our list, the tendency toward transcendence. Nonetheless, I think enough has been said to justify the conclusion that for the long haul through life, a person, generally speaking, needs some mission in life for a *raison d'etre.* I expect that empirical research in psychology will confirm, in essential agreement with William James, that for one who has fulfilled all the precursors to growth, a necessary condition for a life to be significant in the highest degree (and therefore attain to maximal importance) is that it be dedicated to some end to which one is idiosyncratically suited. (Recall that "significant" is one of the synonyms for "important.") I think we may conclude, pending such research, that the norm of dedication is valid.

3.5 Transcendence

Finally, we come to *transcendence.* Recall that we said that the need for transcendence is "the need to free oneself from the here and now and the prison of the self." It follows that the development of a self is a precursor to transcendence. As already indicated in Chapter One, transcendence may be identified, in part, with what Russell [1979] calls "maintaining a wide horizon." People maintain a wide horizon, according to Russell, if they expand and enlarge their thoughts and sympathies and—instead of focussing narrowly on themselves or their own times and places—think universally. Having noted that it is not easy to maintain a wide horizon, particularly when things go "wrong," we also noted in Chapter One that certain techniques may be used to do this even under the worst of circumstances. One technique suggested by Russell [1930] himself is as follows:

> When misfortune threatens, consider seriously and deliberately what is the very worst that could possibly happen. Having looked this possible misfortune in the face, give yourself sound reasons for thinking that after all it would be no such very terrible disaster. Such reasons always exist, since at the worst nothing that happens to oneself has any cosmic importance. When you have looked for some time steadily at the worst possibility and have said to yourself with real conviction, "Well, after all, that would not matter so very much," you will find that your worry diminishes to a quite extraordinary extent.

Russell remarks that the main obstacle to using this technique is the conviction, usually unconscious, that what happens to oneself *is* of cosmic importance, and any threat is therefore to be treated with utmost seriousness. Therefore this technique is only available to those who have already developed some measure of self-transcendence. One needs therefore another technique of greater depth and power if one is to attain this measure of self-transcendence.

There actually is such a technique. This is the practice of meditation. Meditation tends to produce healing not only in the physical sense, but in the spiritual sense. The daily practice of meditation is used by religious teachers all over the world, especially Buddhists and Hindus, to produce what they call "enlightenment." Moreover, meditation is used increasingly by practitioners of alternative medicine as a relaxation (as well as healing) technique (cf. Weil [1995]). The healing effect of meditation acts primarily by bringing about what Maslow [ibid.] calls a *peak experience*. Peak experiences satisfy more than anything else the need for transcendence. It is well known that meditation is credited by mystics for bringing about knowledge that transcends everyday experience. I say no more about this here since matters that impinge on the dogmas of religion are explored in all the detail they deserve in the next chapter (*Beauty, Mysticism, and Religion*). Suffice it to say here that though one may entertain a healthy skepticism about the knowledge achieved through meditation, I think that, of all experiences, a peak experience has the greatest life-justifying potential. A peak experience combines, in a mysterious way, the ultimate of excitement with the ultimate of relaxation. What is more, it seems that peak experiences tend not only to make one glad that one is alive, but to make one no longer fear death. Whitehead [1933] calls it "the union of Zest with Peace." A peak experience of this kind assumes an enormous importance in the life of one who has had it since it brings with it a profound sense of acceptance of—as Milton puts it—"the ways of God with Man." It is no wonder, then, that Aristotle regarded contemplation as the *summum bonum.*

Transcendence not only has a life-justifying potential, but can be used to justify morality as well. As already reported in Chapters One and Three, Whitehead [1938] relates morality to importance and importance to perspective. He says: *Morality consists in the control of process in such a way as to maximize importance.* In the same lectures, he also says: *The two notions of importance and of perspective are closely intertwined.* As we said in Chapter One, the view I defend here is similar: morality demands that people do not remain imprisoned in a particular perspective but adhere to what is truly important. Put another way, morality is basically self-transcendence. Consequently, one can show if the principle of transcendence is valid (and maximum importance in life is tenable), it follows that the imperative for transcendence implies the imperative to be moral. For further explication the reader is referred to Chapter Eleven in PART THREE.

The principle of transcendence invokes something like the spirit of Kant's "good will" rather than sentiments of affection or benevolence. This is, I think, appropriate since sentiments are least subject to voluntary control. The resemblance to Kant stops there, however; for the main argument for the principle of transcendence depends on an analysis of the *consequences* of a lack of transcendence on the attainment of importance. In Chapter One, I argued for self-transcendence or widening one's horizon on three grounds: physical health, aesthetic sensitivity, and the importance of the social bond. I need not repeat here the arguments based on the social bond and the norm of sensitivity, but I want to expand on what was said there about the importance of widening one's horizon for the sake of physical health and vitality.

Before proceeding, however, let us turn to a possible objection. Some philosophers may consider it strange to argue for a higher need (or "value") in terms of a lower one, but this objection is based on a misunderstanding of the status of higher needs (or "values"). As we said in Chapter Three, though they tend in time to become more highly prized than the needs out of which they have arisen, higher needs only emerge because they are useful in the service of lower ones. (The classic example is the need for money, which began as a mere means in the pursuit of economic well-being and ends with money becoming an end-in-itself.) The emergence of higher needs is analogous to the emergence of life itself out of lifeless matter and just as mysterious. The alternative to arguing for higher needs on the basis of the fulfillment of lower ones is to abandon all argument and turn to intuition or the "mystic vision." As stated in Chapter One, those who find the mystic vision persuasive have no need for argument. On the other hand, if higher needs had no earthly use, they would indeed be "useless," and human beings would not adopt them, and if they had, they would abandon them. With this in mind, let us return to the matter at hand, which, as I indicated, is to expand on the benefits, to health and vitality, of spiritual growth.

For the immature it is natural to see things from a narrow and egotistic perspective, but some individuals find it easier to relax their overweening self-concern than others. According to the theory of

Apter (cf. Chapter Three), individuals come in two varieties: one, those who are characteristically under tension and find challenge distressing; the other, those who are characteristically relaxed, and have an appetite for excitement. In this theory, those who are characteristically in the relaxed mode are called *paratelic dominant* and those who are characteristically tense, *telic dominant.* The paratelic dominant are full of vitality, laugh off troubles, and meet challenge with zest. There is sufficient medical evidence to conclude that the paratelic dominant not only enjoy life more but tend to live longer than the telic dominant (Cecil [1985], Cousins [1979], Apter [1989, p.168], Kohn [1999]). The telic dominant are anxious and preoccupied with success, and consequently never have enough of security, power, or wealth. Preoccupation with success—which William James calls "that bitch goddess" (cf. Price [1954])—causes them at times to feel that life is empty. Having no time for others, and being self-centered and incapable of love, their personal troubles range from loneliness to separation, broken homes, and divorce—all of which results in more stress. While the paratelic dominant enjoy challenge, challenge is pathologically stressful to the telic dominant. That is to say, their autonomic nervous systems are in a constant turmoil, making them prone, if not to suicide, then to psychosomatic illnesses such as alcoholism, high blood pressure, and ulcers.

The paratelic dominant do not narrowly restrict their interests to getting ahead in life, but having a relaxed state of mind, they find other things of more interest than their own welfare. For the paratelic dominant it is natural to follow the gospel message (Mathew 6.34): *Take therefore no thought for the morrow: for the morrow shall take thought for the things of itself.*

As we noted in Chapter One, there are at least two ways to health and vitality: first, let your dominant mode be one of relaxation; second, widen your horizon and develop interests that are outside your own skin. For those who already govern themselves by the principles of higher development that we have discussed previously—sensitivity, sapience, self-reliance, and dedication—this is not too difficult to do. I conclude that not only the norm of transcendence, but the requirements for sensitivity, creativity, self-reliance, and dedication may be supported on the ground of health and vitality.

As a final argument for transcendence, consider transcendence as a means to identity. We have already quoted the eminent neurologist Goldstein to the effect that the feeling of unity characteristic of transcendence (or what Goldstein calls "the sphere of immediacy") is the deepest foundation for the experience of well-being and for "self-realization." All explorers, including inventors and philosophers, owe their identity, at least in part, to a desire for transcendence. Russell, out of a desire to establish for all time the transcendental world of mathematics, including the completed infinite, became for a time the pre-eminent philosopher of the modern world. On the other hand, Wittgenstein's desire to transcend the limits of language and reason led him to a distrust of philosophy (at least as ordinarily practiced), and to respect Kierkegaard's "leap of faith" (cf. Monk [1990, p. 310]).

In Chapter Three, it is claimed that transcendence can be used as a means of furthering the impersonal ends with which one is identified. (Note the word "impersonal" here.) There are several reasons why this is true. First of all, anything that improves health and vitality will further whatever other ends a person might have, provided those ends are consistent with health and vitality. (Some personal ends such as vengeance are not in my opinion consistent with health and vitality.) Second, transcendence reduces narrowness and improves the thought process. Many mystics think that transcendence is related to genius, but I say no more about this here since it is considered in the next chapter (Chapter Seven: *Beauty, Mysticism, and Religion*). Finally, those who transcend their old selves, attain to an authenticity which cannot be had by those who follow the easy path of going along with whatever is in fashion or whatever their peers say. They become, as Maslow [ibid.] says, "independent of their culture and environment" and as such they become "themselves."

To sum up: Here and in Chapter One, I give reason to believe that what is important in life can best be obtained by escaping from the "prison of the self." On the other hand, I have argued that the way of transcendence is a way of becoming uniquely "oneself," no longer completely molded by culture and environment. Moreover, I anticipate the conclusion of the next chapter, that transcendence of the kind found in mystical or peak experience or in what Spinoza calls "the intellectual love of God" yields the attainment of importance to the highest degree. There is, I believe, considerable physical and psychological evidence to back up these contentions. Finally, having defined "morality" in such a way that one is moral if one treats living things as ends in themselves rather than as mere means, I conclude that if you are a person who has satisfied the precursors to growth, the norm of transcendence is valid for you, and, what is more, in order to fully satisfy your deepest needs, it is required that it shall be the case that you are moral.

4 Self-actualization

Most moralists would agree that the highest development of a human being involves what I call "self-transcendence." Maslow, nonetheless, calls the style of life of the kind characterized by growth, "self-actualization." It might be conjectured that growth, as defined here, is identical to self-actualization. I intend to make as strong a case for this thesis as I can. All references to Maslow, unless otherwise stated, are to Maslow [1970a].

The characteristics of a self-actualizer may be encapsulated as follows: creativeness, problem-centering, breadth (*Gemeinschaftsge-fühl)*, kindness, integrity, autonomy, resistance to enculturation, detachment, continued freshness of appreciation, spontaneity, realistic orientation, acceptance (self, others, nature), democratic character, oceanic feeling, "philosophical" sense of humor, and integration (of polar opposites). These characteristics will be explained in more detail as we go along.

Suppose that a person whom I call "Ev"—short for "EVery woman or man"—is a person who conforms faithfully to all five principles of higher development. For convenience I use the feminine pronoun. The problem then is to show that Ev has all, or nearly all, of the traits of a self-actualizer.

Many of these traits follow from self-reliance of which we may mention *autonomy, resistance to enculturation, self-acceptance* and *the need for privacy.* Obviously the first two traits follow almost by definition, so we need only consider the remainder. As Emerson [ibid.] notes in a famous essay, self-reliance depends on self-trust—and therefore on *self-acceptance.* He writes, *Trust thyself, every heart vibrates to that iron string.* Moreover, it is easy to see that privacy is a virtual necessity for a self-reliant person. Housewives, for example, whose privacy has been destroyed by the retirement of their husbands, find it difficult to remain self-reliant and autonomous.

Because of a need for detachment and privacy, self-reliant individuals tend to restrict their circle of intimate friends and this may give them an appearance of unsociability. According to Maslow, attachment results from "deficiency-motivation—i.e. motivation based on lower and intermediate needs, especially love, safety, respect, prestige, and belongingness, rather than on the higher needs. He adds (and I quote): *Deficiency-motivated people **must** have other people available.* [emphasis in original]. Nonetheless self-reliant people may have deep personal relations and be kind in their dealing with their fellow men and women. In illustration of this point, I know a self-reliant man—one of the kindest and most genial persons one can imagine—whose daughter, when asked to assemble a group of the man's friends to celebrate his fiftieth birthday said, "He has no friends!".

I conclude that self-acceptance, together with a need for privacy and a modicum of detachment—as well as autonomy and resistance to enculturation—are virtual necessities for Ev.

We come next to another cluster of traits of the self-actualizer: *breadth, kindness, integrity, creativity, "continued freshness of appreciation," spontaneity*, and *"problem-centering."* Maslow does not use the word "integrity," but I think the trait of integrity, as a characteristic of a self-actualizer, is implied when he, Maslow, writes about his subjects as follows: *...they rarely showed...the confusion, the inconsistency, or the conflict that are so common in the average person's ethical dealings.* Since Ev conforms to the principles of higher development and since "continued freshness of appreciation" is practically a definition of aesthetic sensitivity, it is easily seen that she conforms to all of the traits in this cluster. I confine my comments, however, to the last two traits listed, i.e. "problem-centering" and spontaneity.

We have already noted that people who have an authentic identity are, in Maslow's sense, problem-centered; that is to say, they are involved in a cause outside their own skin. Since Ev obeys the dedication principle, she has an authentic identity and is therefore "problem-centered." Moreover, it does not avail one much to be "problem-centered" if one is not spontaneous. Russell [1916b] observes: *A life governed by purposes and desires, to the exclusion of impulse, is a tiring life; it exhausts vitality, and leaves a man, in the end, indifferent to the very purposes which he has been trying to achieve.* Accordingly, as a mentally and physically healthy person who satisfies the precursors to growth and has a "continuous freshness of appreciation," Ev must be spontaneous as well as "problem-centered." I conclude that our paragon is not only problem-centered, but spontaneous.

I turn next to what Maslow, following Freud [ibid.], calls "*oceanic feeling*." Maslow divides his subjects into "peakers" and "nonpeakers," depending on the presence or absence of what he calls "peak experience." The peakers are those prone to mystical experience or at least to what Freud calls "the oceanic feeling." For those who have sufficient aesthetic sensitivity, peak experience is perhaps almost a foregone conclusion. Suffice it to say here that without mystical or peak experience it may be difficult to be sensitive, creative, self-reliant, *and* self-transcendent. I assume it is likely therefore that our paragon, Ev, is a "peaker."

Consider next the characteristic Maslow calls "*realistic orientation.*" Maslow explains this as not only having a "more efficient perception of reality and more comfortable relations with it," but being "generally unthreatened and unfrightened by the unknown". Maslow goes on to say:

> They [self-actualizers] do not cling to the familiar, nor is their quest for truth a catastrophic need for certainty, safety, definiteness, and order...They can be, when the total objective situation calls for it, comfortably disorderly, sloppy, anarchic, chaotic, vague, doubtful, uncertain, indefinite, approximate, inexact, or inaccurate...Thus it comes about that doubt, tentativeness, uncertainty, with the consequent necessity for abeyance of decision, which is for most a torture, can be for some a pleasantly stimulating challenge, a high spot in life rather than a low.

Russell calls this realistic orientation a "robust sense of reality." Does our paragon, Ev, have these characteristics? Some mystics do not have them, but I believe these characteristics are indispensable for those who are serious in their quest for the attainment of importance—especially that aspect of importance that involves perspective, significance and understanding. One must put blinders on oneself—because of some fear of the truth—if one lives in a world of illusion, and I conclude that it is narrowness and self-interest, together with its allies dogma and self-illusion, that keep one from having a realistic orientation. I assume therefore that Ev exercises sufficient control over her mystical tendencies to achieve what Maslow calls a "realistic orientation."

The next trait to be considered is the one that Maslow calls "*acceptance*." I have already contended that Ev has self-acceptance; consequently, I need only show that acceptance of nature and of others may be shown to follow from one or more of the following traits which we have accepted as established: realistic orientation, detachment, autonomy, resistance to enculturation, and aesthetic sensitivity. This may be shown as follows.

First of all, because of detachment, autonomy, and resistance to enculturation, Ev would not be influenced unduly by others, and could therefore judge people and things, not by repute, but by their own qualities. Second, because of a realistic orientation, Ev would not tend to be disappointed in herself or others and would not be put off by the strange and unfamiliar. On the contrary, because of freshness of appreciation and spontaneity our paragon is likely to find something interesting in everything and everybody. Thus, I claim, Ev would feel, like Whitman or Blake, that absolutely nothing is "ordinary." Whitman's statement that even a common mouse "...is miracle enough to stagger sextillions of infidels" is, I claim, typical of the aesthetically sensitive person.

Moreover, I conjecture that a feeling of the wonder of things would cause Ev—to use a phrase of H. G. Wells [1891]—to "rediscover the unique" in all things, and therefore not only accept, but appreciate, not only others but self and nature too. I conclude that it is reasonable to assume that Ev has acceptance of self, nature, and others.

Next let us consider "*democratic character structure*." By this Maslow means essentially not basing one's personal relationships on class, education, political belief, race, or color. This means that a self-actualizer is not—in Whitman's phrase—"a respecter of persons." Transcendence, or lack of narrowness, would tend to give our paragon, Ev, a character structure of this kind. It may be supposed therefore that she has not only acceptance of others, but a democratic character structure.

One fascinating feature that Maslow found in his subjects is that—in a way that may seem mystical—they have an *ability to accept ambiguity.* Maslow writes,

> The dichotomy between selfishness and unselfishness disappears...because in principle every act is *both* selfish and unselfish...Duty cannot be contrasted with pleasure, nor work with play when duty *is* pleasure, when work *is* play, and the person doing his duty and being virtuous is simultaneously seeking his pleasure and being happy. If the most socially identified people are themselves also the most individualistic people, of what use is it to retain the polarity? If the most mature is the most childlike? And if the most ethical and moral people are also the lustiest and most animal? [emphasis in original]

It is easy to see how some of these dichotomies are resolved for those who abide by the five requirements for higher development. Consider the dichotomy of duty versus pleasure. It is through love that a duty can be made into a pleasure. For example, the mother's duty of nursing her child is a pleasure if she loves the child. Ev loves more widely and deeply than others. Were it not for the fear of being thought mawkishly sentimental, I would write: love is, as it were, the philosopher's stone that transmutes the baser metal of "duty" into the gold of joy.

Another whole class of Maslow's dichotomies may be resolved through the use of what may be called "the skeptical principle." This principle, which may be expressed as "Be skeptical of extremes," is a helpful one to keep in mind in any type of exploration, whether in the study, laboratory, or in nature. A sapient person like Ev is apt to accept this principle. The skeptical principle is actually a form of the principle of the golden mean. If we apply the skeptical principle to skepticism itself, it follows that we can have either too much or too little skepticism. The very skeptical fall prey to cynicism and lose their zest for life and freshness of appreciation, but those who are entirely lacking in skepticism suffer from a lack of realistic orientation.

Let us attempt to apply the skeptical principle to several dichotomies: say, asceticism and hedonism, on one hand, and mysticism and logic, on the other. If one accepts the skeptical principle, one is not likely to come down on one side or other. But can one be both hedonistic and ascetic, and both logical and mystical? Consider first, the dichotomy between mysticism and logic. Since the whole of an infinite set is equal to the part, the concept of infinity appears to be contradictory and mystical: yet the mathematics of infinity is logical. The mystical principle that says the part contains the whole even appears in physics nowadays, e.g. in holography. As another example, consider Gödel's incompleteness theorem. In this example, Gödel skirts paradox and contradiction; yet the result is logical and valid. Moreover, mathematics, the most certain of our reasoning, is riddled with paradox and contradiction (cf. Coughlin [1983]). Now consider the dichotomy between hedonism and asceticism. The gap between the two is not absolute, for the intelligent and aware person knows intuitively that deprivation can lead to great joy. Thus food tastes better if you are hungry, sex is better after abstinence, a great play often integrates tragedy with comedy, fasting brings pleasure by renewing the system, nonsmoking brings more pleasure to living by clearing the lungs, exertion in exercise releases endorphins and brings on a high, sensory deprivation leads to peak experience and greater energy. (According to St. Teresa [1961 ed., p.137] *only God could bring such delicious pain!*) The joy of discovery is usually preceded by tremendous effort and pain. Moreover, as Maslow points out, one may be both ethical and moral and yet lusty and animal. One may suspect that the best strategy for maximizing the attainment of importance is to learn to integrate many of the polar opposites.

Finally, Maslow reports that all his subjects have a sense of humor that is not of the ordinary type. He writes:

> They do not consider funny what the average man considers funny. Thus they do not laugh at hostile humor (making people laugh by hurting someone) or superiority humor (laughing at someone's inferiority) or authority-rebellion humor (the unfunny, smutty joke). Characteristically what they consider humor is more closely allied to philosophy than to anything else.

It is easy to see why our paragon, Ev, who conforms to the basic principles of growth, might not have a sense of humor of the common type. A vulgar sense of humor is, as Maslow points out, often unkind, and is often motivated by a repressed need for rebellion. In the next chapter (*Beauty, Mysticism, and Religion*) we return to this subject and, in that chapter, I give reasons for believing that one of the signs of growth is a sense of humor allied to philosophy. Suffice it to remark here that philosophy (and logic) is often so funny that I, along with a colleague, have throughout the years considered writing a book entitled "Laughing our way through philosophy." Anticipating the discussion in the next chapter, I conclude here that it is probable that our paragon, Ev, would have a sense of humor that one might call "philosophical."

To sum up: from the character structure of those who are governed by the five norms of higher development, we have been able to derive most, but perhaps not all, of the traits that, according to Maslow, characterize self-actualization.

5 Concluding remarks

The importance of unity in the attainment of importance and growing toward perfection is emphasized by Whitehead [1938], who writes as follows: *...there are two aspects to Importance, one based on the unity of the Universe, the other on the individuality of the details. The word "Interest" suggests the latter aspect; the word "Importance" leans toward the former.* Moreover, in his essay, *MATHEMATICS AND THE GOOD*, Whitehead [1941, p. 693] writes: *...the world of Value* [i.e. Importance] *emphasizes the essential unity of the many...*

He goes on to say: *Morality is always the aim at that union of harmony, intensity, and vividness which involves the perfection of importance for that occasion.* In essential agreement with Whitehead, I think it can be said that fragmentation is the enemy of those who want to maximize the attainment of importance. This is the case, for, on one hand, importance emerges intellectually when beneath the apparent fragmentation of nature there is a fundamental unity; on the other, importance emerges emotionally when beneath an apparent isolation, the individual finds himself or herself identifying with the whole. But unity itself is not enough for the attainment of maximal importance in life. Because the higher needs include sapience, I believe that higher development, or growth, favors not only unity over fragmentation and depth over triviality, but rationality over senselessness.

One might conclude that "goodness" may be identified with *importance, unity, and rationality.* I neither affirm nor deny the latter assertion, for I wish to avoid disputes about the meaning of words. As I explain in Chapter Two, I intend to avoid meta-ethical questions. For those interested in my reasons I refer them to that chapter.

It should be clear from the present study that it is not easy to maximize importance in one's life. (Recall the oft-quoted remark of Spinoza: *...all excellent things are as difficult as they are rare.*) Moreover, those who embark on a journey of self-development cannot expect immediate understanding from others. According to Maslow, the self- actualizer *...is like an alien in a strange land. Very few really understand him, however much they may like him.* The isolation from others that growth entails cannot but bring a pain to a sensitive soul. I am reminded of the anguish of Shelley, the tribulations of Bertrand Russell, the lonely melancholy of Lincoln, and the isolation—though thoroughly respected—of Einstein. Growth, like anything else that people desire, exacts a price; but a major lesson of this chapter is that lonely isolation in which one does not—to quote Russell's favorite text—"follow the multitude to do evil," tends make one's life more significant than easy conformity.

To conclude this chapter, I make one final remark. Suppose I were asked which of the commands implied by the philosophy of life in this chapter is to be singled out as the most important. If I am allowed just one it is this: *find out what—in your depths—really matters, and use this information as a guide to expressing the excellence that is within you.* This command is in accord with the biblical injunction: *Let your light so shine before men, that they may see your good works...*

Chapter Seven

Beauty, Mysticism, and Religion

1 Introductory remark

So far in this work we have failed to explore the ecclesiastical system, a deficit which, as promised in Chapter Four, is made up in this chapter. Moreover, this chapter fulfills a commitment (given in the previous chapter) to a full discussion of the nature and importance of aesthetic appreciation. The connections among beauty, mysticism, and religion are adumbrated in the previous chapter where it is remarked that there is a "transcendent beauty to life that goes beyond dedication and striving." In this chapter, we make use of concepts like importance and transcendence not only to attempt to understand beauty and its relation to mysticism and religion, but to explore the function of the ecclesiastical system in an age of science and skepticism. We begin the discussion with an exploration of the concept of beauty.

2 Beauty

Beauty, in contrast to religion, only becomes a desideratum for those who have risen above the survival level. On the other hand, being the lowest of the higher needs, beauty is one of the first things that people turn to when they have risen above the survival level.

What makes a thing beautiful? Most men find women to be

beautiful, but others prefer men. Standards of beauty seem to vary not only from observer to observer but from tradition to tradition. According to ancient Chinese tradition, women are considered beautiful only if they have tiny feet, but in the West the size of feet is a minor consideration. On the other hand, certain things are considered beautiful in all ages and nearly all cultures. Chinese paintings are admired in the West and Occidental music is played by orchestras in China, and from such examples one might draw the idea that there are objective norms for beauty.

There is a simple way to increase the apparent beauty, significance, and importance of an object; this is by deprivation. For example, sexual abstinence, on the part of the male, can increase the beauty of a wife in the eyes of her husband. Another technique for increasing beauty, that of meditation, is a species of sensory deprivation. This may not be an argument for asceticism, however; for it is probable that the optimal way to obtain what Maslow calls "the continuous freshness of appreciation" is by temperance where temperance is defined as the golden mean between feast and famine, or as Aristotle put it, between licentiousness and insensibility.

A capacity, like that of Walt Whitman, for the appreciation of beauty is one of the criteria that Maslow uses for classifying people as self-actualizers and may therefore be regarded as one of the measures of spiritual growth. A capacity of this kind is not entirely innate, however; on the contrary, it often has to be taught. There are at least four different domains in which a capacity for the appreciation of beauty can be developed—religion, science, morals, and art. No person who fails to develop a capacity for aesthetic appreciation in any of the areas can be considered fully matured. In the remainder of this section, I examine the domains of science, morals, and art to determine the role of beauty in each of them. (At this point I omit religion since it is a topic for a later section.)

2.1 Science

Science has been blamed for the production of ugliness and the degradation of nature. Blake held science responsible for England's "dark Satanic mills." Science certainly has a dark side. In the present age, for example, science has produced horrors like the atomic bomb. But the true scientist is more akin to a poet than to the creators of the industrial revolution. Shelley [1821] wrote that one of the functions of the poetic faculty is to "create new materials of knowledge." According to Shelley, poetry is the "root and blossom of all other systems of thought." If this is so then it is possible to defend to some extent Keats' famous phrase: "beauty is truth, truth beauty." Shelley [1817] praises the beauty of the poetical imagination—and derivatively the beauty of truth in science and philosophy—when he writes in his *HYMN TO INTELLECTUAL BEAUTY*:

> Thy light alone, like mist o'er mountains driven,
> Or music by the night wind sent
> Through strings of some still instrument,
> Or moonlight on a midnight stream,
> Gives grace and truth to life's unquiet dream.

As typical of the sciences consider mathematics. Plato regards mathematics as not only the touchstone of science but of all intellectual activity. Because of the mathematician's concern for intellectual beauty, mathematics has that "unity in variety" that according to classical aesthetics (Cf. Stolnitz [1967]) heralds the presence of beauty. According to Russell [1917]:

> Mathematics, rightly viewed, possesses not only truth, but supreme beauty—a beauty cold and austere, like that of sculpture, without appeal to any part of our weaker nature, without the gorgeous trappings of painting or music, yet sublimely pure, and capable of a stern perfection such as only the greatest art can show.

Beauty in mathematics, insofar as it exists, is generated by mathematicians, who, as artists, are guided as much by aesthetical considerations as by truth. Moreover, I claim aesthetic considerations are present in the work of all scientists. For example, the beauty in Einstein's equations convinced him of their truth even before observations on the paths of light rays near massive bodies confirmed it. Art has been traditionally thought of as an imitation of nature, but the pure mathematician is an artist who produces equations not necessarily modelled on nature. When nature conforms to the pure mathematician's imagination—for example, noneuclidean geometry—it almost seems that nature imitates art. What nature shares with mathematics is sublime beauty, and therefore beauty offers a clue to the structure of nature. This suggests that the Platonic theory that there is an underlying reality of which both art and nature partake may be correct.

2.2 Morals

Is there a connection between beauty and morals, the next item on our list? Kant [1788] evidently thought so. Recall the famous passage in which he says: *Two things fill the mind with ever-increasing wonder and awe...the starry heavens above me and the moral law within me.* And according to Plato in *SYMPOSIUM* [201] things which are good are also beautiful. Moreover, in *PHILEBUS* [64] Plato has Socrates identify virtue with measure and symmetry, and therefore concludes that "the power of the good has retired into the region of the beautiful." For the Greeks the word for beautiful, *kalos*, also stood for qualities of excellence in general, including excellence of moral character. Nowadays we say that a person may be beautiful on the outside but ugly on the inside and when a beautiful woman has "inner beauty" we say that her beauty is more than "skin deep." For example, the beauty of the movie star, Ingrid Bergman, was at first thought to be more than skin deep, but opinion changed when she had a lapse of "morality." "Goodness" may therefore be regarded as a species of "beauty."

2.3 Art

Art has two major ends: one is *beauty* and the other is *function.* According to Hume [*TREATISE (*III, iii, 1)], function and beauty are related to each other in the sense that "most works of art are esteem'd beautiful, in proportion to their fitness for the use of man, and even many of the productions of nature derive their beauty from that source." Works of art in music or pure mathematics, however, which the vulgar might regard as "useless," are regarded as useful by Hume because they give pleasure. The eminent psychiatrist, Freud [1930], writes: "People who are receptive to the influence of art cannot set too high a value on it as a source of pleasure and consolation in life." Sometimes beauty does not derive from usefulness, but usefulness derives from beauty. For example, beauty is used to attract and in modern culture beauty is used to sell everything from automobiles to hotel rooms. Attractiveness is obviously important in sex: the attractiveness of flowers causes bees to pollinate; the attractiveness of peacock feathers is crucial in the mating of peacocks; lovelight in the eyes of a female human increases her attractiveness; etc. Pleasure in the appearance of the beloved produces a strong bond and the illusion of beauty—if it is an illusion—has therefore great social utility.

Complaining of art in his own day, Shelley [1821] writes: *Comedy loses its universality: wit succeeds to humour; we laugh from self-complacency and triumph, instead of pleasure; malignity, sarcasm, and contempt, succeed to sympathetic merriment...* Shelley, in the same passage, writes that humor which is of the vulgar type is *ever blasphemy against the divine beauty of life.* For sensitive people—and perhaps prudes—nothing destroys aesthetic quality in art more than vulgarity.

Why is this? I believe it is because vulgarity, though it may please the multitude, ultimately fails as art because it is not only hurtful, but there is nothing significant in it: it does not stimulate the imagination nor expand anyone's horizon. As already mentioned in the previous chapter, Maslow considers as one of the marks of growth the development of what he calls a "philosophical sense of humor" as exemplified by such men as Lincoln, Lewis Carroll, Twain, or Shaw. Such humor is gentle and refined rather than hurtful; moreover it does stimulate the imagination. It is for this reason that one can say that the comedies of playwrights of genius such as George Bernard Shaw are generally of a higher aesthetic quality than most comedies.

More generally, in the fine arts, I think aesthetic quality has some relationship to information content. The concept of information, as used in the theory of random processes in statistics, is a measure of "surprise." Surprise, in turn, is a measure of what, in earlier chapters of this work, we called *significance.* There is thus an obvious connection between information and significance since a message that contains no information in the technical sense is insignificant in our sense. In my opinion, however, a subject or an art is trivial—and therefore of lower aesthetic quality—if it is merely surprising or shocking, but has no significance in any deeper sense.

For example, consider poetry. According to Shelley [ibid.]:

> A poet participates in the eternal, the infinite, and the one; as far as relates to his conceptions, time and place and number are not....Poetry lifts the veil from the hidden beauty of the world, and makes familiar objects as if they were not familiar; it reproduces all that it represents, and the impersonations clothed in its Elysian light stand thenceforward in the minds of those who have once contemplated them, as memorials of that gentle and exalted content which extends itself over the thoughts and actions with which it coexists.

I conclude that sublime poetry, like that of Homer, Sappho, Lucretius, Shakespeare, Dante, Milton, Goethe, Blake, Shelley, Whitman,

Dickinson, and Eliot, does not merely excite by means of rhythm and striking metaphor, but has a deeper significance.

2.4 What Is Beauty?

Twain, always the jokester, said of Wagner's music that it is "not as bad as it sounds." The moral is that whether something is beautiful depends on how you classify it (or how you look at it). Mary Shelley's story of Frankenstein, classified as fiction, is beautiful, but to the child the story—at least in the movie version—is real and horrible. A man that does not fit the prototype of what a man looks like is not regarded as "handsome" but as "ugly." Moreover, a bird that looks like a swan cannot be a beautiful duck. In order for something to be recognized as beautiful an object must be a reasonable representative of its class; that is to say, we must be able to say of the object, "that is a beautiful _______" where the blank is filled in by some noun; for example, that is a beautiful painting, that is a beautiful production of Hamlet; that is a beautiful horse, that is a beautiful statue, that graffiti is beautiful, etc.

As we already noted in the previous section, for something to be beautiful it must have some *significance* or *importance*, where I use the terms "significance" and "importance" in the sense they were given in Chapters Two and Three. Objects of beauty range from the meretricious to the sublime depending on their significance. Beauty derives, on one hand, from measure and harmony, and, on the other, from surprise. An object that has a high degree of real significance astonishes—or takes one's breath away—perhaps (as Hartman [1967] would say) because of its rarity. A painting by (say) Picasso is exciting and attention getting because it flirts with chaos and yet, in a mysterious way, it has measure and order. Like all higher ends, beauty delights because it extracts out of the world something of sense where otherwise there might be only frustration and chaos. A work of art (like the Arab's rug in *OF HUMAN BONDAGE*) is a metaphor for life, and just as one hopes to extract sense out of life, it does the same in its limited domain.

Like all talk of "values" it is easy to talk nonsense when talking about beauty. According to Freud [ibid.]: *The science of aesthetics...has been unable to give an explanation of the nature and origin of beauty, and, as usually happens, lack of success is concealed beneath a flood of resounding and empty words.* Despite this, and in spite of Popper's injunction against "what is" questions, I think the identifying characteristics of *aesthetic quality* are such that one might hazard a definition. If I were asked, what is beauty?, I would say (as

Hartman [ibid.] might say) *a beautiful x is an x that not only satisfies the class-concept of an x, but an x that tends to generate an experience of enduring significance or importance in the sense of Chapters Two and Three.* There is an element of idiosyncracy in beauty since not only does the paradigm for what is a proper representative of a class-concept vary from culture to culture, but as we pointed out in Chapter Three what is significant (or astonishing) depends to a great extent on the individual. Walt Whitman was surprised by everything: recall his line about a mouse, quoted in the previous chapter. We have noted that Shelley claims that poetry has the ability to foster such a freshness of appreciation as well as other qualities of the kind that occur in highly developed individuals.

In Chapter Three we discussed the role of arousal in human motivation. We said that an individual is either in a mode of arousal seeking or arousal avoiding, and in the arousal seeking state there is an optimal level of arousal. And art, particularly music, owes some of its aesthetic quality to arousal. The beat in rock music, which is close to the human heart beat, is both arousing and soothing, at least to the young. Light, which arouses us in the morning, and gives us the bright color of flowers and the glitter of jewelry, exercises a mysterious fascination. I believe the arousal factor of light and color is a major source of beauty in art, decoration, and landscape. The blind are deprived of this source of beauty, but for those among them who are mystical there is still the inner light. Mystics report being blinded by an inner light and often regard it as a favor of God.

It follows from our definition of beauty that for an object to be beautiful it must satisfy some paradigm. Paradigms have a disturbing drawback; they foster prejudice and make one blind to other forms of beauty. New music or new art tends to be rejected because it does not fit the current paradigm. A racially prejudiced white American is an American who recognizes only other white Americans as proper representatives of the class of Americans. A journal article that is unusual or unfashionable is rejected for "triviality." As in politics, morals, and science, old paradigms die hard.

Fashions change, however, and one can pinpoint an era by pointing to its art as well as its science, politics, and morality. Of course one may question whether a change in an artistic paradigm is an advance. Nonetheless, as we already have had many occasions to point out, paradigms are subject to critical examination. And art, unlike other domains, is only limited by a failure of imagination and, as Shelley points out, it is the function of the poets, in the broadest sense, to enlarge the imagination of humankind.

2.5 A note on language

So far we have said nothing about language and art. Imperatives enter into art through requirements and admissibilities. Function (or utility) has to do with both requirement and admissibility, but beauty has mainly to do with admissibility. For example, consider the art of bridge building. That it is required that it shall be the case that a bridge is built is determined by the utility of the bridge; that it is admissible that it shall be the case that the bridge is built *according to such and such specifications* is determined partly by function, partly by resources, and partly by aesthetic considerations.

3 Mysticism

The misunderstanding and distrust of peak experience, meditation, and mysticism by even "the great" is astonishing. For example, Russell is said to have said: "I see no reason why the religious mystics should not be placed under the category of drunkards rather than prophets." (This is a curious statement by the author of *IN PRAISE OF IDLENESS*—cf. Russell [1932].) Russell, however, was by no means impervious to religious sentiment. He is said to have refused to enter a Catholic Church when a choir was singing, saying: "The hymns and incense and colored glasses make me confess to feelings my intellect does not approve; I want to keep my mental sky clear of the mystic clouds." (cf. Dilip Kumar Roy [1950, pp. 117, 129-130]). Freud (the great psychiatrist) is an even more astonishing example. Freud [ibid.] apparently thinks the practitioner of yoga and meditation achieves only quietness and an inhibition of the instincts. The truth is that the practitioner achieves not only peace but ecstasy. A peak experience not only tends to bring serenity into a life otherwise characterized (as we noted in Chapter Two) by what Sacks calls *...that eager and anxious tension of anticipation, of intention that drives* [people] *through life*, but tends to bring a surge of energy. Inebriation, on the other hand, not only destroys the body, but produces in the end a zombie. It is true that when the attempt to develop mystical states through meditation becomes the whole of life, it can become a selfish indulgence and therefore to be condemned on moral grounds; but when meditation is not taken as an end itself, and when it does not interfere with the full functioning of a human being, there is no reason, outside of prejudice, to disparage it.

Russell [1917], in another mood, and under the influence of Lady

Ottoline Morell, writes:

> ...the greatest men who have been philosophers have felt the need for both of science and of mysticism: the attempt to harmonise the two was what made their life, and what always must, for all its arduous uncertainty, make philosophy, to some minds, a greater thing than either science or religion.

Great as mysticism's contribution to philosophy may be, it is to religion and art that mysticism makes its greatest contribution. Mystical experience comes in two varieties: bidden or unbidden. A spectacular example of the unbidden kind is that of St. Paul. Everyone has heard the story of how Paul of Tarsus, on the way to Damascus, was suddenly changed from a persecutor of Christians into a devout believer as the result of a mystical experience so powerful that he was knocked off his horse and left temporarily blinded. Mystical experience of the bidden kind is the result of the cultivation of meditation techniques. For a fascinating example, consider the report of Gopi Krishna [1970]:

> Suddenly, with a roar like that of a waterfall, I felt a stream of liquid light entering my brain...The illumination grew brighter and brighter, the roaring louder, I experienced a rocking sensation and then felt myself slipping out of my body, entirely enveloped in a halo of light. It is impossible to describe the experience accurately. I felt the point of consciousness that was myself growing wider, surrounded by waves of light...I...no longer knew myself to be...a small point of awareness confined to a body, but instead was a vast circle of consciousness...bathed in light and in a state of exaltation and happiness impossible to describe.

Other mystics have written of similar experiences; for example, St. Teresa [1588]. She writes, in regard to "transport of the spirit" or ecstasy, as follows:

> ...he is awakened...as though by a rushing comet or a thunderclap...the soul really seems to have left the body; on the other hand, it is clear that the person is not dead, although for a few moments he cannot be sure that the soul is in the body or no. He feels he has been in another world very different from this in which we live, and has been shown a fresh light there, so much unlike any found in this life that, if he had been imagining it, and similar things, all his life long, it would be impossible for him to obtain any idea of them.

For a final example, consider Bucke [1901]:

> It was in the early spring...He and two friends had spent the evening reading [poetry]...They parted at midnight, and he had a long drive in a hansom...He was in a state of quiet, almost passive enjoyment. All at once, without warning..., he found himself wrapped around as it were by a flame colored cloud...Directly afterwards came upon him a sense of exultation, of immense joyousness accompanied or immediately followed by an intellectual illumination quite impossible to describe.

According to Wittgenstein [1922], "Whereof one cannot speak, thereof one must be silent." Nonetheless, the mystic attempts to speak the unspeakable, using language varying according to his or her own background. For example, while mystical experience occurs to both Christians and Buddhists, the Christian finds God, but the Buddhist, for whom neither self nor God exists, finds only "the void." Is it any wonder therefore that skeptics caution against accepting the message of the mystics? And how plausible is it anyway that the truth about life and the world can be established suddenly in a flash of inspiration? Russell [1945], referring to the Platonic Socrates and his mystical tendencies, says that though Socrates may be admitted to the communion of saints, he needs "...a long residence in a scientific purgatory." And yet, despite his skepticism, Russell [1917] writes:

> ... by sufficient restraint, there is an element of wisdom to be learned from the mystical way of feeling, which does not seem to be attainable in any other manner. If this is true, mysticism is to be commended as an attitude toward life, not as a creed about the world.

Maslow [1968], summing up his discussion of the effect of mystic experience, says: *In other words, any person in any of the peak experiences takes on temporarily many of the characteristics that I found in self-actualizing persons.* As we already remarked in the previous chapter, Maslow [1954, 1970a] notes elsewhere that many self-actualizers have had mystical experiences, and divides self-actualizers into "peakers" and "nonpeakers" depending on whether or not they have had significant mystical experience. If we can show that mysticism is in accord with the five principles of growth, it would appear that to become what Maslow calls a "self-actualizer" one "need only" have a mystical experience. In order to follow up on this insight, let us examine each of the five principles of growth, as presented in

Chapter Six, in the light of mystical experience. I begin with the principle of sensitivity.

First of all, note that a mystical experience often results in a profound awakening of what Shelley [1821] calls the poetic faculty. We have already quoted Shelley to the effect that one of the functions of the poetic faculty is to "create new materials of knowledge." The full quote is as follows:

> The functions of the poetic faculty are two-fold; by one it creates new materials of knowledge, and power and pleasure; by the other it engenders in the mind a desire to reproduce and arrange them according to a certain rhythm and order which may be called the beautiful and the good.

The evidence that mystical experience arouses the poetic faculty can only be gleaned by studying the biographies of artists, scientists, philosophers, and prophets. Probably the most complete study of this kind was made by Bucke [op. cit.]. Gopi Krishna describes an enormous increase in his own poetic faculty after he had his first mystical experience. Bertrand Russell [1967] testifies that as a result of a single mystical experience, he changed his entire way of looking at life in a short space of time. He writes:

> Having been an Imperialist, I became during those five minutes a pro-Boer and a Pacifist. Having for years cared only for exactness and analysis, I found myself filled with semi-mystical feelings about beauty, with an intense interest in children, and with a desire almost as profound as that of the Buddha to find some philosophy which should make human life endurable.

Maslow [1970b] describes the result of mystical experience as "resacralization." This means that the universe no longer appears ordinary, but as Wordsworth says, the mystic sees

> The earth and every common sight, ...
> Apparelled in celestial light.

According to William Blake [1946 ed., p. 657], *If the doors of perception were cleansed everything would appear to man as it is, infinite.* No matter the literal truth of Blake's statement, it is clear from the testimony of all mystics that mystical experience has a remarkable power to renew and strengthen one's aesthetic appreciation of the world—even to the extent of making the world and all the things in it appear worthy of worship. Hence, it may be concluded that mysticism is in accord with the aesthetic sensitivity requirement.

The requirement of sapience, the next need to be considered, enjoins one to be clever, curious, and creative. History is replete with examples in which mysticism leads to an increase in sapience. To cite one example, Russell [1967] discovered his well-known paradox and wrote his famous essay *A FREE MAN'S WORSHIP* after a mystical experience in 1901. If, on one hand, the mystic's search for sapience is often wrong-headed and even a little crazy, on the other, recall that Kepler was an astrologer and, moreover, it was the alchemist's search for the philosopher's stone that led to chemistry. What is more, the doctrine of the mystics Pythagoras, Parmenides, and Plato that appearance is not to be taken seriously, for reality is eternal and mathematical, not only contributed to mathematical research, but encouraged a mathematical interpretation of time, space, and matter. Even Democritus, with his appeal to occult entities, must have been encouraged by the idea that there is a reality beneath appearance. The certainty, which the scientist shares with the mystic, that the universe makes sense, that—in the words of Einstein—*God is subtle but not malicious,* has driven men like Kepler, Newton, and Einstein to our present understanding of the large scale features of the physical universe. Moreover, the poetic faculty, which mysticism stimulates, supplies the material without which the scientist cannot create his or her grand theories of the universe.

According to the mystic view, the universe has the following four interrelated characteristics: one, beneath appearance there is a reality to be explored; two, there is—as Einstein [1931] says—intelligence manifested in the universe; three, the world is not an absurd chaos, but an intelligible whole; four, life itself has significance. It follows that the mystic's view of reality encourages the sense that the universe though mysterious is comprehensible. Einstein [ibid.] calls the sense of wonder and mystery about the cosmos, which reaches its acme in the mystic, "cosmic emotion." Although the beliefs of the mystic are not always in accordance with science, the belief that the universe has the aforementioned four characteristics cannot but help inspire inquiry.

Nonetheless, the examples of India and China show that mysticism may not inspire science of the Western kind. In India, exploration, inspired by mysticism, is the investigation of inner reality together with an investigation of the mind's ability to control the body as in the practice of yoga. Exploration of this kind is not to be discounted, however, and, what is more, meditation is itself a kind of exploration. In China, Taoism has led to the study of *chi*, which the Chinese medical people use to control the body and cure disease (Cf. Eisenberg [1987]). We may conclude therefore that the mystic vision is consistent with the principle of sapience. Let us therefore pass on to the third higher need, self-reliance.

One of the great benefits of mystical experience is that the mystic is no longer driven by fear and desire. Having lost all fear of the future, particularly the fear of death, the genuine mystic—as distinct from bogus ones such as Bagwan Rajneesh—tend to follow the biblical injunction (quoted in Chapter Six and elsewhere): *Take therefore no thought for the morrow: for the morrow shall take thought for the things of itself. Sufficient unto the day is the evil thereof.* Some mystics speak of an inner voice. Socrates calls his inner voice a "daimon," some mystics call it "God." I conclude that the mystic (who follows his or her own inner voice) lives in accord with the self-reliance requirement.

It is otherwise with *followers* of mystics. The milieu of those who seek for spiritual enlightenment tends to be cultish. Maslow [1970a] warns that guru-disciple relationship leads straight to "cultishness and a retreat from rationality." (See also Maslow [1970b, pp. viii-xi].) Cult members, in the hope of attaining to nirvana, tend to follow the master like lemmings plunging into the sea. *Mystics, themselves, are at fault here, for they tend to encourage dependence.*

Dedication, the next higher need on our list, is defined as being dedicated to some cause or calling. There is little doubt that the mystic enjoys this property to the highest degree. The mystic follows Russell's maxim (already quoted): *Aim to create, not to possess.* Some mystics become religious teachers, some, like Kepler, devote themselves to science. Others, as we have already noted, are philosophers or poets.

We turn now to the fifth and final higher need, the need for transcendence. The principle of transcendence enjoins one to avoid narrowness and, in the phrase of Russell, "maintain a wide-horizon." The ability of mystic experience to expand consciousness is attested to by many authorities. Moreover the transforming power is tremendous. Maslow [1968] reports the cure "forever after" of neurotic symptoms in at least two cases as the result of mystical experience. What is more, according to Russell [1917]: *...if we have the mystic vision, we find the whole world worthy of love and worship...* If this is the case, one way to have a change of consciousness, and bring about universal love throughout the world is to induce a mystical experience on a mass scale. A dramatic tale of how this might be brought about has already been envisioned by the science-fiction writer, H. G. Wells [1906]. In his tale, *IN THE DAYS OF THE COMET*, a comet passing through the earth's atmosphere produces just such a change of consciousness.

Science-fiction aside, I claim that mystical experience not only works as a kind of psychotherapy, but is in many ways enlightening. For example, as explained in Chapter Six, what before seemed to be polar opposites now appear to be reconcilable. Dichotomies such as those between friend and foe, self and nonself, duty and pleasure, sense and nonsense, good and evil, etc. are seen as aspects of a unified whole. For example, for many of us the world appears at bottom absurd, but the mystic sees a unified whole. (For a discussion of the contradiction between a world of chaos (or nonsense) and a world of order (or sense) see the next section.) The mystic achieves an integrity or wholeness, an integrity that comes from within and not for the sake of reputation, honors, or fame. This wholeness is the result of purity of motive, a steadfastness of purpose, a lack of desire for possessions, and a knack for avoiding being trapped into unwise commitments. Since, as Russell [ibid.]in his *A FREE MAN'S WORSHIP* puts it, mystics have expelled all eagerness of temporary desire, they have little or no reason to lie or cheat. In the mystic vision, the world is seen as a unified whole and the concerns of the mystic therefore broaden to encompass the world. Since the mystic vision leads one to believe with Leibniz that "All is for the best in the best of all possible worlds," it might be expected that the mystic would be unconcerned about human suffering. Yet the facts show that this is not the case; consequently, we may conclude that mystics live in accord with the principle of transcendence.

Though the mystic may be a self-actualizer, mysticism has both a positive and a negative aspect. On the positive side mystical experience constitutes, as we have seen, a kind of psychotherapy. On the negative side, mystical experience brings many disturbances in its train. In particular, the change of personality that results from mystical experience is often disruptive and may have a negative effect on social relations. A typical example of such an effect is the downhill course of Russell's marriage soon after his mystical experience of 1901. In my opinion, however, the greatest danger from mysticism is that it tends to encourage a master-followership mentality.

Mystics, in general, are not noted for their robust sense of reality and some achievements attributed to them must be taken with a grain of salt. Sad to say, a display of supposed paranormal powers is, in all probability, the sign of a charlatan rather than the sign of a genuine transformation of character.

There are two major motives that lead people to seek mystical experience: one, the desire for an experience whose ecstasy exceeds in depth that found in love; the other, the desire for what is called "secret knowledge." For both of these reasons, many people have tried meditation in the hope of becoming mystics. Except when spontaneously induced, however, the road to becoming a mystic is a rough one. Not only do many find meditation boring, but the success rate for attaining the mystic vision through meditation is not high. According to Maslow [1970a]: *Instant self-actualization, in a moment of conversion, or insight, or awakening* ***does*** *happen, but it is extremely rare and should not be counted on* [emphasis in original]. Worse still, it is said that meditation may pose a risk of psychological damage. For a graphic description of risks of a personal type one might read Gopi Krishna [ibid.]. On the other hand, as we have noted, meditation has impressive therapeutic benefits. Medical scientists are more and more realizing the therapeutic value of meditation in treating conditions such as high blood pressure [20/20, 2/19/93]. Meditation, at the very least, offers a time of quiet in which, not only is bodily energy impressively increased, but the seething mind restores itself into balance. (The new source of energy is called "*prana*" by the Indian swamis, "*chi*" in Chinese medicine, and "orgone" by the therapist, Reich.) Though it does not necessarily cure the mind of all of its ills or produce a full-blown mystical experience, the regular practice of meditation is a technique that is helpful in enriching life and bringing about a genuine transformation of character.

3.1 Enlightenment

There are two kinds of enlightenment of which the world is in need: one is in the realm of the spirit, and the other is in the realm of practice. The need for enlightenment in the latter domain is apparent. Even God, according to the ancient myth in the Torah, has, at least once, yielded to despair for the human race. Separation between the two realms—spiritual and mundane—hardly seems possible, however; for without the development of spiritual enlightenment, it is extremely unlikely that human beings can make it in the realm of practice. Though human beings may save themselves temporarily by prudence or enlightened self-interest, I do not believe that a stable world order can be built on prudence alone. The world needs the teachings of mystics. The mystic does not obey moral rules out of prudence but out of love. Sages have always contended that what is destructive of human society is not so much the violation of moral rules as the hatred of others. As the sayings attributed to Jesus in the Gospels tell us, moral principles are subject to scrutiny, and when they do not serve humanity they may be discarded. Mystics form a nucleus for the enlightenment of the human race.

In the previous chapter, we have in effect already shown that a person who has undergone the ultimate in personal growth has no need to cheat for personal gain; for he or she has lost all self-concern and is serene in the conviction that whatever happens cannot matter that much in the ultimate scheme of things. One of the purposes of philosophy is to help people develop in such a way as to reach that state of mind—(or at least that was its purpose before it became a profession!) After all, the attitude of people like Socrates is the one that we still call "philosophical." As long as it remains true that philosophy helps people reach a higher state of development, society would stand to benefit from an encouragement of the pursuit of philosophy.

In the history of philosophy, perhaps Spinoza is the one who exhibits in purest form the attitude that we call philosophical. Spinoza teaches us to regard the world from the standpoint, not of here and now, but eternity; moreover, he claims that with a sufficient amount of understanding nothing which happens to one need matter. According to him, it is not truth, but one's *comprehension* of the truth that makes you free. Spinoza believes that if one is concerned with the world of truth—as Plato was—rather than the phenomenal world, then one is no longer a victim of the ups and downs of this world since the world of truth is immortal and imperishable. Bertrand Russell [ibid.] echoes Spinoza when he urges us in *A FREEMAN'S WORSHIP* to "burn with passion for eternal things." Spinoza, in agreement with Aristotle, thinks that the highest form of life is contemplation of the eternal world. He calls it the *intellectual love of God.* According to Watts [1970], *..."happiness" in the Aristotelian and Thomistic sense of man's true end or destiny...means union with God, or in Oriental terms, harmony with the Tao, or moksha, or nirvana.* Even the Gospel of Luke anticipates Spinoza when it says, in agreement with Plato and Aristotle, that t*he kingdom of God lies within* and urges the faithful not to lay up treasures on earth.

The lack of enlightenment in *homo sapiens* is perhaps not surprising since human beings have evolved, in large part, because of the Darwinian struggle for existence. But what is "enlightenment"? For my part I believe that "enlightenment" is nothing but a transformation of the spirit or psyche that leads one to live in accordance with the principles of higher development.

As we have already noted in many places in this book, Maslow writes of a hierarchy of needs, and recognizes that although people do "not live on bread alone," bread is a prerequisite for the development of the human potential for the spiritual life. But we hardly need anthropologists and psychologists to tell us that the higher things in life are lost on those who have starving children and do not have the basic things of life like food, water, clothing, and shelter. Aristotle [11773b33] recognizes this, and therefore counsels us first to fulfill in an optimum way the nature humans share with animals before embarking on a task, essential for real happiness, which is to develop our "divine" nature (the intellect) and "make ourselves immortal," i.e. become like gods.

Aristotle regarded wisdom as being the divine part of human nature and as such may be identified with what in the *GOSPELS* is called the "kingdom of God." According to *MATTHEW* [7.33] if you first seek the kingdom of God, all the other needs will be taken care of. Wisdom is not sought for the sake of pleasure, however; yet one may argue that in a mysterious way, the highest pleasure for a human being is not possible without wisdom. I believe that mystical experience, in spite of being ineffable, is an indispensable source of that wisdom.

4 Religion

Primitive people tend to worship power. The Pharaohs of Egypt were god-kings. The Babylonian kings annually received their royal authority from the priests of Marduk, the chief Babylonian deity. Even in the democratic Athens of antiquity, there was an established religion based on the worship of the goddess Athena; moreover, as witnessed by the fate of Socrates, impiety was a crime punishable by death. The Jews from the time of Abraham onward were defined by their relation to the god *JHVH*—the very name "Israel" being derived from the Hebrew word meaning "wrestler with God"; moreover, even today Israel is a theocratic state. In imperial Rome, the Caesars were semi-divine and, moreover, refusal to sacrifice to the gods was punishable by death; what is more, Rome became even more of a theocracy when Constantine made Christianity into a state religion. Some three hundred years after Constantine, around 622 C.E., Muhammed founded Islam, which, like Christianity, was spread by the sword. Jesus, himself, whom some called "The King of the Jews," is said to have said [Matt:10.34-35]:

> Think not that I am come to send peace on earth; I came not to bring peace, but a sword.

> For I am come to set a man at variance against his father, and the daughter against her mother, and the daughter in law against her mother in law.

[So much for filial piety!] The relation of religion in the Far East to power is more complicated, but even here it is rare to find a separation of church and state. Tibetans, for example, have traditionally lived in a theocratic state (although "theocratic" may not be quite appropriate in this case). Even the Chinese emperors received their mandate "from heaven." Moreover Indians to this day are largely governed by Brahmanism and the Japanese have until recently lived in accordance with a religion (Shintoism) in which the emperor is treated as a god.

In every state the dominant religion attempts to take over the reins of power. In the United States, for example, as well as in modern India, there is a continuous struggle to maintain the separation of church and state. In the marvelous phrase of Norman Mailer [1998]), "...the fundamentalists search for an all-purpose power grip over other humans." Although many countries (like Sweden and Great Britain) have functioned quite well with a state religion, organized religion, especially when allied to the state, remains a significant threat to human liberty.

Religion, at least in the West, acts like a combination of the good cop and the bad cop. The bad cop arouses fears, but the good cop overwhelms you with kindness. Freud [1930] describes the Christian religion as the system of doctrines and promises which, on one hand, explains the riddles of this world with enviable completeness and, on the other, assures its adherents that a careful Providence will watch over their lives and compensate them for any frustrations that they suffer here. Freud goes on to say: *The whole thing is so patently infantile, so foreign to reality, that to anyone with a friendly attitude to humanity it is painful to think the great majority of mortals will never be able to rise above this view of life.*

The motives that lead the mass of people to accept religion are, I am afraid, much the same as mine were as a child. These motives are clear and simple: filial piety, fear of damnation, and the need for acceptance. Filial piety led me to believe what my parents wanted me to believe even when it became apparent (e.g., when they lied about Santa Claus) that they were not infallible. Like many in the West, I feared that if I did not believe strongly enough I would surely die, or worse yet, roast in hell. What is more, the need for acceptance made me fear not only being thought wicked but of *being* wicked.

The search for power is the source of much in human life including religion. Russell [1917] writes: *Brief and powerless is Man's life; on him and all his race the slow, sure doom falls pitiless and dark. Blind to good and evil, reckless of destruction, omnipotent matter rolls on its relentless way...* What Russell says is, in essence, true, and therefore it is little wonder that human beings reach out to "higher powers." Human beings, being relatively powerless, seek either to obtain power or, failing that, propitiate the powers that be. Belief in God is, at a minimum, a belief in a "higher power." On earth, the source of power in nature is primarily the sun; as a consequence, Amenhotepp IV (Ikhnaton) claims that the sun, whom the Egyptians called *Aton*, is the only god. The claim of Jesus of Nazareth to divinity is based at least in part on his god-like powers; for example, the ability to walk on water and turn water into wine. The belief in the West that there is a *single* higher power, which leaves "evil" unaccounted for, is an embarrassment to theologians. To account for "evil," the Jews, perhaps influenced by Zoroastrianism, invented a rival power they called *Satan*. Satan may not seem as credible in the West as he once did, but many Christians continue to believe in more than one higher power (e.g., Mary, "the mother of God," as well as a host of angels, dominations, and powers. Mary is thought to intercede in behalf of the faithful).

While in the West people tend to worship external sources of power, people in the Far East tend to respect the power within. Yogis claim that by turning inward one can obtain supernatural powers, powers like levitation and teleportation. Buddhists claim that meditation gives them the power to vanquish craving. The Taoists emphasize yielding and renounce all strife after power, while the goal of the Hindu is to achieve oneness with the source of all power.

Freud's criticism of religion, though correct as far it goes, is narrowly based. If this were all there is to say about religion there would be little point in discussing it further. But religion has an important role in society which, as described in Chapter Four, is to help bring "meaning," beauty, and fellowhip into the lives of people. It follows that religion not only helps people to make sense out of their lives, but in bringing people together in a common faith helps strengthen the social bond. Moreover, in bringing "meaning" and beauty into life, religion allies itself with art and psychotherapy in an effort to uplift the human spirit. I conclude that religion has two main functions (one, social; the other, individual): the first is to strengthen the social bond; the other to uplift the human spirit and to help people make sense out of life. Let us consider these two functions separately.

First the social function.

Religion strengthens the social bond not only through the sharing of a sacred message and sacred symbols and ceremonies, but by the warmth and fellowship in occupying a sacred place of worship. Moreover, through its role in establishing institutions for healing the sick, tending to the wounded, and educating the young, religion tends not only to remove obstacles to personal growth but to act as a break on the harshness of society. The message of a religion, to which the young are exposed, transmits the prevailing culture and reinforces the prevailing beliefs. Through holy works or other means, the prevailing socio-political paradigm becomes codified in lists of counsels of socio-political wisdom such as the Jewish *DECALOGUE* or the Hindu *LAWS OF MANU.* This means of social bonding is almost always effective, but one may doubt whether it is always wise.

Though religion unites the faithful, it also tends to set one group against another. We find Gentile against Jew in Nazi Germany, Protestant against Catholic in Northern Ireland, Muslim against Hindu in India and Pakistan, etc. Moreover, as Lucretius saw, religion is often perverted into a force that tends to rouse fears and restrict human liberty. It follows that the use of religion, especially in its more primitive forms, is often counterproductive when it is used as a force for strengthening the social bond. I turn now to consider the other primary function of religion, already mentioned: *to uplift the human spirit and to help people make sense out of life.*

Religion helps people make sense out of life in at least three ways: one, it supplies a scheme for explaining life and the universe and tends to reduce the bewilderment and the feeling of strangeness which might otherwise engulf a person who found the universe unintelligible; two, it explains to people how to conduct themselves so that they can get the most out of life; three, it supplies drama, fellowship, and other accoutrements which reconcile people to life. One way that religion enhances importance in life is through its aesthetic dimension. The pageantry, the glory of ritual, the music of Bach and Handel, are highly significant in the lives of many people. What is more, great cathedrals are breath-taking reminders that one does not live by bread alone. It is the aesthetic aspect of Catholicism that attracts many people, among them, according to Russell [1956], Santayana.

As has been suggested, religion, beauty, and mysticism are intimately connected. A mystical experience is either so terrifying as to cause derangement or so sublimely beautiful as to arouse the consciousness of a divine presence. The words "mystic" and "mystery" derive from the same Greek root, *mystes*, and as we have noted elsewhere, Einstein said that the most beautiful thing that we can experience is the mysterious. On the other hand, anything really beautiful, whether in art or nature, is mysterious and therefore tends to evoke feelings of worship. Great music, masterworks of art, and awesome cathedrals are therefore exploited in religion to evoke feelings of awe, wonder, and worship.

The purest religion is perhaps to be found in monasteries and ashrams. Although a lot of hocus-pocus goes on there, at least in theory people are there *voluntarily* for spiritual purposes. Monasteries provide a much needed haven for spiritual rest and quiet contemplation. Moreover, monasteries are also at times havens for learning. Some monasteries in the *dark ages*, like those in Ireland, did their best to preserve classical learning by treasuring and copying manuscripts. Moreover, the monastic tradition fosters mystics, for example St. Teresa and Vivekananda.

For most religious people, nothing in life enhances importance and makes life intelligible as much as the revelations of the mystics. As we have already stated, according to the mystics there is a hidden and mysterious reality underneath the world of appearance. Knowledge of this reality—which does not make sense in the same way that empirical science does and cannot be explained to the hard-headed rationalist—is conveyed through rites, sacraments, and reading sacred works as well as through prayer, meditation, and worship. Christians regard this reality as God; Buddhists regard it as "pure, objectless, consciousness." According to the Christian mystic such as St. Teresa, in mystical experience, one has a sense of God's presence as distinct from oneself. On the other hand, the Hindus believe that in mystical experience one sees that the distinction between God and oneself is an illusion.

Religion maintains its sway by various means. In the past the rack and screw were used, but nowadays, at least in the West, the authority of religion is maintained by means of three doctrines: the first of these, the doctrine of sanctity, assures the believer that the dogmas of the church are unquestionable; another doctrine, the salvation myth, instructs the believer on how to find salvation or freedom from suffering; the third, the doctrine of divine retribution, a doctrine not often invoked in the present but which hangs over the nonbeliever like the *sword of Damocles*.

According to the doctrine of sanctity everything in the universe runs perfectly in accordance with the divine plan—except for human beings who have free will. This view of the universe, namely that the universe runs in accordance with an intelligible plan, is shared by some scientists such as Einstein. We may contrast that view with the skeptical view. The skeptical view is well-expressed by Macbeth: *Life's but...a tale told by an idiot, full of sound and fury, signifying nothing.* The ultimate skeptic says there is no reality and there is no meaning or truth.

From physics there is evidence for both points of view. On one hand, according to relativity theory, the universe unfolds in a predictable and orderly way, with a beginning and perhaps an end; on the other hand, quantum mechanics gives an impression of chaos such as described by Milton: *Chaos umpire sits and by decision more embroils the fray by which he reigns: next him high arbiter chance governs all.* Though quantum mechanics asserts that nature is nondeterministic at the submicroscopic level, it is true that there are statistical laws, conservation laws, symmetries, and other structural aspects of the universe that imply a limitation on chaos. Thus the grand design of the universe, if it exists, is achieved in such a subtle fashion that the truth about nature involves a Hegelian synthesis of opposites: that is to say, although the universe is at bottom absurd, nonetheless in a mysterious way the universe manifests a guiding intelligence. The rational course therefore is to believe in neither order nor chaos, but to approach the universe with a full panoply of emotions, including awe, fear, and laughter. Organized religion is largely based on the first two—awe and fear—and tends to neglect laughter.

The various churches have different doctrines of salvation, each of which being contained in a holy book. History shows that these doctrines tend to get more rigid as time goes on. It seems that as a religion ages it accumulates doctrines in a manner that is reminiscent of the manner in which calcium accumulates in plumbing or cholesterol accumulates in human arteries. The authorities in each church believe that the faithful must not presume to criticize any of the doctrines, but allow themselves to be guided by the wiser (and usually male and older) heads. Each church claims to have the whole truth not only about the origin and purpose of life, but what God (or salvation) requires of human beings. Religions, in addition to making truth claims, generally lay down rules of requirement in the conduct of life. That is to say, religion answers to a need for an authority to determine, not only matters of ***truth***, but matters of ***requirement.*** Since the day of Galileo, science has largely replaced religion in the realm of truth, but so far there is no acknowledged authority to replace religion in the realm of requirement.

There is no doubt that authorities are sometimes useful. No layman can be trusted to know whether a gall-bladder operation is required or not. The besetting sin of authorities, however—whether they are clergy, scientists, newspaper pundits, or government bureaucrats—is arrogance and cocksureness. Governments, wielding the death-penalty, presume to know who is guilty and who is innocent. Scientists, railing against obscurantists, presume to know what is possible in this world and what isn't. According to Whitehead (Cf. Price [1954, p. 7]):

> Nothing is more curious than the self-satisfied dogmatism with which mankind at each period of its history cherishes the delusion of the finality of its existing mode of knowledge. ***Sceptics and believers are all alike.*** At this moment scientists and sceptics are the leading dogmatists. Advance in detail is admitted; fundamental novelty is barred. This dogmatic finality is the death of philosophic adventure. The universe is vast. [emphasis added]

Blind faith in authorities belongs to the childhood of the human race and a mature race will develop beyond this need, especially in matters of ultimate importance. According to Humboldt [1854, p. 92]:

> ...doubt is torture to the believer only, and not to him who follows the results of his own inquiries; for, to the latter, results are in general far less important. During the process of inquiry, he becomes conscious of his soul's activity, its inherent strength; he feels that his perfection, his happiness, depends on this strength; and instead of being depressed by his doubts concerning the principles believed to be true, he congratulates himself that his increasing force of thought enables him to see clearly through errors that till now remained hidden.

Religious authorities—Catholic, Protestant, Jewish, Islamic, and others—no matter how intelligent, tolerate no ambiguity or uncertainty. According to Whitehead [1933], *the folly of intelligent people, clear-headed and narrow-visioned, has precipitated many catastrophes*. The doctors of the church, with learned assurance, interpret for a credulous laity the myths of humanity in its infancy. *Where is Socrates now that we need him?*

The most effective tools of criticism of the doctrines of religion, whether in matters of truth or requirement, are of two kinds: one, critical scholarship; the other, *laughter*. Critical scholarship reveals the Bible (Christian or Jewish) as the product of fallible scribes whose writings, in many respects, are an insult to the deity (Cf. Mark Twain

[1974 ed.]). Laughter is a corrective, not only to religion, but to all folly. Without the freedom to laugh, critical scholarship and science—the other method of attack on the doctrine of sanctity—is in great difficulty; for rational inquiry is hindered when the revealed TRUTH is so sacred that those who mock it are put to death. (Remember Rushdie!)

The myth of salvation states in part that "sin" is an obstacle to being saved. In Christianity this myth is reinforced by the belief that all human beings are tainted by the "original sin" of Adam. As everyone knows, the doctrine of original sin is based on the account of the creation in *GENESIS.* According to this account, Adam and Eve are tempted into disobedience of the "Lord God's" edict not to eat of the tree of "the knowledge of good and evil" by a "serpent" who "was more subtil than any beast of the field." Adam and Eve are told there is a death penalty for disobedience; nonetheless, in their eagerness to become "as gods, knowing good and evil," they ignore the "Lord God's" warning and eat "of the tree which is in the midst of the garden." Two points about this story may be noted. First, if Adam and Eve were really like the animals and didn't know right from wrong, one cannot in good conscience attach any blame to their actions. The second point is that one may infer from the story that "evil" in the minds of the writers of Genesis had something to do with sex since Adam and Eve, after their "eyes were opened," knew they were naked and were ashamed. This priestly view of sex, which is still held among many holy men today, has put a pall on the relations between men and women, and only now are we (at least in the West) slowly emerging from its restricting influence.

Theologians agree that it is freedom of the will that makes "sin" possible. As few theologians acknowledge, however, without the knowledge of "good and evil," neither "sin" nor freedom of the will make any sense at all. Freedom of the will may be defined as freedom from instinct together with freedom to choose based on reason. Note that, according to *GENESIS*, the emergence of human freedom to choose guided by the knowledge of good and evil was not the result of the "Lord God's" intention but came about almost by accident.

The view of Kant [1785] that instinct is a surer guide to happiness than reason is debatable. Be that as it may, it is obvious that freedom from instinct magnifies the possibility of making mistakes. Mistakes are not always to be deplored, however. The marvelous ability of human beings to change would not be possible if human beings followed instructions like a robot and had no capacity to make mistakes.

As Whitehead [1938] points out: *Great advances of thought are often the result of fortunate errors.* Moreover, according to Lewis Thomas [1979], "the capacity to blunder slightly" is the "real marvel" of living material: specifically Thomas points out that mistakes in the replication of the DNA molecules, out of which genetic material is constructed, accounts for all of evolution.

One may interpret *GENESIS* as saying that the genus homo emerged from its animal state by choosing to be free of instinct (or the "Lord God") and charting its own course. The "serpent" may be interpreted as the spirit of knowledge and reason. Accordingly, I suggest that "original sin" is the exercise of freedom *and* being guided by knowledge rather than instinct. This freedom is both a *freedom to* pursue and be guided by knowledge and a *freedom from* being dictated to by God or nature.

The exercise of human freedom and human reason—i.e., I suggest, "original sin"—is both the glory and curse of humankind. From human freedom and human reason follows everything that makes humankind human—all art and all science. Moreover, freedom and the power of human reason give humanity a glimpse of the eternal. On the other hand, the exercise of human freedom and human reason sets humankind apart from God and nature and encourages the tragic sense of life. It makes people long for salvation and sanctity. It shuts people out of the paradise of the *Garden of Eden* and brings to every human all the troubles of *Pandora's box.*

The idea of divine retribution must have haunted humanity ever since the notion of gods and demons was first hatched in the overactive imaginations of early peoples. Early gods, like *JHVH,* were jealous and inclined toward harshness and cruelty; however, believers are generally undisturbed by stories of divine cruelty. Once, when I tried to make an eleven year old (who is generally not without sympathy) realize the full horror of the Biblical story in which the Lord God drowned almost all humanity (not to mention most land animals) in the Flood, she dismissed my concerns with the remark, "These people were not Christians." Some of the talk of Jesus in the New Testament about forgiveness sounds hollow after reading the following passage (explaining the parable of the tares in the wheat field) [Matt:13.37-43]:

> He that soweth the good seed is the Son of Man;
> The field is the world; the good seed are the children of the kingdom;
> but the tares are the children of the wicked one;
> The enemy that sowed them is the devil; the harvest is the end of the

> world [close of the age]; and the reapers are the angels.
> As therefore the tares are gathered and burned in the fire; so shall it be at the end of the world.
> The Son of Man shall send forth his angels, and they shall gather out of his kingdom all things that offend, and them which do iniquity;
> And shall cast them in a furnace of fire; there shall be wailing and gnashing of teeth.
> Then shall the righteous shine forth as the sun in the kingdom of their Father. Who hath ears to hear, let him [or her] hear.

Islam and many other religions seem to have similar harsh doctrines. The persuasiveness of all religious doctrines—including the doctrine of divine retribution—depends on the doctrine of sanctity, and hence its demise has had repercussions for all tenets of faith. Ever since Galileo, science has steadily won the fight against the doctrine of sanctity. In the East, the Buddhist view—that life is suffering and that only the enlightened can escape the supposed round of births and deaths—is no longer as plausible as it once was. Moreover, the Western view that one must choose between eternal torment in hell and eternal bliss in heaven has also lost most of its credibility. The power of the myth of salvation has crumbled in part because (as Lucretius foresaw), owing to the demise of superstition and the rise of scientific thinking, survival after death is no longer thought to be a realistic option.

As we have already noted, religion, especially in its more primitive forms, tends to arouse fear and restrict human liberty. On the other hand, many religious people are in the forefront in the fight for peace, distributive justice, and human freedom. There are no greater opponents of capital punishment than religious nuns. (It must be emphasized, however, that many, if not most atheists, share these ideals.) An enlightened religion will not only fight to relieve suffering, work for political justice, and bring peace and harmony in the world, but enable its adherents to develop freely and without spiritual fetters. Religions of the past have focussed on human flaws and the tragic sense of life. For the Buddhist, life is suffering and salvation is escape from the endless round of births and deaths. For the Christian, life is a vale of tears and humanity is tainted by original sin and *hubris* and needs cleansing. There is no doubt that life is often tragic and human beings are flawed. But human beings are also god-like. Even the ancient religions out of which Christianity and Buddhism grew testify to that. Jesus [John:10.34] quotes Psalms 82.6— (*I have said, Ye **are** gods; and all of you **are** children of the most High* [emphasis in original])—in defense of his claim to be the son of God. Moreover, Hinduism, out of which Buddhism developed, makes the claim that the soul of a human and Brahman (God) are one and the same. Instead of feeling guilt in trying to—as *GENESIS* says—"be as gods" (and for constructing ever more glorious towers of Babel "whose top may reach unto heaven,") the pious of the future may yet celebrate the attempts of humankind to be god-like. Instead of grovelling before gods, could not the religion of the future focus, not on "original sin," but on "original purity"?

Organized religion sways more people nowadays by means of raw power than most governments. *At least in form*, aside from the heirs of Mao-Tse-Tung, the largest totalitarian power remaining in the world is the Roman Catholic Church. Fortunately, the papacy, unlike some of the Islamic clergy, cannot (generally) get governments to enforce its pronouncements. Moreover, the United States has more to fear from the fundamentalism of Protestant sects than it does from the Catholic Church, which is graced by vast resources of internal criticism and scholarship. It may be noted that the Church is besieged by its own membership, and I predict that sometime in the next century the Pope will become a figurehead much like the monarchs in Great Britain and Sweden. Reforms will be made such as giving power to women—who make up the majority of the Church—as well as allowing priests to marry. What is more, I predict that the hierarchy will be elected by the priesthood instead of appointed, more local autonomy will be granted, and the work of deciding official church doctrine will fall to a permanent council, perhaps along the lines of Vatican II but representing laity as well as clergy.

Religion, like all else in this world, has either to adapt to changing conditions or else perish. People will continue to look for guidance from an institution only if the advice that they receive helps them in their daily lives. We see springing up around us new groups and cults to fill the vacuum when the old answers no longer seem to work. Existing institutions can only survive if they respond to the needs of those who are no longer satisfied with the dead formulas of the past.

A greater problem than the dominance in the Catholic Church of moribund institutions, is the problem of fundamentalism in both East and West. Fundamentalism, as we have noted, stalks many governments in the world—and none worse than in the countries dominated by Islam. It will perhaps take many generations until free thinkers in Moslem countries can enjoy the freedoms that are taken for granted in much of the West.

If the fight against fundamentalism in both the East and the West is ultimately won, the question will still remain: can religion continue to serve either of the two functions mentioned earlier: one, strengthen the social bond; the other, uplift the spirit and help people make sense out of life? In order for this to happen the ecclesiastical system must expand its horizons and become less dogmatic. The Quakers have shown that it is possible to have a religion without religious authorities and creeds. The Buddhists and the Taoists, unlike their Western counterparts, have shown that it is possible to have a religion that remains agnostic on ultimate questions. The Buddha is reported to have said (cf. Burt [1955, p.38]): *No dogmatist can win, by self-concocted views, the way to purity.*

Religion has always looked to mysticism for guidance. In the quietness of meditation, when people listen to what is deepest inside themselves, the mystic vision comes and tends to rend asunder the harsh doctrines of the past. It is to mystics we owe the most profound of counsels whether it is the golden rule from the *NEW TESTAMENT*, the counsel of nonattachment in the *BHAGAVAD GITA,* or the doctrine of forgiveness in the *DHAMMAPADA* where it is written (Cf. Yutang [1942, p. 327]):

> "He abused me, he beat me, he robbed me"—in those who harbor such thoughts hatred will never cease. "He abused me, he beat me, he robbed me"—in those who *do not* harbor such thoughts hatred *will* cease. For hatred does not cease by hatred at any time: hatred ceases by love—this is an old rule. [emphasis added]

Technical innovation has brought many improvements to modern life, but the accompanying problems (to which we turn in the next chapter) show that intellect alone cannot save civilization. Perhaps we can once more look to mysticism to produce the vision necessary to produce a religion that satisfies the intellect, enhances the sense of the importance of life, and urges humanity toward a higher state of perfection.

Chapter Eight

Several Suggestions for the Salvation of the Planet

1 Introduction

According to Marty Feldman [1980], the meek shall inherit the earth only after the strong are through with it. On the other hand, as *ECCLESIASTES* reminds us, *...the race* ***is*** *not to the swift, nor the battle to the strong...*[emphasis in original]. When Bertrand Russell [1962] wrote his book, *HAS MAN A FUTURE?,* the strong both in the Soviet Union and the United States threatened to render the earth worthless through a nuclear interchange; but fortunately owing to Gorbachev and Yeltsin the strong in the Soviet Union disappeared. On the other hand, the Soviet nuclear weapons *have not disappeared* and whether the world is actually safer remains problematic (see *FRONTLINE* (*PBS)*: *RUSSIAN ROULETTE)*. Many problems remain and Russell's question is still open. In this chapter I examine some of these problems, beginning with the problems of the global economy.

The world requires a political system that enables men and women—as well as all other life on this planet—to live and flourish inpeace and harmony. For such a political system to exist there is needed a political economy for the world that satisfies four requirements: one, it must be a political economy that ceases to plunder and pollute the planet; two, it must be a political economy that ensures an abundance sufficient to house and feed everyone; three, it must be a political economy that shares the wealth and power fairly among all human beings on the planet; four, and not least, it must be a political economy the fosters the free development of all human beings. Why don't our economists and politicians invent such a system? According to H. G. Wells it is because, like the people in Plato's myth, they see the shadows on the wall and are unaware of the underlying realities. Wells [1935, p. 49] writes:

> The besetting vice of economic science, orthodox and unorthodox alike, has been the vice of beginning in the air, with current practice and current conviction, with questions of wages, prices, values, and possession, when the profounder issues of human association are not to be found at all on these levels. The primary issues of human association are biological and psychological, and the essentials of economics are problems in applied physics and chemistry. The first thing we should examine is what we want to do with natural resources, and the next, how to get men [and women] to do what has to be done as pleasurably and effectively as possible. Then we should have a standard by which to judge the methods of today.

There are economists who tend to agree with Wells. According to Heilbroner [1999, 1980], modern economic models are too simplistic and he says that in trying to be scientific, modern economists favor two-dimensional models that leave out too much, and consequently are left "without a true understanding of how the system works." Heilbroner quotes Keynes [1951] as saying, "No part of man's nature or his institutions must be entirely outside of his [the economist's] regard." He might have added, with Wells, that in a modern world, which is in large part dependent on technology and in danger of exhausting or polluting natural resources, economics is, in part, "applied physics and chemistry."

Those strictures against modern economics by Wells, Heilbroner, Keynes, and others go a long distance toward explaining why no really workable political economy has been created, and barring the appearance of a new Einstein of economics, the future of the world economy must be relatively bleak.

2 The global market

The so-called developed nations, whose development is widely attributed to the market economy, are full of paradox. They are rich and yet approaching bankruptcy. They are highly productive, capable of producing more food, goods, and services than anyone needs, and yet there are homeless people on the streets and children go underhoused, underfed, and undereducated. We are told that competition makes us more productive and hence "better off," and yet we are also told that owing to competition, we cannot afford what we used to be able to afford. In America the trend toward concentration of wealth in few hands marches on unabated (cf. New York Times [9/30/97]); moreover, except for the fortunate few, it is only a matter of time before the trends toward robotics and free trade will mean the end of work in advanced countries (cf. Rifkin [1995]). The effective unemployment rate in the United States, though it is officially around 5% or less, may be actually closer to 14%; the unemployment rate in the inner city is much higher. Europe's unemployment rate is in the double digits and idleness in Japan, though unofficial, is almost as great (cf. Thurow [1996]). Thurow [1999] points out that "...median family incomes [in the United States]...haven't risen since the early 1970's, yet the average wife is working 15 more weeks a year than she did back then. Real wages for 80 percent of the male labor force are below where they used to be." This does not worry the elite, but can the boom last forever? The financier, George Soros [1998], writes: *...I cannot believe that the present boom will not be followed by a bust until history proves me wrong.*

As was noted in Chapter Four, all the world is mired in debt. As a warning about what may happen in the future consider the causes of the *great depression* of the 30s and the great war that followed it. According to H. G. Wells [1931], after World War I, "All the world was burthened with debt and it probably would have eased the world situation immensely if there could be a concerted simultaneous inflation of all currencies..." World War II was almost a direct consequence of debt; for without the debt of Germany after World War I it is doubtful that Hitler would have come to power. Third World debt may have similar effects at the present time. Moreover the growing international indebtedness of the United States will end in impoverishing Americans. The commanding position of the American economy has already been eroded and the dollar is bound to weaken. Eventually the world, including Japan, will tire of getting nothing but worthless dollars for their products.

Monetary policy (together with the global market) is the culprit, however. It is spinning out of control, driving up interest rates, pitting not only the young against the old but the workers of each country against workers in all other countries. Keynes [1933] warned that in a global economy, capital shortage in developing countries would drive up interest rates and produce widespread unemployment. Each nation keeps its interest rates high both to finance its national debt and to shore up the price of the national unit of currency. Taxes on capital are kept low, in part, to prevent capital flight. In the United States the needed revenue is increasingly squeezed out of the class of people who can least afford it—the working poor and the lower middle class—who are already undergoing a financial crunch as business everywhere downsizes, cuts wages and pensions, and wherever possible, replaces people by robots (cf. Rifkin [ibid.]). According to Soros [ibid.], *What has happened...is that taxes on capital have come down while taxes on labor have kept increasing.* Labor has been unable to do anything about these practices for at least two reasons: one, if labor protests and wants to put a stop to them the multinationals (and other companies) simply move their factories to other places on the globe; the other, with the demise of socialism, business no longer fears labor.

The truth is that as a result of global competition there is a mad scramble by investors pursuing the highest return on capital and a correlative mad scramble of governments raising their interest rates in a vain effort to pay for the welfare state. Soros [ibid.] quotes the international economist Dani Rodrik to the effect that "globalization increases the demand on the state to produce social insurance while it reduces its ability to do so." For the sake of all peoples these mad scrambles must be brought to an end. The only way this can be done is for the developed nations to get together and stop being pawns in the international chess game that capital plays against labor. This will require reining in corporations and setting up an international authority with the power to reform the monetary system, end speculation in currency, end exploitation of the poor in all nations, monitor industrial waste and overexploitation of natural resources, and encourage full employment.

As long as there is a global economy, the current situation is not going to be reversed by letting things drift. A global economy cannot be run by Adam Smith's "invisible hand." On the contrary, it can only be effectively managed by planning—and central planning at that. For "better" or "worse," the world is one unit and its economic well-being cannot be achieved without the intervention of a central agency—a *de facto* world government if you will. Some mainstream economists agree. According to Thurow, for example, [1996, p. 138]: *Cooperative policies are needed to make the global economy work...Interest rates and budget balances have to be agreed upon collectively for some group of countries.* Moreover, even some international financiers agree. Soros [ibid.] points out that: *There is no international authority for financial markets, and there is not enough international cooperation for the taxation of capital.* There is already a world bank, but unfortunately it is part of the problem instead of part of the solution (see Clarke [1996]). Unless there is a reform of the international monetary system, the competition between the United States and Europe will drive interest rates ever higher. The closest thing to a global currency is the dollar, but the dollar is a shaky currency, no longer backed up by the preponderance of economic power. Europe, if it can get its act together, will soon outstrip the United States and the future currency of choice may well be the European *euro.* A battle of these two currencies, fueled by speculation, does not make sense, however.

The problem is global because capital nowadays is highly mobile. If the government thinks of regulating business or raising taxes, business needs only to threaten to move the base of their operations overseas to stop any such nonsense. Patriotism ends where the pocket book begins. According to Thurow [ibid., p. 138]:

> The era of national government regulation of business is over. Activities go to where they are unregulated and often that relocation can happen without anyone physically moving. Insurance and financial activities are electronically performed in Bermuda or the Bahamas, while almost all of those doing the activities are still sitting in their offices in New York or London.

I conclude that in the modern world, *even if subregions of the world retain a greater or lesser degree of economic autonomy,* there is a need for a global government or a central agency to control the money supply and the expansion of credit, and end—or at least moderate—speculation in currency.

Concerning control of this kind Thurow [ibid., p. 138] writes: *It could only be democratic if there were an* ***elected democratic world government,*** *yet the Left and the Right would be the first ones to object to any such government* [emphasis mine]. Some liberals may well unite with conservatives against the institution of a world government. Liberals are not all of the same stripe. Liberals divide into two types: imperialists, on one hand, and what may be called internationalists, on the other. The imperialists are patriots who agree with liberals about the welfare state and with moderate conservatives about nearly everything else. An example of such a liberal is Michael Lind [1996]. He writes:

> Against pacifists on the left and isolationists on the right, a new generation of national liberals would defend the interests of the United States as a great power in a dangerous world, and support a strong and appropriate (though not bloated...) military-industrial sector. National liberals would unite their defense of the American national-security state...with a defense of the American welfare state...

The national security state of the United States, which Lind would leave in place, has at least two preoccupations: one, at home, the other, abroad. Domestically its preoccupation is the invention, production, and stock-piling of more and more powerful weapons, beefing up the FBI, and expanding the police force; abroad, its preoccupation is world hegemony. A powerful national security state is able to maintain its leadership in the world because the world respects power. (There is that old joke: Where does an 800 pound gorilla sleep? The answer, "Anywhere he wants to!")

Liberals have fought for internationalism for two basic reasons: one, no one nation can solve the problems of peace and war; the other, no one nation can solve its economic and environmental problems alone. The need to solve world economic problems is not a recent problem emerging out of a sudden development in the late twentieth century of a "global economy." In order to see that, consider what happened after World War I. The United States and France cornered the market on gold and therefore the United States and France were at the time the only nations with cash reserves—the world being at that time on the gold standard. It was the tariff and fiscal policies of these two countries that triggered the *great depression* in the twenties and thirties. Writing at the time, H.G.Wells [1931] reports "France and America still remain great absorbers of gold..." According to Wells, if in 1926—before the great depression—America had eased up on its hold on gold and credit and lowered its tariffs "prosperity might have spread across the Atlantic and the balance restored." The point of the story is that prosperity depends on the money supply and even in an era of more or less autonomous nations there is a need for a centralized agency to come to the aid of ailing regions, regulate financial relations, and keep the peace.

3 Conservative versus liberal

A liberal—as well as a conservative—wants only as much government as is necessary. In a way almost everybody is a "liberal" nowadays since—though the conservative talks of small government—historically it is the liberal who began the call for small government. For example, the liberal Thomas Jefferson, who hated merchants (cf. Jefferson [1810]), is often quoted—probably inaccurately: see Mieder, Kingsbury, & Harder [1992]—as saying "That government is best that governs least." The stance of the liberal—and many conservatives as well—is that enunciated by Abraham Lincoln: *The legitimate object of government is to do for a community of people whatever they need to have done, but cannot do at all, or cannot do as well, for themselves, in their separate and individual capacities.* (Cf. Current [1996]) Conservatives and liberals generally differ, however, on what people can do for themselves.

The conservative professes to hate government but the reality is that the conservative only hates government when it regulates commerce or spends money on the poor and the environment. But perhaps the greatest divide between liberals and conservatives is their respective attitudes toward public money. It is *not* the case, however, that it is only liberals who want to spend public money. The liberal wants to spend money on the economic and social infra-structure and on the poor; the conservative (by and large) wants to spend money only on corporate welfare and the military.

Utopia for the conservatives would be to eliminate unions, increase competition, and let wages sink to the bottom. The inevitable result would be a return to a two-tier society consisting in a wealthy elite and a largely uneducated underclass. As a harbinger of such a society, many of the rich already live in enclaves resembling the walled cities of the middle ages. If trends continue, some jobs requiring skill—such as surgery—might still be performed by well-paid professionals, but the rest would be performed by robots, immigrants, and people in factories overseas. In the two-tiered society of the conservatives, the underclass need not know unemployment since, wages being low, those with low-tech skills will—like in the good old days—find a berth serving the elite. In the conservatives' dream there is no class warfare because the poor are happy to serve their masters whose position in life all right-thinking persons acknowledge is a richly deserved reward for pluck, superior knowledge, and breeding.

The liberal utopia, on the other hand, is a one-class society. It is the position of the liberal that though socialism may err when everything is run by government, capitalism errs when everything is run by private enterprise. Liberals will settle for (as the Poles say—cf. Greider [1997b]) "capitalism with a human face." Under the threat of communism there had been some amelioration of the fate of the poor, but in recent times capitalism has turned ugly once more, especially in the United States. Many people, especially blacks, are in danger of being thrown away on the scrap heap of history. What do I mean by this? Let us review the record.

After emancipation, African-Americans—having been brought to America against their will—first, eked out a poor existence on the land; then, as farming became mechanized and work scarce, Southern blacks went north in search of jobs. Though they were given these jobs, the joker is that these jobs have now disappeared. Having escaped slavery, African-Americans in the inner cities are still bound by invisible chains that bind them to poverty and degradation. Many people in the inner cities do not possess the skills to hold down a job in a modern factory (see William Julius Wilson [1996]). In this hopeless environment, the young, seeking some escape and revenge, descend into drug dealing and other criminal acts and wind up in prison. It is shocking to observe that the response of white people often seems to be "Am I my brother's keeper?" (Americans forget that they *became* their brothers' keeper when they imported slaves.)

The root cause of the plight of the inner city is the decline of labor. The Achilles heel of capital, however, is that it needs both customers *and* investment. Unless most people become pensioners or stockholders, the largest pool of customers must come from the ranks of labor, whether from the inner city or elsewhere. But the working class is getting poorer as business gets richer; consequently there is a decrease in the proportion of the world's workers who have the wealth to buy the products that capital and labor produce together. Moreover, as business downsizes and the global economy puts a downward pressure on wages, a potent source of investment—the pension fund—declines. The problems that are posed for capital by the shrinking assets of labor and the end of work (cf. Rifkin [ibid.], Wilson [ibid.], Greider [1997a,b]) are not immediate, but they will no doubt become obvious as the pool of customers shrinks and the level of savings continues to decline in the years to come. Eventually one of two things is bound to happen: one, the corporations, acting in their own self-interest, decide to share more equally the fruits of capital and labor so that people can not only buy the goods that are being produced but can save some for future investment; the other, willy-nilly, set the stage for a world-wide revolt. There are some faint signs that business may yet see the light and be able to avoid a revolution which no one wants. Modern industry requires brains and what is now called "the knowledge industry" is in the driver's seat. H. G. Wells had the dream that someday people with brains would unite and take over the world. Could it be that the chief executive officers in the knowledge industry, Bill Gates, for example, together with financial wizards like George Soros, might decide, in their own interests, to get together to see what could be done to make sure that workers all over the world get a fair shake? (This may be an idea whose time has come. I understand that Ralph Nader is trying to get something like this going.) Liberalism first got its start in Europe when merchants and bankers got together to overthrow the old feudal system. *Perhaps the use of money in fighting for liberal ideas is not quite finished.* Before it is too late, organized labor can make use of its vast pension funds to buy corporate stock and take its place beside capital in corporate boardrooms (cf., e.g., Hedrick Smith [1998]). Some of the rich must know that they cannot continue to take a greater and greater share of wealth for themselves without subverting democracy and killing the ultimate source of wealth—which is the people.

There are many things that can be done. I mention a few of them here, beginning with the most important. First—in conformity with suggestions already given in the previous section—reform the monetary and tax systems in such a way that cities, states, and countries are no longer pawns in a national and global chess game played by capital against labor. Part of this reform (from the United States point of view) would have the aim of ensuring that multinational corporations who sell their products in the United States are forced (or encouraged) to locate their factories in places in the United States where they are needed. When environmental costs are factored in, together with the continual depletion of natural resources, the toll that transportation costs take from the real world economy must eventually make it evident that local production is more economical than importation. (I return to this point later.) Moreover, the ability to pay for social welfare depends in part on full employment, taxes on the rich and on wealthy corporations, and perhaps taxes on currency transactions. Much of the money that is needed can be taken from such places as the bloated military budget, the CIA—an anachronism as well as a laughing stock in a post cold-war period—and from corporate welfare. Reforms of this kind, if adopted, would ensure that the workers once again have the clout they need to ensure a fair shake at the bargaining table. Second, reverse the trend toward downsizing. Everywhere in the Western World the work has been spread more uniformly among the population by shortening the work day, decreasing the number of days per year that employees work, and giving early retirement. There is no reason that this could not continue. Third, I suggest a massive education program targeted to the inner cities. Funding at present is mostly local. This has to change to ensure that all regions share in the responsibility to educate all citizens, regardless of whether they live in affluent suburbs, in the relatively impoverished countryside, or in the inner city. Fourth, drop the current drug program, which encourages illegality among inner city residents and is largely responsible for the incarceration of fully one-fourth of African-American youths. Fifth, as Rifkin [ibid.] suggests, expand the service sector of the economy. Sixth, add to the ability of people to participate in the economy by expanding the number of people on pensions or owning stock. In order to do this not only would the trend toward early retirement have to be continued, but the trend toward raising the age for social security would have to be reversed. (For further discussion of this matter see Section 5.) If all the above working together do not work there is one final option: institute a massive public works program.

The conservative argues against a radical program to help the poor on the basis that we cannot afford it. But this argument rests, as we have seen in Chapter Four, on the need of governments to keep interest rates high so as to attract investment and to reduce inflation. As we have already noted, these reforms, particularly control of interest rates, cannot be done without both national and global cooperation.

4 Toward regionalism?

The free trade battle has entered a new phase, however. We have noted that in order to escape taxes and high labor costs, productive capacity is being moved from places where taxes and wages are high to places where taxes and wages are low. Countries that previously were only a source of food or of raw materials are now sources of cheap labor. Some writers in the West see a bright side to all of this. For example, according to Greider [1997b], "Even given its ugliest exploitive aspects, the process of productive dispersal has the potential to produce a dramatic departure from the past, an opening for greater equality on a global scale." But Greider is well aware of the dark side as well. He adds, "Yet this same opening feeds the economic loss and growing social distress of the older wealthier nations, where inequalities deepen and marginalized people forfeit control of their own lives."

A locality that does not produce all it needs has a weakness. In part, Vietnam was able to hold out against America so long because it produced all it needed to survive. As productive capacity is moved overseas and beyond borders, so-called developed countries are becoming less self-sufficient. According to Sale [1980]:

> Self-sufficiency may perhaps seem a foolish anachronism in a world of interconnectedness, an economic cul-de-sac in an era of mutual dependence. And indeed it may not be obtainable in its most absolute form...But as a goal and an ideal...self-sufficiency has inestimable virtues.

In the modern global economy, nations are as much threatened by debt as they were previously by predators. The international bankers are in control and nations are being squeezed like Germany was in the 1920s and like farmers were in nineteenth and twentieth century America. Today (1999) Thailand, Brazil, Indonesia, Russia, and South Korea, to mention a few, are in a debt crisis; in a previous year it was Mexico, and before that Argentina. Developing nations complain that "there must be something wrong with a system that wipes out years of hard-won development" because of global cash flows (Cf. Cowell [1999]). As already mentioned, even the United States of America, which now depends on imports, is threatened by international indebtedness. According to Sale [ibid.]: *Any nation that comes to depend on imports will find itself perpetually vulnerable, scrambling always to create exports that will be sent off in return or else living in ever-deepening enthrallment.*

The doctrine of free trade is based mainly on two arguments. One is based on the self-fulfilling prophecy that the global economy is inevitable and there is nothing a mere mortal can do about it. This argument—if one can call it an argument—has the same intellectual respectability as the thesis of Marx that capitalism is destined to be replaced by communism. The other argument, which has considerable intellectual prestige behind it, is based on the theory of "comparative advantage." The idea is that everybody profits if people do what they do best. David Ricardo, the venerable British economist, explains it as follows (and here I quote David Morris [1996]):

> Two men can both make shoes and hats and one is superior to the other in both employments; but in making hats he can only exceed

> his competitor by one-fifth or 20 percent, and in making shoes he can exceed him by one-third or 33 percent. Will it not be for the interest of both that the superior man should employ himself exclusively in making shoes and the inferior man in making hats?

Aside from the fact that the developed nations produce neither shoes nor hats, what is wrong with this argument? According to Thurow [ibid.] and Rifkin [ibid.] the knowledge industry is the only source of long-run sustainable competitive advantage for the United States. Thurow, however, envisions a two-class society in which those with "first-world" skills will live in luxury while those with third-world skills will resort to crime, live in poverty, or perish. (Thurow [1995] worries that such a society may not be sustainable and that there may be a massive revolt.) Whether the United States can long sustain even an advantage in brain power in a global economy remains to be seen.

There is no need for me to go into a detailed refutation of the free trade arguments here, especially when this has been done by other people whose brain power exceeds my own. The interested reader can consult David Morris, for example, who gives a masterly refutation of the full panoply of arguments for free trade. He says in part (cf. Morris [ibid., p. 226]):

> The theory of comparative advantage is fast losing its credibility. Time was when technology spread slowly...Today, technology transfer is simple. According to Dataquest, a market research firm, it takes only three weeks after a new U.S. made product is introduced before it is copied, manufactured, and shipped back to the U.S. from Asia. So much for comparative advantage.

For a fuller analysis of the fallacies of the free trade arguments I refer the reader to the full text. Perhaps the best argument against free trade, however, is the practical one. The supposed opulence to which free trade leads can be refuted by walking down the streets of places like Youngstown, Ohio or by visiting the *Maquiladora* region of Mexico. (Cp. Daly [1993]).

If capitalism is to work as envisioned by Adam Smith the size of the market has to be large enough to make it possible to encourage specialization and consequently keep down the cost of production. It might be thought therefore that local production could be too costly a practice for a subregion to maintain. Nevertheless, there must be a limit to the size of the market needed for a degree of specialization sufficient to bring the costs of production down to a profitable level. At the present time the world is threatened more by overcapacity than underproduction (cf. Greider [1997a]). The basic question that needs an answer is how large a market is necessary to supply the dynamics necessary for an adequate standard of living? Nowadays, with modern techniques such as computers and robotics, this could be rather small (Cf. Rifkin [ibid.]). In the modern world what is more important than free trade in goods is free trade in information; for if information about how to produce goods cheaply is shared there is no reason that a large subregion cannot have both the talent and capital to produce goods as efficiently as any other place. According to Keynes [ibid.]:

> Ideas, art, hospitality, travel—these are things which should of their nature be international. But let goods be homespun whenever it is reasonably and conveniently possible; and, above all, let finance be primarily national.

In the future, one would expect that there would be a mass of information and techniques (including reverse engineering) available for producing the necessities of life, and even luxuries, at low cost. Even now the inefficiencies of local production tend to be exaggerated. One glaring omission from the equation is the cost in energy depletion as well as costs to the environment. Undeveloped nations, where production tends more and more to be relocated, are notorious polluters of both air and water. Moreover, the energy used in transportation must inevitably hasten the depletion of the world's oil reserves, driving up the cost of energy, and forcing a return to local production after it is too late. Sale [ibid., p. 70], reporting on the work of Stein [1974], writes:

> ...as Barry A. Stein, the leading scale-economist, has shown, "studies of productivity, as measured by value of shipments per manufacturing employee, show that the highest efficiencies tend to occur not in the largest size plant, but in those of moderate size," *that is, with fewer than forty five people.* There is a myth that big companies make more profit than small ones—but in fact Stein's figures show that "there is a declining relationship between profitability and size," that "smaller firms are more efficient users of capital," and that as firms grow bigger they show a "decreasing efficiency of asset use." [emphasis in original]

It is to be expected that local autonomy eventually would lead to an economy less subject to disruption from outside forces and this should make it possible more easily to solve the unemployment problem. This is the case since, on one hand, if local production turns out to be inefficient, there would be a need for a large pool of workers and the problem of unemployment would not arise; on the other hand, if the local production turns out to be efficient and there is a need for few workers, there are simple remedies like a shorter work week. If more workers are needed, however, and workers are recruited from outside the local region, care needs to be taken that immigrants are not paid a substandard wage—for this would tend to erode the local standard of living.

One of the arguments for free trade is that jobs are lost when interregional trade is reduced significantly by means of tariffs or otherwise. Though disruptions of this kind would certainly occur in the early stages on the road to independence, some jobs would also be gained in producing the goods that were formerly made by cheap labor or by more efficient industries abroad. Even under free trade agreements such as GATT, trade is less free than the rhetoric of political and editorial writers might lead one to suppose (cf. Greider [1997b]). Moreover, trade restrictions such as tariffs do not cut off foreign competition completely since a buyer is often willing to pay a higher price for an item of higher quality. There are two classes of items for which there is no reason to restrict interregional trade: one, items otherwise unavailable such as essential minerals, art and information produced abroad, and crucial medical supplies; the other, luxury items such as jewelry, exotic fruits, and precious stones. In the case where import of such items results in unemployment, however, those who are rendered jobless by foreign competition must be retrained for other jobs.

I conclude that it is foolish for developed regions to sacrifice autonomy for the sake of trade, especially when it means a lower standard of living for the average individual. Moreover, it is not only foolish, but dangerous, for developing regions to allow themselves to become helpless clients of international bankers and multinational corporations. Let us suppose, therefore, that the world is divided into unified and autonomous subregions and, moreover, let us suppose they are stripped of advanced weapons of destruction such as nuclear bombs. Disputes over the uses of the seas, outer space, or remote places like Antarctica would still arise of course, but these are matters to be settled by world government and interregional law. Given that autonomous subregions are in the main responsible for their own security and not armed against one another, one would expect that the most troublesome disputes to be within subregions. It is to be hoped therefore that within the various autonomous subregions freedom and democracy arise and that eventually disputes internal to a subregion would be settled locally within the courtroom and not (as in the Balkans) on the streets and in the fields.

The autonomous subregions of the globe might be North America (including Central America and the Caribbean), South America, Europe (including much of the former Soviet Union, Iceland, Greenland, and possibly Israel), Islam, Africa (perhaps excluding North Africa which is Islamic), India, China, and Southeast Asia (including Japan). The world government might be located in Israel, with Israel under the special protection of the world government. Special protection might also be offered indigenous populations and cultures throughout the world such as Eskimos, American Indians, Ainus, Tibetans, Australian Aborigines, Pygmies, Lapps, the tribes of Borneo, etc. Australia and New Zealand might also need special protection, and might therefore be used as bases for world government armaments and troops. Alternatively Australia and New Zealand might decide to belong to Southeast Asia or (say) North America.

In order for peace and prosperity to come to the world it is not only necessary that no region of the globe be in the grip of poverty, but it is necessary that it is no longer the case that some regions of the globe have what the others want and can't get. It is hardly to be supposed, however, that the subregions of the world, such as North America, cannot produce for themselves the basic necessities of life. A region can only be autonomous if it has renewable resources including food, energy, and minerals and it is sufficiently populous to develop a reasonable division of labor. Conservation, recycling, and the finding of substitutes is a necessary condition for self-sufficiency. Besides certain precious minerals, the main lack in North America is an energy supply sufficient for its needs. But North America (especially the United States) is enormously wasteful, and this lack could, I am told, be entirely made up by conservation. If North American oil becomes insufficient, fusion power may eventually fill the gap, but until that time, sources such as coal, natural gas, solar power, wind power, water power, geothermal power, power from the jet stream or the gulf stream, etc. would have to be phased in. (Because of its polluting potential, fission power is to be avoided if at all possible.)

For the sake of development, a region will inevitably engage in deficit financing of the kind that financed the war effort in the United States during World War II. Unlike the financing of armaments, however, deficit financing for the purpose of development does not impoverish a region: on the contrary, the capital developed when money is spent on education and industry increases wealth. According to Soros [ibid.], the state has played an important role in the economic development in countries deficient in local capital. In these countries "it has allied itself with local business interests and helped them to accumulate capital. This strategy has proved successful in Japan, Korea, and the now wounded tigers of Southeast Asia." It is helpful, moreover, when underdeveloped regions such as India and China adopt free market policies internally and, what is more, have free trade within the region such as is found in the European Union. This should not be more difficult in India, where the education level is high (though not yet sufficiently well-spread), than it is in China (cf. Amartya Sen [1997]).

I expect a time will come when the world is stable and there is established some parity between the wealth of various regions of the world. In the meantime, however, the need for capital drives developing subregions, such as South America and India, to welcome investments by foreigners. But they need investments *and not loans that make them vulnerable to high interest rates*. Moreover, it will not help a subregion if the countries in the region are forced to relinquish local ownership and control of their own economy. (At the present time, foreign aid to Haiti, for example, is held hostage to the demand that the government allow international corporations and/or the local elite to take over its infrastructure.) Investments are to be encouraged, however, not only because developing regions need capital, but because they promote peace; for a foreign power is less likely to bomb a country in which it has vested interests. It would not be too much to ask a developed region to spend one-percent of its gross domestic product on subsidizing industry in those developing regions of the world that are making progress toward freedom and human "rights." (The developed world is now taxed to support wars such as the war against Iraq in 1991.) But an investment of this kind is not to be encouraged if the purpose of the investment is to exploit the willingness of people in undeveloped countries to work for virtually nothing and to flood the world market with imports that could just as well be produced domestically.

Though I argue here for regionalism, I do not retract my previous call for world government. There are at least three areas in which central control is needed. One is in the area of finance, another is in the area of interregional conflict, the third is in the area of the environment. The need for central control in the matter of financial matters we already discussed, however inadequately, in the second section. I turn now to consider the other two, beginning with interregional conflict.

With regard to interregional conflict there are several points to be made. The first point is that there is a need for a world government to help settle local disputes. The second point is that in order to prevent disputes from escalating it is necessary that those not directly involved in a dispute remain neutral. The argument for neutrality is obvious. The danger of entangling alliances, against which George Washington warned, is that they tend to make a local conflict into a global one. We have seen such alliances break out into two world wars during the past century. It is important therefore that regions not directly involved in a dispute remain neutral. The argument for a world government is equally simple. First note that to reduce the potential for destruction it is necessary that no weapons be imported into a region of potential conflict, and that there be machinery set up to settle local disputes peaceably. But these things need policing, and therefore it is necessary that there be a world government which not only has sufficient clout to stop arms shipments and to bring about world disarmament, but sufficient prestige and skill in diplomacy to bring the parties to the negotiating table.

A world government is a necessity for the development of a workable system of economically self-sufficient blocs that can live in peace with one another. Economic self-sufficiency should help ensure that the world government does not become completely dominant. The eventual goal is to do away with the ability of one region to dominate another and yet to reach H. G. Wells's dream of one world. What I have in mind is a kind of federation of the world in which the component states have considerably more control over their local economy than the States in the United States have over theirs.

The environment, the third area to be considered, is an area of concern if the whole world is to undergo further economic development. The planet's ecosystem is already overburdened. Global warming is a major threat. The resources of earth are being rapidly used up. For example, top soil is being eroded at an alarming rate and if allowed to continue will turn much of the world into desert. Regions need to get together to control pollution, which is simply another factor pointing toward the need for—if not world government—at least some measure of central control.

Communist countries have been major contributors to pollution, but in the West and since the demise of communism, pollution is a biproduct of the market economy. This brings us to the topic of the next section.

5 The market economy and wage slavery

Though it is to be expected that a version of the free-market economy will be developed in each region of the globe, the free-market economy has some severe drawbacks, and therefore it is inevitable that there be some movement to ameliorate it.

The most discussed drawback of the market economy is that it is subject to cyclic recessions. There are parts of the economy that are not allowed to fail, however, including major industrial and financial industries. If a bank is big enough, at least in America, it can be sure to get a bailout from the *FEDERAL RESERVE* (cf. Greider [1987]) or from the American taxpayer. It is apparently the case that the United States of America is convinced that socialism works—not for the individual but—for large corporations (like Chrysler— now Daimler-Chrysler) and the major financial institutions. Recessions are caused largely by artificial restrictions in the money supply, on one hand, and overproduction, on the other. As we have already indicated, while in the old days the money supply was restricted by the gold standard, nowadays it is restricted primarily by the effort to fight inflation. Thanks are due the gods—or whoever is in control—that the world is no longer "crucified on a cross of gold."

Let us suppose that the major problems of the market economy such as unemployment and cyclical recessions were solved. There would still remain work as a "necessary evil." As one who has worked in industry, I know that, by and large, the employee's lot is not a happy one. I intend here to discuss what can be done to improve the lot of the working man and woman. I suggest that there are three reforms that are necessary if what has been called "wage slavery" is to be a thing of the past. The first is the elimination of the worst aspects of capitalism by the free availability to all of the basic necessities of life, the second is the reform of management through more democracy in the work place, and the third is a removal of the threat to the well being of both humanity and nature by the present emphasis on economic considerations. I discuss these in order.

John Kenneth Galbraith comments somewhere that according to the conventional wisdom, money works in two different ways on rich and poor: it is thought that rewarding the rich makes them work harder, but giving money to the poor takes away their incentive to work. Even if this were true there is an easy solution: eliminate the poor by making them rich—then everybody will have the same incentive for work. I believe, however, that the desire for increased opulence is a sufficient incentive even among the poor. If the basic necessities of life were freely available, the incentive for work would remain since things like jewelry, movies, television, an automobile, or a finer house, or even alcohol, tobacco, or drugs, would not be available without labor. Of course, even today, people can obtain luxuries without work if they are good at borrowing or begging. Nonetheless, one would expect that the desire for challenge as well as a desire for luxuries would cause most people to work hard. Idleness is not attractive to most people; for, as pointed out in Chapter Three, after a glut of idleness, healthy people go out looking for challenge. What is more, by the application of intelligence, most boring jobs can be eliminated, or failing that, made attractive by large incentives.

The ideal of the anarchists like Krotpotkin [1902] of free food, clothing, warmth, safety, shelter, medical services, waste disposal, a healthful environment, education, and libraries and museums is already within reach in advanced countries such as the United States and Sweden. Until recently food stamps and earned income credit have been available for those whose income and assets are below a fixed level, and there is reason to hope that eventually all the basic services will be available to all who need it. When this becomes a reality, no one would have to sleep on the streets of New York, London, or Calcutta; nor forego medical or dental treatment. There is no reason to suppose that the towns and cities in advanced countries such as the United States cannot be made as attractive as those one generally finds in Sweden or Germany. When this happens, the incentive for robbery would be lessened, and there would be a vast improvement in the health and safety of the cities.

I come now to the second reform to be discussed: give workers more say about what happens in the work place. According to Chomsky [1997, p. 20]: *A business or big corporation is a fascist structure internally. Power is at the top. Orders go from top to bottom. You either follow the orders or get out.* Without totally abandoning capitalism, there are many ways to change this. (For early proposals see, for example, Russell [1919].) One proposal is to allow workers to buy out the owners, the feasibility of which has already been demonstrated in thousands of companies. In the United States there are *Employee Stock Ownership Plans,* supported by the Federal Government, which are already in operation. In Quebec, labor makes use of its pension funds to buy up stock and take a place in corporate boardrooms (cf. Hedrick Smith [ibid.]). Though plans like these can help immensely in bridging the distance between workers and management, most of them do not go so far as to give workers a say in how the business is run (cf. Sale [ibid.]). Caution is needed here; I do not suggest that democracy be carried to the extreme that the bulk of administration is to be done by committee. Spare the workers long boring hours in interminable discussions, but let the administrators be accountable for what they do. Until there is some accountability of the management to the workers for the decisions they make and some correlative assurance that the workers are to share fairly in the fruits of their labor, workers cannot be expected to have the same enthusiasm for work as their employers nor can they be expected to behave responsibly. Greider [1997b] points out that if capital were spread more uniformly among the people, the class struggle between capital and labor could be brought to an end. There are at least two areas of concern in which management can be made accountable to the workers: one, wages and working conditions; the other, the expansion or contraction of production. The latter brings us to the third way of eliminating wage slavery to be discussed: a removal of the present threat to the well being of both humanity and nature by the present emphasis on economic considerations.

The modern emphasis on wealth and power may be viewed as an overreaction of the human animal to the need for the basic necessities of life. As John Kenneth Galbraith [1967] points out:

> If people are hungry, ill-clad, unsheltered or diseased, nothing is so important as to remedy their condition. Higher income is the basic remedy; their problem is thus an economic problem...With high income, questions beyond the reach of economics obtrude. These require consideration of how much beauty should be sacrificed for increased output.

The world's overemphasis on economic conditions has a negative effect on the quality of life. Consider, for example, how the burgeoning production of motor cars affects the environment. The environment is filled with huge affluent suburbs in which you cannot buy a newspaper or a loaf of bread, huge expressways that are choked to capacity soon after they are built, and impoverished central cities that look like war zones. The early romance of travelling by motor car is turning into a nightmare. A return to rail transport and the bicycle would do much to decrease congestion as well as result in a huge increase in health. I am as reluctant as anyone to give up the privacy and freedom of the automobile, yet the sacrifice must be made. Let us hark back to the days of the corner grocery store, neighborhood shopping, and local delivery. This is already being done in progressive places like Denmark (see Walljasper [1998]). For a modern twist, let us encourage shopping via the television, computer, and phone. This should not be too difficult if, as recommended by the sages (see Chapter Six: *What Are We To Do With Our lives?*), the emphasis were on creativity rather than possessions.

Though wealth is in some sense important for happiness, there is no reason to suppose that life in a more primitive time and place could not be on the whole happier than life in modern times. People enjoy roughing it, and in the future we may find that Schumaker [1973] is right when he says "small is beautiful." Moreover, what is worth money is a matter of opinion, e.g. van Gogh sold only one painting and hence produced little or no wealth during his lifetime, but after his death his work is worth millions. If van Gogh had been convinced of the economic dogma he would not have painted one picture. In one of two ways, austerity is the wave of the future: either life will become more austere owing to crowding and pollution, or we shall adopt a more austere lifestyle in order to save the environment. The latter alternative is more pleasant to contemplate than the former.

In much of the world the problem is to produce a sufficient quantity of goods so that people have enough of the basic necessities of life, but in the developed world one finds not only a huge unused capacity for production, but an expansion of what are thought to be basic necessities. As spiritual leaders are fond of pointing out, the real standard of living of humanity does not depend on bread alone. At some point we need to get beyond the notion that the more goods one has, the better off one is. Wordsworth complained:

Getting and spending, we lay waste our powers:
Little we see in nature that is ours.

What is needed is to reconcile people to a limitation of the production of both goods and people. The main benefit of such a limitation would be a great improvement in the physical environment, but a whole range of other benefits would accrue. To take one example, not crucial perhaps but important: if there were a lesser emphasis on production, it would no longer appear an evil that women on welfare stay home to take care of their own babies. Other benefits include reducing crime, and—owing to a moderation in competition—harmonizing different races and ethnic groups and fostering world peace. Most important of all, if people could be released from concern for their own welfare, they would be freed for unlimited spiritual adventures of the kind discussed in earlier chapters of this work.

6 Toward devolution

If smallness is to be preferred to bigness, one need not stop at a division of the world into subregions. Why not further divide each subregion *ad infinitum*? For the sake of administration, one already expects complexes such as the former British empire, the former Soviet Union, the United States of America, Australia, and India, to be divided into provinces having some measure of local autonomy. A decentralized system, with its greater local autonomy, allows for local leadership which is more capable of responding to the people. As noted in Section 4, there are advantages to local autonomy but also dangers such as local conflict. Here I consider some of the ways—outside of those having to do with the economy—in which local autonomy confers advantage (and sometimes disadvantage) on a region. One has to do with defense; the other with culture. I consider these in order.

A state is like an organism: the more highly organized the easier it is to destroy. Consider, for example, the earthworm; if you cut an earthworm in two you have two earthworms. The moral is that if an entity is decentralized and constructed like a hologram—or the human brain—there is no way to destroy the whole by destroying a part. For example, consider Vietnam. Vietnam is decentralized and therefore all the military might of the United States was powerless to destroy it. It follows from this argument that if the vital services needed for survival were to be found at the local level throughout the country it would be relatively invulnerable, and consequently from the military standpoint it may be prudent to decentralize.

I turn now to culture. The impoverishment of local regions as against the centers of power is not only true in a material sense but also in a cultural sense. Art and science travel to where the money is. If wealth were more evenly distributed it is likely that artistic talent and scientific acumen would also be more equally dispersed. Were such a dispersal to be realized, it is probable that we would see a cultural renascence such as was found in Athens in 500 B.C.E. As it is, talent is rejected by local regions not only because there is no facility to exercise it, but because it tends, on one hand, to be stifled by local ignorance and prejudice and, on the other, finds its haven only in the more tolerant quarters of the big cities. Even in tolerant Athens, Socrates was tried and put to death, and today there is a spirit abroad in the land, especially in the so-called hinterlands, that stifles dissent and discourages creativity. The need to frustrate the poisonous network of Bible thumpers and fundamentalists, whether in Rome, Cairo, Moscow,

Teheran or in Dayton, Tennessee, is as pressing now as it was when Galileo was silenced and Bruno was put to death. Free spirits, whether they are apologists for privilege like Socrates or revolutionaries like Shelley, need not only the protection of a constitution such as that of the United States, but vigorous protagonists and protectors of liberty such as Jefferson, Lincoln, John Stuart Mill and the Warren Court. For this to happen the wall of ignorance that insulates small localities from the larger currents of thought must be penetrated not only by large-minded individuals, but by an open communication system that reaches into small towns and hamlets and an education system that tends to open the eyes of young people to the splendors of which they might be a part.

On the other hand, owing to the fact that the larger the unit the more impersonal and machine-like it tends to become, the cult of bigness everywhere cuts down the quality of life. This cult has even penetrated education. In the United States the young are herded in large schools in which they are treated like raw material to be processed. It is well said that a fitting punishment for any crime would be four years in high school.

It seems hopeless to try to raise the cultural level of art and entertainment if this level is determined by mass culture. The appetite of the masses for images of destruction, whether it be the gladiatorial contests of Rome, human sacrifice in religious services, bull fights in Spain, modern wrestling matches, mass media, etc. seems to be insatiable. Plato thought that the popular culture of his day had to be banned. When attempts are made to regulate morality by repressive means, however, it is soon found that the cure is worse than the disease. Instead of repression, the general culture level might be raised if somehow big money no longer dominated art and entertainment. If art and culture were locally produced it is possible that people would develop greater discrimination of judgment.

History shows that masses of people are not needed to produce quality. Athens in its ancient glory did not contain more than a few hundred thousand persons. At the present time, local creativity is inhibited because certain people get national or world attention, often irrespective of the quality of what is produced, while people who are often of equal or greater talent languish for lack of recognition. According to Bertrand Russell [1949, p. 59]:

> The inferiority of our age in some respects is an inevitable result of the fact that society is centralized and organized to such a degree that individual initiative is reduced to a minimum. Where art has

> flourished in the past it has flourished as a rule amongst small communities which had rivals among their neighbors, such as the Greek City States, the little Principalities of the Italian Renaissance, and the petty Courts of German eighteenth century rulers. Each of these rulers had to have his musician, and once in a way he was Johann Sebastian Bach, but even if he is not he was still free to do his best. There is something about local rivalry that is essential in these matters.

I think Russell is correct in this and that it is the near invisibility of local people that has an inhibiting effect on art and scholarship. If a locality of the size of (say) Toledo (Ohio) were separated from the rest of the world, and if the inhabitants thereof pay as much attention to what is done locally as they formerly did to the doings of celebrities, there is little reason to believe the talent developed would not begin to be raised to world class level. Of course, in today's world of instant communication and ease of movement it is difficult to amass, in a restricted locality, a critical mass of talent not only because local people are in awe of celebrities and share the skepticism of the Israelite who asked, "Can there any good thing come out of Nazareth?", but because of the constant drain of talent and brains toward the centers of power and influence such as New York, London, Paris, or Tokyo—or Harvard, Oxford, or Sorbonne. Moreover, according to Russell [ibid.]:

> In those who might otherwise have worthy ambitions, the effect of centralization is to bring them into competition with too large a number of rivals, and into subjection to an unduly uniform standard of taste.

As a way to save humankind from the "boredom relieved only by disaster" which he regards as resulting from the "unbearably rapid growth and centralization" of organization, Russell suggests that we restore to localities their traditional importance and that cities "...develop an artistic pride leading them to mutual rivalry," and that each city develop "its own school of music and painting, not without vigorous contempt for the school of the next city." More generally, he recommends that *organization should be made more flexible, more relieved by local autonomy, and less oppressive to the human spirit through its impersonal vastness.*

Whether the trend toward bigness can be stemmed remains to be seen. Yet I think it can be concluded that the argument for decentralization based on culture is equally as persuasive as the arguments based on politics and economics.

7 Other obstacles to personal growth

As Plato observes in *THE REPUBLIC*, without the illumination of the idea we are like prisoners in a cell without a window to the world. In this section I want to make a plea for greater imagination in handling some of the problems that face humanity at the present time. I start with the drug problem.

It should be obvious that in the case of drugs, what is being done is not working and we need the illumination of a new idea. As already pointed out in Chapter Four, we pay a large price for the belief that people are not adult enough to decide what to put in their own bodies. At one time there was prohibition of alcohol in the United States, but now alcohol is freely available and one finds that the consumption of alcohol is decreasing. It is argued that if drugs were as freely available as alcohol, the results would be disastrous. For example, it is said there would be more damaged babies born to mothers addicted to crack. It turns out that this danger has been greatly exaggerated. The decriminalization of the use of the drugs need not have such an untoward effect.

No country now makes drugs freely available without prescription because such a policy would have severe repercussions on its foreign relations. This is the case since drugs that are freely available in one country can and will be smuggled into others. In Great Britain drugs are available, but only by prescription, and as a consequence the British have managed to reduce criminal activity associated with the selling of illegal drugs to a minimum. It is the British position that people are not being killed by drugs but by impurities in the drugs, and for this reason they allow certified addicts to obtain, by prescription, the amount of the drug necessary to keep the addict free from symptoms of withdrawal. (For common sense about drugs consult Goodman and Gilman [1941] (and later editions)). The street addict dies either from violence or through infection, contaminated needles, or drug impurities. According to British physicians, most addicts in the British program will eventually get off of drugs of their own accord, but in the meantime a maintenance program saves the addict from criminal activity and may save his or her life.

The current policy on drugs is the result of an absolutist conception of morality. The social penalties imposed by an absolutist conception of morality—which condemns as immoral a long list of activities including taking drugs or alcohol, card playing, smoking, nudity, fornication, masturbation, homoerotic behavior, euthanasia, abortion, etc.—often result in more suffering than is the case when people are given the liberty to choose for themselves. The world over, a standard such as the Bible or Hindu scripture is taken as the last word on morality, no matter the consequences. For example, injunctions, such as those found in the *NEW TESTAMENT*, that husbands be the masters of their wives, have stunted the development of at least half the human race. But freedom is of one piece. Unless women and minorities are free, no one is free, and freedom to live is not complete unless it includes the freedom to die.

If, instead of an absolutist morality, the method of rational inquiry were the preferred approach to human problems, there would be a tendency toward less arrogance when it comes to judging the behavior of others. Moreover, since punishment reinforces harsh behavior and only kindness has the potential to reform, I believe that psychologists like Skinner [1953] are correct when they say that adages such as "give a dog a bad name and it will live up to it" and "honey will attract more flies than vinegar" express the proper spirit when it comes to controlling antisocial behavior. Recently, I have heard on CNN of experimental prisons in Spain that have the promise of demonstrating the truth of this position.

8 The planet in peril

In this section I turn to consider the peril that cattle ranches and eating habits pose to the planet. Singer [1992] writes: *That the growth of human population is one of the greatest risks to the survival of all species on this planet is now a commonplace; but do we ever hear of the threat from the world's cattle population?* According to Rifkin [1992]:

> There are approximately 1.28 billion cattle populating the earth. They take up nearly 24 percent of the landmass of the planet and consume enough grain to feed hundreds of millions of people. Their combined weight exceeds that of the human population on earth.

Along the same lines there is another statistic quoted by Singer [ibid.] (who credits Durning and Brough [1992] for his information): *There are now three times as many domestic animals on this planet as there are human beings.*

The threat to the human survival and human happiness because of eating beef and other meat products is at least fourfold: one, the use of grain to feed livestock reduces rather than augments the food supply;

two, large herds of cattle and other livestock have a ruinous effect on the environment; three, cattle farming separates people from the land and from their traditional source of livelihood and forces them either to emigrate or to swell the ranks of the poor in the already overcrowded cities; fourth, meat is not only inherently unhealthy to eat, but people are being made sick by unsanitary conditions in the markets. The unhealthiness of eating meat is not hard to document. As shown by investigative reporters of the ABC network program (*PRIMETIME LIVE),* despite stringent rules, it is difficult to keep the meat markets in grocery stores sanitary. Moreover, meat tends to get infected by the animals own feces and, what is more, there are deadly parasites in meat that affect people the world over, such as *trichina spiralis*, the cause of trichinosis. But even where meat is rendered sanitary and free of parasites, the deleterious effect of meat eating on the coronary arteries, which has lately been much discussed in the news media, remains an insuperable obstacle. Changing to a diet of fish, which is a superior source of protein, may only be a temporary stopgap since there are not enough fish in the sea to feed everyone. The appetite for meat cannot be justified on the ground of the need for food since feeding grain to cattle destroys much more food than it produces. According to Singer, "Some 38 percent of the world's grain crop is now fed to animals, as well as large quantities of soybeans." There is no doubt that more people could be fed if grain were fed to people rather than to cattle. This fact has long been known. Shelley [1811] wrote:

> The quantity of nutritious vegetal matter consumed in fattening the carcase of an ox would afford ten times the sustenance...if gathered immediately from the bosom of the earth. The most fertile districts of the habitable globe are now cultivated by men for animals at a delay and waste of aliment absolutely incapable of calculation.

Modern research backs up Shelley. Singer writes: "The inefficiency of feeding grain to cattle has been well known since Frances Moore Lappe's *DIET FOR A SMALL PLANET* was first published in 1971." According to Wilson [1993],

> ...our liking for meat causes us to use the sun's energy at low efficiency. It is a general rule of ecology that (very roughly) only about 10 percent of the sun's energy captured by photosynthesis to produce plant tissue is converted into energy in the tissue of herbivores, the animals that eat the plants. Of that amount, 10 percent reaches the carnivores feeding on the herbivores.

If the appetite for meat were given up, vast stretches of land would be opened up for wildlife and recreation, for living space to relieve the overcrowded cities, and for the production of crops to feed the hungry. A country that ate its grain rather than feeding it to cattle, and turned the land saved into forest and cropland, could, more than likely, produce all its own food. Shelley [ibid.], arguing that land is much better used for growing grain or vegetables rather than for grazing sheep, wrote:

> Of what consequence would it be that they [i.e. other countries] refused to take her [Britain's] woolen manufactures, when large and fertile tracts of the island ceased to be allotted to the waste of pasturage?

Throughout history (beginning perhaps with Cain and Abel) people were driven off the land, *banished*, as Karl Marx [1873] says, *by sheep and ox*. Considered in this light, the phrase *man versus beast* takes on a new meaning. The snobbery of the elite in this regard is reflected in a witticism of Samuel Johnson [1755], which is found in his famous dictionary: *oats. A grain, which in England is generally given to horses, but in Scotland supports the people.* As an example of the devastation that pasturage does to people, consider Ireland. Marx notes that before 1846, the English Corn Laws secured for Ireland a monopoly of free importation of corn into Britain. With the imposition of the doctrine of free trade, this monopoly was suddenly removed. Marx [ibid.] writes:

> Apart from all other circumstances, this event alone was sufficient to give a great impulse to the turning of Irish arable land into pasture land, to the concentration of farms, and the eviction of small cultivators. After the fruitfulness of Irish soil had been praised from 1815 to 1846, and proclaimed loudly as by nature destined for the cultivation of wheat, English agronomists, economists, politicians, discover suddenly that it is good for nothing but to produce forage...Therefore her depopulation must go yet further, that thus she may fulfill her true destiny, that of an English sheep walk and cattle pasture.

As a final remark on the Irish question, Marx [ibid.] notes that the Irishman, "banished by sheep and ox" reappears in America as a "Fenian" (i.e. an Irish revolutionary) and the young giant Republic rises to threaten the "old queen of the sea."

In Scotland, the highlands were cleared of both forests and people, to be replaced by sheep and deer. In England and Wales, enclosure laws eliminated communal property and resulted in what Alfred Daniel Hall [1929] calls "a race of landless men." Those who did not emigrate, either starved or filled the already overcrowded cities. The clearing of the land for sheep and ox began very early as we find if we read More's *UTOPIA* [1516]:

> ... your shepe that were wont to be so meke and tame, and so smal eaters, now, as I heare saye, be become so great devourers and wylde that they eate up, and swallow downe, the very men themselfes. They consume, destroye, and devour whole fieldes, howses, and cities.

Things were not substantially different in antiquity. Marx quotes Appian (Civil Wars, I.7) as follows:

> The rich had got possession of the greater part of the undivided land. They trusted in the conditions of the time, that these possessions would not be again taken from them, and bought, therefore, some of the pieces of land lying near theirs, and belonging to the poor, with the acquiescence of their owners, and took some by force, so that they were now cultivating widely extended domains, instead of isolated fields. Then they employed slaves in agriculture and cattle-breeding, because freedmen would have been taken for military service. The possession of slaves brought them great gain, inasmuch as these, on account of their immunity from military service, could freely multiply and have a multitude of children. Thus the powerful men drew all wealth to themselves, and all the land swarmed with slaves. The Italians, on the other hand, were always decreasing in number, destroyed as they were by poverty, taxes, and military service. Even when times of peace came they were doomed to complete inactivity, because the rich were in possession of the soil, and used slaves instead of free men in the tilling of it.

Perhaps the most important threat that the beef industry poses to the well-being of human beings on this planet is the destructive effect of cattle on the environment. In Florida, Lake Okeechobee has become a sewer (cf. *THE ST. PETERSBURG TIMES*, May 18, 1998). But for a full explanation of the destructive effects of cattle ranching on the environment the reader is referred to Rifkin's book, *BEYOND BEEF*. Suffice it to say here that if the beef industry were to disappear tomorrow, the rate of loss of top soil would be vastly reduced, slashing and burning of the Amazon forest would be stopped, the rivers of the world would run purer, dead zones in river estuaries would recede (see *NEW YORK TIMES*, Jan. 20, 1998), and the rate of emission of greenhouse gasses such as methane, carbon dioxide, and nitrous oxide would also be significantly cut back.

Nonetheless, for me, the most telling charge against the beef industry is that it represents harshness on a scale almost unimaginable. Shelley [ibid.], referring specifically to hunting, wrote that for a person "unvitiated by the contagion of the world":

> ...it will be a contemplation full of horror and disappointment to his mind that beings fully capable of the gentlest and most admirable sympathies should take delight in the death-pangs and last convulsions of dying animals.

Why is it that it is rare that people let their hearts go out to the animals that are daily bludgeoned or having their throats cut in the slaughter houses throughout the world? Why is it that most of those who otherwise shrink from violence, do not choke as they reflect on the grisly source of their hamburgers? The taste for beef, like the taste for sugar, is an unhealthy appetite, but unlike an animal in the feedlots and slaughter houses, sugar cane, so far as is known, feels no pain.

It is of course obvious to practical people that to rail against meat eating is like railing against the wind. Whitehead [1933] reminds us, however, that *impractical ideals are a program for reform...not to be criticized by immediate possibilities.* I close this section with a quotation from Shelley's *QUEEN MAB*, written when Shelley was a mere lad of eighteen:

> And man, once fleeting o'er the transient scene
> Swift as an unremembered vision, stands
> Immortal upon the earth: no longer now
> He slays the lamb that looks him in the face,...
> No longer the winged habitants,
> Flee from the form of man; but gather round,
> Which little children stretch in friendly sport
> toward the dreadless partners of their play.

9 Liberalism

We noted at the end of Chapter Four, that the socio-economic doctrines of this work might be fairly characterized as "liberal." Liberalism is an outgrowth of two traditions, one from religion, the other from commerce. From commerce comes the realization that you can get something that you want from all people provided you treat them well. From religion comes a concern for those less well off than oneself. In Deuteronomy [15.11], for example, one reads: *Thou shalt open thine hand wide unto thy brother, to thy poor, and to the needy, in thy land.* These two traditions inspired two ideals, both of which are in a sense enshrined in the *DECLARATION OF INDEPENDENCE.* One of these, the ideal of ***equality***, is embodied in the principle that people are not to be divided into different classes and treated differently according to their class; the other, the ideal of ***freedom***, is embodied in the principle that each person is entitled to an environment in which people can, under their own initiative, flourish. Liberals, in general, have argued for these ideals on the basis that the community as a whole benefits if people are free and equal. For example, Adam Smith argued that if personal economic interests are freely pursued, then—as if by an "invisible hand"—the interests of the general public are served. As has already been pointed out, this argument has its limitations. Nonetheless, with suitable modifications, the argument for liberalism is apt to be much the same today.

As we have indicated, the aim of liberalism is to promote individual development and freedom. Historically, this aim was first enunciated, *explicitly,* by Wilhelm von Humboldt [1854] who, as we noted in Chapter Two, was inspired by the French Revolution. (Of course Humboldt had forerunners like Kant.) In political terms, a liberal is one who believes that the state functions well insofar as it promotes individual development and freedom. In agreement with the principles set forth in Chapter Four, let us interpret this to mean that the proper measure of how well a society functions is the degree to which the society tends to satisfy the needs (including the higher needs) of its members. Nowadays it is common to measure the functioning of a society or state by its *gross domestic product* or *GDP.* GDP (the total monetary value of the goods and services produced domestically) measures the well-being of a society in narrow materialistic terms. It measures what I have termed the *precursors to growth* rather than personal growth or development. The aim of liberalism, as conceived here, is to promote personal freedom and growth even at the expense of the gross domestic product.

The key is for government to use its financial power not only to set new priorities but to energize its citizenry. Men and women, whose talent and genius have been amply demonstrated in science, engineering, and art—e.g. splitting the atom, taking us to the moon, producing color television, and building the Taj Mahal—can hardly fail, if given the resources, to produce equally beautiful results in the socio-economic field. For this to happen the society must, through its taxation, monetary and spending policies, set a proper environment. A society through its taxation and monetary policy can encourage the growth of business, large and small. But even more important for the purpose of surviving in the modern world is to stem the tide of lawlessness, disease, and environmental degradation. As has already been pointed out, a government, which routinely grants contracts to the military-industrial establishment for the purposes of war, can grant contracts of other kinds if it wishes to solve its overwhelming socio-economic problems. A variety of different contractors is needed for this purpose including research establishments, the medical establishment, peace corps, endowments for the arts and a free press, and schools. I suggest that governments need to grant contracts for the following purposes: rebuild the infrastructure, clean up the environment, build new homes, make homes and factories more energy efficient, finding new sources of clean energy (e.g. harnessing the gulfstream?), make medicine available to all, find new cures for diseases, find how to make products more cheaply and more efficiently and more attractive, develop new products, reduce the size of schools and make learning more efficient and more attractive, encourage art and architecture, teach people how to have children only when they want them, etc. This list may not be complete, but it is a start.

I do not doubt that the program, outlined above, will be criticized as old-fashioned "liberalism," fiscally irresponsible and out-dated. Inevitably, when an administration proposes to make resources available for a project, whether it be for a super-collider or a poverty program, the nay sayers, sounding the alarm against both debt and taxes, ask where the money is to come from. Taxes, they say, will take away money needed for investment. Deficit financing, they say, will impoverish future generations. And yet—as we already pointed out—it is clear that deficit financing (and a realistic taxation policy) can lead to prosperity as it did in the USA after World War II. The nay sayers neglect to acknowledge that if interest rates are kept at a level not to exceed economic growth, money people borrow from themselves can produce wealth. It is true that this wealth can be realized only if nations (or regions of the world) move toward self-sufficiency and stop trying to live on cheap imports and/or expensive exports. The wealth to be produced is of two kinds: one, hardware; the other, intangible. The hardware includes bridges, highways, parks, clinics, as well as facilities for production. Wealth of the intangible kind includes vast discoveries whose monetary value is as yet unknown, increased skills, better health, a beautiful environment, greater equality, and a reduction in crime and violence. There is a need for a new accounting system that counts as capital, and against the national debt, both kinds of wealth.

Unless the modern world becomes impoverished through overpopulation and pollution, each locality can have the resources for a flowering of culture surpassing that of ancient Greece. The rub is, however, for this to happen the life styles of people must change. The whole structure of the modern world order—or rather chaos—must change.

According to Whitehead [1933]: *The modern evolution of big business involves a closer analogy to feudalism, than does feudalism to slavery...Individualists and socialists are merely debating the details of the neo-feudalism that modern industry requires.* Some steps toward combatting this "neo-feudalism" and ameliorating the strangle-hold that modern industry has over the conduct of human affairs were mentioned earlier in this chapter as well as in Chapter Four. One step (suggested in Chapter Four), that would utterly destroy the ability of businesses and corporations to block needed reforms, is to take away their privileged status as a "person." Two other steps toward reform are mentioned in Section Five: one of these is for organized labor, with its vast holding in pensions plans, to make its presence felt in corporate boardrooms; the other is the use of trusts or employee-stock ownership-plans [*ESOPs*] in order to enable workers to own the businesses in which they work. According to Scale [1980] the plan works like this:

> ...an *ESOP* begins with a corporation setting up a trust for its employees, say with 10,000 dollars worth of stock. The trust then borrows money from the bank against those shares, but it gets special beneficial rates and write-offs that the parent company would not be able to get. The trust then gives the money to the corporation, as payment for the stocks it has been given, and it pays off its low-interest loan with some part of the corporate profits that it is given every year—and that the corporation writes off as a tax deduction. And all the employees get the stock or may cash it in when they retire.

According to Greider [1997b, pp. 419-420], Louis Kelso, a prominent investment banker, invented the idea of employee-ownership trusts nearly fifty years ago. He goes on to say: *When Kelso died in 1993, the obituaries cited the contribution of ESOPs* [employee-ownership trusts], *but did not explain that employee trusts were only one device in his grander plan for rescuing capitalism itself.* Among these devices is the idea of a citizen-ownership trust, which is a trust funded directly by central banks like the *FEDERAL RESERVE*. Like *ESOPs* these trusts buy up companies in the name of the workers (or a community) and hold the title until the loan is paid off by the company's earnings and dividends. Greider [ibid., p. 438] writes:

> The idea of a central bank purchasing the loans of private companies, instead of simply buying government bonds, may sound unorthodox, but it is actually the method the *FEDERAL RESERVE* originally used to manage the nation's money supply, and the practice is still employed among European central banks.

Many on the left and the right are opposed to Kelso's ideas for spreading capital more widely. Scale [ibid.], a leftist, noting that *ESOPs* were sponsored in the Senate by Russell Long, questions whether anything to the benefit of the masses could be sponsored by an *unreconstructed* Senator from Louisiana. Greider, on the other hand, reminds us that the sponsor's father, Huey Long, was a populist who called for a radical redistribution of wealth, and claims that the son thinks "his father attacked the right problem." Be that as it may, it is clear that some such means are the only ones now available for a redistribution of capital if we are to stay within the capitalist system.

The main enemy of the liberal is narrowness. It is narrowness (as well as greed) that makes the chief executive consider nothing but the bottom line. But it is not only the merchant who is narrow. The stockholder who invests in any business so long as it is profitable is narrow. The educator who educates only to prepare students for exams or jobs down the road is narrow. The person is narrow who prates of "family values," but fails to value people regardless of their national origin, the color of their skin, their sex or sexual preference, or their social status. Those who would deny others the privilege of speaking freely and thinking their own thoughts are narrow. Those who would control what goes in, and what comes out of the bodies of others, are narrow. Narrowness rears its ugly head in racism, on one hand, and in sexism and homophobia, on the other.

10 Conclusions

Unfortunately, we have not spent much effort in this work on how society can foster an atmosphere in which spiritual growth becomes the norm. We have concentrated on the lower levels of complaint ("...*what ye shall eat, or what ye shall drink; nor yet for the body, what ye shall put on*") that belong to the dismal science of economics. We need to get beyond this to reach a society in which the complaints are at a higher level, e.g. the complaint, often uttered by sophisticated people who have little else to complain of, that the music written recently is not up to the level of the great masters.

The problems of earth and our present sophistication about "human nature" make us despair of what our sexist ancestors called the "perfectibility of man" and, moreover, they render conceptions of a heaven on earth an impossible dream. Heaven is relative, however. For the starving, heaven is a place where there are plenty of things to eat. In 1987, when I returned from the Soviet Union, Finland seemed like heaven, but as soon I became accustomed to relative freedom my critical faculty returned. As long as we are human we never cease to find disappointments or things that need to be changed. Heaven might be defined as the state in which the grumbles are at a relatively high level, for example: having a manuscript rejected by a publisher or too many ants at a picnic. For this to be possible the minimum achievement is that the ecology of the earth is saved and human beings remain a part of it.

As noted in Chapter Seven, Kant, in his *GROUNDWORK*, weighs reason against instinct as an instrument for the preservation of species and asserts that instinct is superior. Kant's view is at least debatable. Though it is true that human reason is less sure in *practical* fields such as ethics, religion, and politics than it is in theoretical fields such as mathematics and physics, it is apparent that it is only owing to reason that a comparatively weak animal such as the human being has been able to triumph over other animals and thereby fulfill the command we find in *GENESIS*: ***Be fruitful and multiply, and replenish the earth, and subdue it: and have dominion over the fish of the sea, and over the fowl of the air, and over every living thing that moveth upon the earth.*** On the other hand, Kant's view may prove ultimately to be the correct one. Human beings are indeed "too clever by half." According to the sage Aurobindo [1950]:

> Science pursuing its cold and even way has made discoveries which have served on one side a practical humanitarianism, on the other supplied monstrous weapons to egoism and mutual destruction; it has made possible a gigantic efficiency of organization which has been used on one side for the economic and social amelioration of the nations and on the other for turning each into a colossal battering ram of aggression, ruin and slaughter. It has given rise on the one side to

> large rationalistic and altruistic humanitarianism, on the other it has rationalized a godless egoism, vitalism, vulgar will to power and success.

As Aurobindo and other sages make clear, human reason is a two-edged sword. More worrying still, one may note—as we, in effect, did in Chapter Four—humanity is not unlike the sorcerer's apprentice who unleashes forces with which she is unable to cope. For example, scientists have unleashed the force of the atom before learning how to ensure that it does not destroy the earth. It has become obvious that if the world is to be saved the leaders of humanity can no longer rely on measures based on blind instinct or short-sighted cleverness, but must develop a new wisdom of a depth so far not seen in the world.

Where are we to find this wisdom? A hint is given in the last chapter where it is claimed that there is a source of wisdom that transcends reason and instinct. Socrates calls it his daimon. In Chapters One and Seven we call it the "mystic vision." Mathematicians call it "intuition." (The brain seems to work in mysterious ways to deliver the solution to a mathematical problem. See Poincaré [1913].) Though the mystic vision in some sense transcends reason and instinct, it is *in accord* with reason and instinct when reason and instinct are purged of those elements that stand in the way of spiritual growth and what Whitehead [1933] calls *the final Beauty with which the Universe achieves its justification.*

What is to be done? One of the conclusions of this work is that if the earth is to be saved and humans beings are to remain part of it, the world must stop its overweening concern for economics and turn toward inner development. Moreover, we have concluded that in order for the world to be at peace there needs to be world government with the power to broker disputes between different localities, regulate financial relations between localities and promote development where it is needed, protect the global biosphere, and oversee the use of the oceans and outer space. But the activities of this government need to be severely circumscribed; for peace and a clean environment are of little avail unless they enable people to live lives in which their higher needs are fulfilled. The greatest challenge for the future is how to combine the vast organization needed to maintain peace and a liveable environment with the flexibility and devolution necessary for maximum personal development.

There is no guarantee that this challenge will be solved. Aurobindo (Cf. Roy [1950, p. 311]) writes (emphasis in original):

> Men [and women] do not abandon a cause, because they have seen it all or foresee its failure; and they are spiritually right in their stubborn perseverance. *Moreover, we do not live for outward result alone; far more the object of life is the growth of the soul, not outward success of the hour or even of the near future. The soul can grow against or even by a material destiny that is adverse.*

Note:

According to newspaper reports (see New York Times, May 23, 1999) the employment rate in the United States for young black men has surged. It is reported that the unemployment rate among those African-American youths in major metropolitan areas who are not disabled or in prison is only forty nine percent. But we need not worry—Alan Greenspan will soon put things to rights. Already he has (as of 6/31/99) raised interest rates by a quarter point.

Part Two

A LOGICAL BASIS FOR A PRACTICAL PHILOSOPHY

Chapter Nine

The Language of Requirement

1 Introduction

The theory of requirement and admissibility, as presented in this chapter, provides a new foundation for normative systems. Two earlier versions of this theory may be found in the literature, one in *THEORIA* (see Johanson [88]), another in *DEONTIC LOGIC, AGENCY AND NORMATIVE SYSTEMS* (see Brown & Carmo (Eds.) [1996]). The theory is a wide ranging logical system that encompasses in its scope not only deontic logic but a theory of norms whose main rival is the theory of "rights" and "legal positions" of Kanger and Lindahl.

Whenever a new theory arrives on the scene one always hears the plaintive cry: what is it possible to do by this new system which cannot be done better by the old and familiar ones? This may or not be a good question, but it requires an answer. The success of the new system—which I refer to here as the *functorial framework for practical logic* (because of the prominent part played in the theory by what mathematicians call *functors*)—can be measured, in part, by how much it is an improvement over both standard deontic logic and the theory of Kanger and Lindahl. Other systems of deontic logic have appeared in recent times that avoid the pitfalls of previous theories, but these unfortunately are so abstruse and technical as to be practically unusable. My comments in this section are restricted to standard deontic logic—the Kanger-Lindahl theory is discussed at length in

Section 10.

I believe there are three important characteristics that a system should have if it is to replace standard deontic logic. These are: one, it resolves the paradoxes of standard deontic logic; two, it shows clearly the relation of deontic operators to imperatives; three, it is in some sense dynamic rather than static. Other criteria could be added, but I think these are sufficient for our purposes. I consider each in turn.

In standard deontic logic attempts such as that of Chisholm [1963] to bridge the chasm between fact and obligation have led to outright contradiction. On the other hand, though one can deny, as Hume did, that obligation statements can be derived from factual ones, one cannot deny that only by confrontation with the facts can it be determined whether an obligation has been violated. What is needed therefore is a language in which the relation between fact and what Chisholm [1964] calls requirement is made clear. I claim that this is precisely what the present theory does.

As is well known, in standard deontic logic, obligation commutes with conjunction and permission commutes with disjunction. In the view of von Wright [1981] it is the acceptance of rules of this kind that has led to the paradoxes. In the functorial approach, the situation is altogether different. Instead of one kind of simple duty there are two. One is called *irrevocability,* the other, *compulsivity.* (There are other kinds of "compulsions" or "duties" in the theory, but of the basic variety there are only two.) Not all of the basic "duty" operators commute with conjunction and not all of the associated "permission" operators commute with disjunctions. It is in part this richness and flexibility that enables the present theory to allow conflicting requirements and avoid the contradictions of standard deontic logic (cf. Section 8).

The relation of deontic operators to imperatives is left untouched by other approaches to deontic logic, but in the present system imperatives take a central role. Indeed, one of the fruits of the functorial framework for practical logic is a logic of imperatives. There are many reasons why it is both necessary and advantageous for a deontic theory to take imperatives into account. Principles of conduct when expressed in the form of maxims and commands have a directness and clarity that is not possible to achieve in any other way. The moral principles in the great religions are expressed in imperatives. One only has to think of the *Ten Commandments, The Sermon on the Mount,* the *DHAMMAPADA,* the *TAO-TE-CHING,* etc. I believe it is in no small part because the basic principles of the ethical formalism of Kant and the utilitarianism of Bentham and Mill are easily expressible in the form of maxims that these are the doctrines that dominate moral philosophy in modern times. (In the case of utilitarianism the maxim is: *Act in accordance with the policy that in the present case* [*or at least as a general rule*] *will lead to the maximum utility—however defined.*) As mentioned in the ***INTRODUCTORY***, it is central to the present theory that there is a pair of operators—*Let* and *Required*—which stand to each other in the relation that in mathematics is called *adjointness.* (The meaning of "adjointness" is explained in Section 2 below.) *Let* assigns to each statement an imperative and *Required* assigns to each imperative a statement. The deontic operators of the present theory are constructed by means of these two operators. As already mentioned, these operators are of the kind that mathematicians call "functors" (see below) and hence the name "functorial framework." Unless I am very much mistaken, these operators can take over the role of the operators in standard deontic logic without running into any serious paradox.

The dynamic or temporal aspect of a deontic theory, the final characteristic to be considered, is widely recognized as an essential ingredient in any theory that is to replace standard deontic logic, e.g. by Thomason [1981] and Meyer [1988]. The present theory is dynamic in two respects: one, it is time-dependent; the other, it makes use of agent-introducing operators of the kind due to Kanger [1957]. Time dependence is built into the present theory because two of the basic operators of the theory—*Let* and its associated operator *Sustain*—are defined by quantifying over time. Further discussion of this topic is best postponed, however, until after the formal apparatus of the theory is introduced.

To sum up: the functorial framework for practical logic, if judged on the basis of the problems we have been discussing—as well as how well it overcomes the shortcomings of the Kanger-Lindahl theory—is an advance on other systems of deontic logic. In particular, the present system allows for conflicting requirements and, unlike standard deontic logic, does not plunge into inconsistencies when *prima facie* duties have been violated. Moreover, the functorial approach not only relates deontic statements to imperatives, but incorporates time dependence in its formulas. Furthermore, not only does it offer solutions to problems outside the ken of other theories, but it has the merit of assembling in one package all the different strands of this field (except perhaps dynamic logic) without resort to extraneous or *ad hoc* hypotheses.

I return to the topic of the use of the present theory after introducing the formal apparatus. (See Section 5)

2 The formal apparatus

I outline here a framework *K* in which to carry out the investigation. (This is the framework that we have already labelled *the functorial framework for practical logic*, which hereafter will be called simply *the functorial framework.*) As has already been indicated, the formal apparatus of this work — (i.e., the framework *K*)— has been treated by the present author ([1988, 1996]) in two different places. The later version is least unsatisfactory and I freely excerpt from it here. There is a need for a new version, however, since none of the previous treatments do full justice to the theory.

As the name "functorial framework" suggests, the present theory is based on the *mathematical theory of categories* which was invented in mid-century by Eilenberg and MacLane (see Chapter Two). As a concession to those philosophers who are not familiar with category theory, I make every effort in this chapter to make the theory self-contained. (That is less so in the next chapter; for there the full apparatus of category theory is assumed.) Since our attention in this work is confined to logic and lattice theory, there is no need to give the standard definitions of categorical terms when the terms, restricted to logic and lattice theory, may be more simply defined. The procedure adopted in the present chapter is that terms such as "functor" and "adjoint" are defined by properties that apply only to lattices and logic, and accordingly the properties used in this paper as "definitions" are actually provable from the standard categorical definitions in their full generality. Why the categorical terminology then? My excuse is that standard categorical results on adjoint functors are appealed to; moreover, since the theory is a simple application of category theory, it would not be proper to disguise the source from which the theory derives its concepts

As is well known, when logically equivalent formulas in a formal system are identified, the underlying structure may be regarded as *a lattice* ***ordered by entailment*** (cf. Birkhoff [1948]). Such a lattice is either a *Heyting algebra* or a *Boolean algebra*, depending on whether or not the law of excluded middle holds. In *K* there are two logics (or formal systems): one, **PROP** (a first-order predicate logic); the other, **IMP** (a logic of imperatives). A formula in **PROP** may (or may not) have one or more free occurrences of a variable *t* interpreted as the time variable.

The purpose of the theory *K* is to serve as a foundation for normative systems. Informally, a "normative system" is a system of norms, rules, or imperatives such as one finds in law, religion or personal ethics. Formally, by a *normative system* ***N***, I understand a logical system, axiomatized over ***K***, with the following property: ***N*** *contains, as subsystems, two theories (formal languages): one, a first-order theory axiomatized over* ***PROP****; the other, a theory (not necessarily first-order) axiomatized over* ***IMP****.*

As in any theory, in a normative system there are two kinds of axioms: one, the *logical* axioms, common to all normative systems; the other, the *proper* axioms that are over and above the logical axioms. (Cp. Mendelson [1987])

As already indicated, the main action in the present theory occurs by means of operators called "functors" which establish a relationship between the two logics somewhat as described by Mill. A functor is defined as follows: Let Ξ and Ψ be lattices in which the order relation is written $<$; then a mapping $f\colon \Xi \longrightarrow \Psi$ is a (lattice) ***functor*** provided that $fp < fq$ in Ψ whenever p and q are elements of Ξ such that $p < q$. There are basically two pairs of functors in ***K*** as shown in the following diagram:

$$\mathbf{PROP}\ \underset{\boldsymbol{R}}{\overset{\boldsymbol{L}}{\rightleftarrows}}\ \mathbf{IMP}\ \underset{\boldsymbol{S}}{\overset{\boldsymbol{A}}{\rightleftarrows}}\ \mathbf{PROP}$$

The functor R is called *requirement* and the functor A is called *admissibility*. The mapping L is called *"Let"* and the mapping S is called *"Sustain."* The application of the mapping R to an imperative U yields the expression RU (read: *it is required that U*); the application of the mapping A to U yields the expression AU (read: *it is admissible that U*); the application of L to an element F in **PROP** yields an imperative LF (read: *it **shall** be the case that at some time t, F is true*); finally, the application of S to an element F in **PROP** yields an imperative SF (read: *it **shall** be the case that for all time t, F is true*).

It should be apparent from the preceding discussion that L and S are built up from more primitive operations, one of which is the *imperative operator* ! (read: *it **shall** be the case that*). It is assumed that the imperative operator is a functor that (*so to speak*) commutes with negation and therefore, by rule of interchange of quantifiers, it follows that L and S are interdefinable (see below). Moreover we interpret negation and admissibility in such a way that the phrase *it is admissible that* is synonymous with *it is not required that not,* and therefore admissibility and requirement are also interdefinable.

In the theory ***K***, a time variable *t* is assumed. A time variable, like any coordinate variable, is basically a map from events (or spatial locations) into the set of real numbers. (Cf. Munroe [1963]) This means that in interpreting the theory , we must assume, at a minimum, a theory of real numbers. Hence, for the sake of applications, I assume that *the basic formal apparatus of* ***K*** *includes a set theory (or category theory) adequate for arithmetic, geometry, lattice theory and real analysis.*

The interpretation of the time variable *t* depends, of course, on the time frame adopted (i.e. whether time is measured in minutes, hours, or seconds, etc. and whether we are using local time or (say) Greenwich Mean Time). It is often convenient, however, to choose a *moving* frame in which *now* is assigned the number 0 (cf. Munroe [ibid.]).

The use of the time variable *t* is crucial in practical logic since it makes no sense to issue commands intended to affect the past. In the interpretation of commands the tacit assumption is that in the phrases "for all *t*" and "for some *t*," the variable *t* is restricted to the present or the future. One must admit, however, that it is not always possible to restrict the variable *t* to the present and future since, generally speaking, the commands found in law or legislation apply to past actions as well as future ones and, moreover, people are accountable for past actions in courts of law.

Note: In order to avoid excessive pedantry, linguistic expressions will very often be treated *autonymously*, i.e. as names of themselves. Moreover, sentential expressions such as U_j (or *F, G, U, V etc.)* are metavariables, i.e. replaceable by *names* of sentences instead of sentences. It follows that readings such as "*It* ***shall*** *be the case that F!*" (in the case of *!F*) or "*It is required that U*" (in the case of *RU*) are not actually correct as they stand for they abuse language by conflating use and mention. The correct reading is obtained by substituting the actual sentence for the name of the sentence. For example, if *F* is *"The door is closed,"* *!F* is read: *It* ***shall*** *be the case that the door is closed!*—not: *It* ***shall*** *be the case that* "The door is closed"*!* The case of *RU* is similar, except that, in this case, *U* must be put in "proper form" before a reading of *RU* makes sense—where by "proper form" I mean a sentence of the form "It *shall* be the case that *F*." For example, if *U* is the imperative "Shut the door!"; then a suitable replacement for U is the sentence "It *shall* be the case that the door is shut!" If a replacement like this is made, then *RU* can be read: *It is required that it* ***shall*** *be the case that the door is shut* (**not:** *It is required that* "Shut the door!" or *It is required that* "It *shall* be the case that the door is shut!").

The operators in *K* are of two kinds: one, operators that are internal to the subsystems **PROP** and **IMP** such as the connectives, predicate letters, the quantifiers, and the "agent-introduction operator" *d*; the other, mappings such as *R* and *L* between the two logics. The structure of neither **PROP** nor **IMP** is independent of the mappings between the logics, however; for not only do expressions such as *RF*, where *F* designates an element of **IMP**, designate "well-formed" elements of **PROP,** but all imperatives (belonging to **IMP**) are formed by applying the imperative operator (or equivalent) to an element of **PROP.**

2.1 Symbols

The symbols *internal* to **PROP** and **IMP** are as follows:

(i)	parentheses:	(,)	
	brackets:	[,]	
(ii)	connectives:	$\neg$	(negation)
		&	(conjunction)
		$\vee$	(disjunction)
		$\rightarrow$	(the conditional)
		$\leftrightarrow$	(the biconditional)
(iii)	statement letters:	F, G, H, F_j, j = 1, 2, 3, ...	
	n-ary predicate letters:	$F(n, j)$, $n = 1, 2, 3, ...; j = 1, 2, 3, ...$	
	imperative letters	U, V, U_j, j =1, 2, 3, ...	
(iv)	individual variables:	t	(time),
		x, y, z, x_j, j = 1, 2, 3,...	
(v)	individual constants:	a, b, c, a_j, j = 1, 2, 3,...	
(vi)	lattice constants:	0*	(smallest element),
		1*	(complement of 0*)
(vii)	function letter:	*d*	(*agent-introduction* operator)

Terms are individual variables or individual constants. The *agent-introduction operator d* applied to a term *x* yields the *agent operator* [*x*,] (read: *x sees to it that*). The agent operator [*x*,] applied to an element *F* in **PROP** yields the expression [*x, F*] (read: x *sees to it that F*). The agent operator is a mapping from **PROP** to **PROP.**

(viii)	ordered couples	(a, b)
(ix)	ordered n-tuples	$(a_1, a_2, a_3, ..., a_n)$
(x)	quantifier symbols:	$\exists, \forall$
(xi)	time constant	*now* ($t = 0$)

Note: For convenience, the same symbols are used in *both* **PROP** and **IMP** to denote the smallest element 0* and its negation 1* (¬0*) respectively.

Other symbols:

(i)	entailment	=>	(the order relation in **PROP** and **IMP)**
(ii)	imperative functor	!	(read: *it shall be the case that)*
(iii)	*Let* functor	*L*	(read: *let it be that*)
(iv)	requirement functor	*R*	(read: *it is required that*)
(v)	*Sustain* functor	*S*	(read: *see to it that it remains the case that*)
(vi)	admissibility functor	*A*	(read: *it is admissible that*)
vii)	agent operator	[*x*,]	(read: *x sees to it that*)
(viii)	if and only if	iff	

2.2 Definitions

D1. The order relation < in **PROP** and **IMP** is defined as *entailment* =>.

By definition therefore (see above) a *functor* in *K* is a mapping that preserves entailment, i.e. a mapping *f* from **PROP** to **IMP** or from **IMP** to **PROP** such that *fp* => *fq* whenever *p* => *q*.

D2. **PROP** is the smallest complete, complemented and distributive lattice that contains *distinct* elements 0* (defined as the smallest element) and 1* (defined as ¬0*, i.e. the complement—or *negation*—of 0*) together with the following:

(i) $J(b_1, b_2, ..., b_n)$ where *J* is an n-ary predicate letter and the b_j, j = 1, 2,..., n are terms.

(ii) *RU* where *U* is an element of **IMP**.

(iii) [*x*, *F*] where *x* is a term and *F* is an element of **PROP**.

It follows that **PROP** is a complete *Boolean Algebra.*

Informally, a "statement" is a sentence in the indicative mood; formally, however, a *statement* is defined as a member of the lattice **PROP**.

D3. A statement that contains no free variable is *closed* and a statement that contains a free variable is *open.* A statement is *t-closed* if it does not contain the time variable free.

D4. The *time variable* t is a mapping from events into real numbers. (Ignoring relativistic considerations we assume events can be linearly ordered.)

D5. **IMP** is the smallest complemented and distributive lattice containing the following:
(i) *distinct* elements 0* and 1* ($0^* \neq 1^*$)
(ii) and the elements of the form $!F$ where F is a t-closed statement in **PROP**.
It follows that **IMP** is a *Boolean Algebra.*

Formally, "imperatives" are sentences in the imperative or optative mood (Cp. Hamblin [1987]); formally, however, an *imperative* is a member of **IMP**.

(A *directive* is an imperative such as "John, shut the door!" which is directed at an agent. A *fiat*, on the other hand, is an imperative such as "Let there be light!" in which the domain of application of the imperative is not made clear. The terms "directive" and "fiat" are used only intuitively; and consequently, in order to avoid excessive pedantry they are not given formal definitions.)

D6. The *connectives* in **PROP** and **IMP** are defined as follows:
(i) $\neg p$ for the complement of p.
(ii) $(p \,\&\, q)$ for the greatest lower bound of p and q.
(iii) $(p \vee q)$ for the least upper bound of p *and* q.
(iv) $(p \rightarrow q)$ for $(\neg p \vee q)$.
(v) $(p \leftrightarrow q)$ for $(p \rightarrow q) \,\&\, (q \rightarrow p)$.

D7. The quantifiers in **PROP** are defined as follows:
(i) Given an element F in **PROP** and a variable x (not t), $(\exists x)F$ is the least upper bound of all elements of **PROP** that are like F except that (possibly) each occurrence of x in F is replaced by another variable or a constant. In a similar way, $(\forall x)F$ is defined as a greatest lower bound.
(ii) Given an element F in **PROP**, $(\exists t)F$ is the least upper bound of all elements of **PROP** that are like F except that each occurrence of the variable t in F is replaced by the name of a real number. $(\forall t)F$ is defined similarly as a greatest lower bound.

Of course, $(\exists t)$ is read "***there exists a time t such that***" and $(\forall t)$ is read "***for all time t.***"

D8. Let Ξ be a lattice and let the order relation in a lattice be written "<"; then if x and y are two elements in the lattice, we write $x = y$ provided $x < y$ and $y < x$. Hence in **PROP** and **IMP** we write $\alpha = \beta$ provided $\alpha \Rightarrow \beta$ and $\beta \Rightarrow \alpha$. (In other contexts, e.g. in D3 above, the equal sign "=" means logical identity.)

D9. A formula α in **PROP** or **IMP** is said to be *valid* if $1^* \Rightarrow \alpha$. Since 1^* is the complement of the least element, it is the largest element, and hence for every α, $\alpha \Rightarrow 1^*$ and $0^* \Rightarrow \alpha$. 0^* is called the *initial element* and 1^* is called the *terminal element.* It follows that α *is valid iff* α *is terminal* (i.e. equal to 1^*).

D10. The *imperative operator* ! is a mapping from the sublattice of t-closed statements in **PROP** to **IMP.** (The symbol ! is not to be confused with "bang" operator in linear logic. See Hofstadter & McKinsey [1939], Kanger [1957].) The function of this mapping is to operate on expressions in **PROP** and turn them into imperatives. That is to say, if the symbol ! is applied to a symbol F that designates an element of **PROP** that does not contain t free, the result is an expression $!F$ that designates an element of **IMP**.

D11. The operators L and S are defined as follows:

(i) $LF = !(\exists t)F$ (read: *It shall be the case that there is a time t such that F*).

(ii) $SF = !(\forall t)F$ (read: *It shall be the case that for all time t, it is the case that F*).

It follows that if the letter L (or S), which designates a mapping, is applied to an expression F that designates an element of **PROP**, the result is an expression LF (or SF) that designates an element of **IMP.**

D12. If $\alpha: \Xi \longrightarrow \Psi$ *and* $\beta: \Psi \longrightarrow \Xi$ are lattice functors, ξ an element of Ξ and ψ an element of Ψ, then (α, β) is a pair of ***adjoint functors*** provided that $\alpha\xi < \psi$ iff $\xi < \beta\psi$; the functor α is ***left adjoint*** to β and β is ***right adjoint*** to α. (See McLarty [1993] or MacLane [1971] for the more general—and much more complex!—categorical definitions.)

The fundamental assumption in the functorial framework is that there is a functor R (called requirement) that is right adjoint to the let functor L. In the next section we spell out what this means.

D13. Two functors α and β are said to be *associate functors* if $\alpha = \neg\beta\neg$ and $\beta = \neg\alpha\neg$. The functor A is defined to be the associate functor of R, i.e. $A = \neg R \neg$ and, consequently, $R = \neg A \neg$.

3 The formal properties of the operators

The imperative operator ! is postulated to have the following properties:

(i) Given a *t*-closed statement *F* in **PROP**, $\neg F$ is an imperative or optative in **IMP** such that $\neg!F = !\neg F$.

(ii) If *F* and *G* are *t*-closed statements such that *F* entails *G* in **PROP**; then $!F$ entails $!G$ in **IMP**.

(iii) $!1^* = 1^*$

Property (i) states that *the imperative operator commutes with negation*, (ii) states that *the imperative operator is a functor* (order-preserving mapping) from the *t*-closed statements of **PROP** to **IMP**, and (iii) states that *the imperative operator preserves validity.*

As a consequence we have:

(1) *S and L are functors that preserve validity; moreover S and L are associate operators in the sense that* $S = \neg L\neg$ *and* $L = \neg S \neg$

Proof: Writing $S = !(\forall t)$ it follows from (i) and an interchange of quantifiers that $S = \neg!(\exists t)\neg = \neg L\neg$, and hence $\neg S\neg = L$. Moreover, one can prove that products of functors that preserve validity are functors that preserve validity, whence *S* and *L* are functors that preserve validity, since they are each products of functors that preserve validity; namely the imperative operator and a quantifier. ♦

As we have stated above, the fundamental assumption of the functorial framework for practical logic says that there is a unique functor *R* (which we call "requirement") that is right adjoint to *L*. In precise language the fundamental assumption says:

(2) *There is a functor R (unique up to logical equivalence) such that*

$$LF \Rightarrow U \text{ (in } \mathbf{IMP}) \text{ iff } F \Rightarrow RU \text{ (in } \mathbf{PROP})$$

By preestablished harmony, as it were, *R is the right adjoint* and *L is the left adjoint.* Now from the fact that *R* is right adjoint to *L*, we easily establish that:

(3) *R is a functor from* **IMP** *to* **PROP** *that preserves validity*

Proof: According to (2), $L1^* \Rightarrow 1^*$ iff $1^* \Rightarrow R1^*$; we need only notethat 1^* is terminal in order to conclude that *R* preserves validity. ♦

It remains a question whether the same is true of *A*. One can easily prove that *A* preserves validity *provided **K** is a complete system.* (See the note at the end of the next chapter.) However that be, it has not been shown that ***K*** is a complete system, and hence we need the axiom:

(4) *A is a functor from* **IMP** *to* **PROP** *that preserves validity.*

The next result is that

(5) *Admissibility A is left adjoint to S (Sustain), i.e. given F in* **PROP** *and U in* **IMP**,

$$U \Rightarrow SF \text{ (in } \mathbf{IMP}) \textit{ iff } AU \Rightarrow F \text{ (in } \mathbf{PROP})$$

This result follows immediately from (2) and the fact that *R* and *A*, as well as *L* and *S*, are associate functors.

As an illustration of adjunction of type (2), consider the following rough example from traffic theory (already given in Chapter Two): "*The light is red*" implies "***It is required** to stop*" if and only if the optative, *Let the light be red!*, implies the imperative, *Stop!* A similar heuristic example from the theory of motor traffic, illustrating adjunctions of type (4), may be stated as follows: The traffic law which says *it is admissible* that you move your vehicle forward into an intersection only if *the traffic light is green* holds if and only if the *imperative* to move forward implies the optative "*Let it be sustained that the light be green!*" The reader will find that these examples are rendered with more precision in Section 4 below where we discuss semantics.

The following are some easy results that follow from adjointness: The first result is that *L* and *A* preserve disjunctions and *R* and *S* preserve conjunctions. In particular, since *L* and *A* are left adjoints, one may infer (cf. McLarty [1992, p.97], MacLane [1971, p. 114 ff.]) that *L* and *A* preserve disjunctions, i.e.

(6) $A(U \vee V) = (AU \vee AV)$ and $L(F \vee G) = (LF \vee LG)$

Because of duality,

(7) $R(U \,\&\, V) = (RU \,\&\, RV)$ and $S(F \,\&\, G) = (SF \,\&\, SG)$

Some further results are also consequences of adjointness; e.g., $LF \Rightarrow LF$ and $SF \Rightarrow SF$ yield

(8) $(F \Rightarrow RLF)$ and $(ASF \Rightarrow F)$

Moreover it follows from (5) together with the fact that the functors *A* and *L* preserve validity that

(9) $RU \Rightarrow AU$ and $SF \Rightarrow LF$

What is more, it follows from (6) and modus ponens that

(10) $R(U \rightarrow V) \Rightarrow (RU \rightarrow RV)$ and $S(F \rightarrow G) \Rightarrow (SF \rightarrow SG)$

As shown in the next chapter, the following is an easy consequence:

(11) $(RU \,\&\, AV) \Rightarrow A(U \,\&\, V)$ and $(SF \,\&\, LG) \Rightarrow L(F \,\&\, G)$

Finally, consider the agent operator [*x*, __]. The purpose of the agent operator [*x*, __] is to bring directives within the scope of the theory. When the operator *L* is applied to [*x*, *F*] the result is the directive $L[x, F]$ (read: *it **shall** be the case that at some time, x sees to it that F*); similarly when S is applied to $[x, F]$ the result is the directive $S[x, F]$ (read: *it **shall** be the case that it is always true that x sees to it that F*).

Contrary to Chellas [1969], Pörn [1970], and Johanson [1996] there is no reason to suppose that the agent operator $[x,\ ]$ is a functor, nor need we assume that [x, 1*] is valid. As a result, formulas such as $RS[x, F] \Rightarrow RS[x, (F \vee G)]$ are not valid.

The agent operator $[x,\ ]$ has the following two properties:

(i) $[x, F] \Rightarrow F$;

(ii) $([x, F] \& [x, G]) = [x, (F \& G)]$

The fundamental properties of the agent operator are exactly like those of the *Do* operator introduced by Kanger [1957, 1966, 1972]. The following are valid:

$$([x, F] \to \neg[x, \neg F]),\ \neg[x, 0^*],\ (([x, F] \& [x, (F \to G)]) \to [x, G]).$$

Notes:

1. As we have already indicated, lattices may be regarded as categories (see McLarty [1992], MacLane [1971]). This fact is exploited in the next chapter.
2. The notion of functor is due to Eilenberg and MacLane [1948] and the notion of adjoint functor is due to Kan [1958]. (See Chapter Two)
3. The mapping L supplies the *syntax* (or the syntactical structure) of the imperatives in **IMP**. Moreover, it may be remarked that the mapping R supplies the *semantics*. This may be seen as follows: Since L preserves validity, (i.e. $L1^* = 1^*$), it follows from (2) that

$$1^* \Rightarrow U \text{ iff } 1^* \Rightarrow RU,$$

whence

$$RU \textit{ is valid iff } U \textit{ is valid.}$$

As a corollary we have:

U entails V (in **IMP**) *iff $R(U \to V)$ is valid* (in **PROP**)

It follows that *requirement R is analogous to truth.*

4 An interpretation of the functors *L, S, R*, and *A*

Why introduce the operators L and S? I claim that these operators are necessary in any deontic logic that is in some sense dynamic rather than static and in which imperatives are not neglected. For if the logic is to be time-dependent, **PROP** must be a quantification logic. Now a statement with a free time variable admits at least four possibilities: always true, sometimes true, sometimes false, and always false. (Of course these four possibilities are neither mutually exclusive nor exhaustive. Other possibilities include the possibility that a statement is true exactly once or is indeterminate.) As a simple illustration, if F is the statement that the door is closed (or "The door is closed at time t")

then there are four possibilities: *sometimes closed, always closed, sometimes open, and always open.* I conclude that the simplest formulas in **PROP** constructed out of a statement F and closed with respect to the time variable are the statements: $(\exists t)F$, $(\forall t)F$, $(\exists t)\neg F$, $(\forall t)\neg F$. Applying the imperative operator to these four forms we arrive at four possible formulas: $!(\exists t)F$, $!(\forall t)F$, $!(\exists t)\neg F$, $!(\forall t)\neg F$. These four forms are abbreviated: LF, SF, $L\neg F$, $S\neg F$.

As examples, let F be the statement considered earlier: *The door is closed at time t.* Then *LF* may be read "*It **shall** be the case that there is a time t such that the door is closed at time t!*" and *SF* may be read "*It **shall** be the case that for all t the door is closed at time t!*", i.e. "*It **shall** be the case that the door is always closed!*" (As abbreviations, *LF* can be read "*Let the door be closed!*" and *SF* can be read "*Let it be sustained that the door is closed!*") If one wishes to express more timely compliance one can ensure this by including a time parameter ***a*** in *F,* e.g. let *F* be "*If the time t =* ***a****, the door is closed at time t*" and let *G* be "*The time t =* ***a*** *and the door is closed at time t.*" If "***a***" is "noon," *SF* says "*It **shall** be the case that for all t, if the time t is noon, the door is closed at time t!,*" i.e. "*It **shall** be the case that the door is always closed at noon!*" and *LG* says "*It **shall** be the case that there exists a time t such that t is noon and the door is closed at t!,*" i.e. "*It **shall** be the case that the door is sometimes closed at noon!*" (It may be noted that in this example, *LF* and *SG* are either vacuous or nonsensical.)

Some requirements and admissibilities are based on fact; others are not. Those that are based on fact may be called *contingent*; those that are independent of the facts *absolute.* I begin with the absolute ones.

The simplest absolute requirements and admissibilities are the trivial ones, i.e. the formulas which like tautologies hold in every possible world. For example if α is $(F \rightarrow F)$ then, since *R, A, L,* and *S* preserve validity, the formulas $RS\alpha$ and $RL\alpha$ are absolute requirements, and the formulas $AS\alpha$ and $AL\alpha$ are absolute admissibilities.

Nontrivial requirements and admissibilities require for their explication what we have called a *normative system.* Nontrivial requirements and admissibilities are requirements and admissibilities based on the proper axioms—as opposed to the logical axioms—of a normative system. Consider, for example, the absolute requirements and admissibilities based on the manual for writers, *ELEMENTS OF STYLE* by Struik and White. This manual is an informal normative system which may be converted into a formal system ***N*** by putting its principles in the form of proper axioms over **IMP**. One such principle, "Omit needless words," can be reformulated as the proper axiom *SF* (read: *it shall be the case that from now on one omits needless words)* where *F* is the sentence *one omits needless words*. Since *R* preserves validity, we may infer from *SF* the absolute requirement *RSF* (read: *it is required that it shall be the case that, from now on, one omits needless words*). Needless to say, the absolute requirement *RLF* and the absolute admissibilities *ASF* and *ALF* also follow.

Of course, not all admissibilities follow from requirements. Indeed, since $(AU \leftrightarrow \neg R \neg U)$ is a valid formula, it follows that given an imperative μ, it is admissible that μ provided $\neg R \neg \mu$ is provable. Caution has to be exercised, however; for given an incomplete normative system one cannot prove $\neg R \neg \mu$ by showing that $R \neg \mu$ is not valid. For example, suppose one is given the normative system based on *STRUIK & WHITE* and assume one cannot prove in that system a sentence logically equivalent to "*It is required that it shall be the case that one does not copy the work of others.*" It would be absurd to conclude that according to *STRUIK & WHITE* it is admissible to copy the work of others. To draw such a conclusion (or the opposite one) one would need to introduce a new axiom into the system.

Nevertheless, the formula $(AU = \neg R \neg U)$, properly interpreted, does give us a criterion for introducing a proper axiom of the form *AU* into a normative system. This is: for it to be appropriate to introduce an axiom of the form *AU* into a normative system for governing activity in a certain field ***W*** there are two conditions to be satisfied: one, the introduction of *AU* does not lead to an inconsistency; the other, it is intuitively clear that in ***W*** it is not required that not *U*. (In other words from the intuitive standpoint, "*it is admissible that*" is treated as interchangeable with "*it is not required that not.*") This criterion is easy to state but is often difficult to apply since there is no general solution to the problem of what is or is not "required" in each field; instead there is a host of possible considerations that enter into the equation, some of which are based on religious and moral scruples, some on precedent, and still others on pragmatism. Let us consider an example.

Take the classic system for writing English written by Fowler. According to Fowler, it is not required that it shall be the case that one always puts a comma after the next to the last term in a list. Accordingly, if we let Fowler be reformulated as a normative system ***N*** in which there is a formula *LF* whose intended interpretation is "*it shall* *be the case that occasionally one omits to put a comma after the next to the last term in a list,*" we may adopt *ALF* (or ¬*R*¬*LF)* (read: *it is admissible that it shall be the case that occasionally one omits to put a comma after the next to the last term in a list*) is a proper axiom of ***N.*** "Admissibility" in this case follows from standard practice in writing English. In other cases it is more difficult to arrive at what is or is what is not required. Take the pro-life position in the matter of abortion, for example. (I cannot help remarking at this point that legislation consists in adding (or deleting) proper axioms to a normative system.)

I turn now to consider contingent requirements and admissibilities. Much of what has been said above about absolute requirements and admissibilities applies also to the contingent ones. For example, some contingent requirements and admissibilities are deducible from the valid formulas of ***K*** and hence ultimately derive from the logical axioms of a normative system. Thus if *F* states a fact, the contingent requirement *RLF* and the contingent admissiblity *ALF* derive not only from the fact *F* but the valid formulae (*F* → *RLF*) and (*RLF* → *ALF*). On the other hand, many contingent requirements and admissibilities must either be assumed as premises or derived from proper axioms.

As an illustration, I return to discuss the traffic examples we have used to illustrate adjointness. Let us assume a normative system ***N*** whose function is to regulate traffic. Suppose one of the proper axioms of this normative system is a law of the form (*LF* → *U*) where *F* is the open statement "*The traffic light is red at time t* = ***a***" and *U* is the command "*It **shall** be the case that no vehicle moves into the intersection at time **a**!*" Informally, the formula (*LF* → *U*) expresses the command, "No vehicle shall move into the intersection when the light is red!" (Literally, (*LF* → *U*) is read: *It **shall** be the case that there is a time t such that the traffic light is red at time t* = ***a*** *only if it **shall** be the case that no vehicle moves into the intersection at time **a!***) Since (*LF* → *U*) is an axiom, we may infer *LF* => *U*; moreover, by adjointness (2), (*LF* => *U*) iff (*F* => *RU*). Consequently, given *F* (that the light is red at time t = a) one may infer the contingent requirement *RU* (read: *it is **required** that it shall be the case that no vehicle moves into the intersection at time **a***). (Cp. Chisholm [1964]) on the subject of requirement.).

As an illustration of a contingent admissibility, assume the same normative system ***N*** and suppose one of the proper axioms of *N is* a law of the form (*U* → *SF*) where U is "*It **shall** be the case that a motor* *vehicle z moves into an intersection y at time* ***a****!*" and *F* is "*If t is* ***a****, the traffic light at y is green at time t.*" Informally, the formula ($U \rightarrow SF$) expresses the command, "*It shall be the case that a vehicle moves forward at a traffic light only if the light is green!*" (Literally ($U \rightarrow SF$) reads: *If it* ***shall*** *be the case that a motor vehicle z moves into the intersection y at time* ***a****, then it* ***shall*** *be the case that for all t, if t is* ***a****, the traffic light at y is green at time t!)* By adjointness (4), ($U \rightarrow SF$) implies ($AU \rightarrow F$), which says it is admissible that a motor vehicle moves into an intersection only if the traffic light is green. (Literally, ($AU \rightarrow F$) is read: *It is admissible that it shall be the case that a motor vehicle z moves into an intersection y at time* ***a*** *only if, if t =* ***a****, the traffic light at y is green at time t.*) Thus we have an admissibility that is contingent on a fact; namely, that the traffic light is green.

I claimed in Chapter Two that whenever there is a problem to be solved it is *prima facie* the case that *it is admissible* that it shall be the case that one does whatever is necessary solve the problem. On the other hand, given that to do *x* will solve the problem, it is not necessarily the case that it is admissible that it shall be the case that one does *x*. For example, it might be that to do *x* is illegal. It follows that in order for one to be sure that it is admissible that it shall be the case that one does *x*, it is necessary to check *x* and its consequences to see if there might be other grounds for rejecting x. The principle here is the familiar one: *the end does not justify the means.* We may call it the "means-ends principle."

This principle is discussed further in Section 3 of Chapter Eleven.

5 Of what use is the formal apparatus?

There should be no need for an apology for a language that enables one to express clearly what could be expressed only clumsily before; that enables one to avoid using controversial terms such as "good," "right," and "ought"; that enables one to reason precisely about imperatives; and, as shown in Section 8, avoids the paradoxes of standard deontic logic. But we want a framework that does more than this, however. What is wanted, after all, is a system that can be used to clarify questions in practical fields.

On the other hand, one must not expect too much. The most that can be hoped for at this stage is to have a framework for practical logic that not only delivers us from bad (i.e. fallacious) modes of reasoning, but enables us to confirm scientifically those results and only those results that have an empirical basis. The use of the present theory is somewhat like the use of formal logic in mathematics: a brilliant mathematician can *see* (without formal proof) that a theorem is valid, but the mathematical community will only accept it as valid if there is a proof. As indicated in the introductory remarks, the problem is not so much that all the old systems of ethics and politics were wrong: the problem is that there has not existed a proper framework from which to establish what is valid in them.

The intended use of the formal apparatus in this work is to scrutinize the adequacy of derivations of results in practical fields like ethics and politics. The method is as follows: first, decide on the field whose results are to be critically examined; second, set up a normative system for that field within the functorial framework ***K***; third, express the results under scrutiny in the formal language of the system; fourth and finally, check the validity of the results using the formal apparatus of *K*. In ***Part Three*** the reader will find normative systems for ethics and politics formulated within *K* as well as suggestions of how to derive within *K* some of the results that were only derived informally in ***Part One***. The adequacy of these systems and the success of the derivations are matters that I trust will be subject to scrutiny in the years ahead. I think what I have done is an advance on what has gone before, but I am sure that my efforts are a mere beginning and they can be much improved upon in the future. The best I can hope for is that I have in some degree forwarded the hopes of Mill (see the concluding remarks or *Epilogue* in Chapter Eleven) of bringing the moral sciences "into a more satisfactory state."

6 Simple norms

Some of the statements in **PROP** may be regarded as expressions of norms. Formally, *norms* are statements in **PROP** that have occurrences of RU (or a logical equivalent) where U is an imperative in **IMP.** Informally, a statement is a norm (or normative statement) if it expresses certain modal properties, e.g. *feasibility, irrevocability, tenability*, and *compulsivity*. In this section I explore the simplest norms based on a single statement F.

The eight norms that may be constructed by first applying the operators S and $\neg S$ to F and $\neg F$ and then applying R and $\neg R$ to the results are called *simple norms*. These norms may be written thus: $RSF, RS\neg F, RLF, RL\neg F, AL\neg F, ALF, AS\neg F, ASF$. Let us call the first four of these—which have the interpretations: *irrevocability* (RSF), *nonfeasibility* ($RS\neg F$), *compulsivity* (RLF), and *nontenability*

($RL\neg F$)—***the primary simple norms***. The remaining four norms, being negations of the first four, are accordingly *contradictability* ($AL\neg F$), *feasibility* (ALF), *avoidability* ($AS\neg F$), and *tenability* (ASF). Why these names? Let us consider each primary norm (and its negation) in turn.

There are two criteria for the statement, *It is irrevocable that F:* one, F is already true; the other, it is required in some sense that it *shall* be the case that F is always true. Since the formula ($RSF \rightarrow F$) is valid, both these features of irrevocability are satisfactorily expressed by the formula RSF (read: *It is required that it shall be the case that F is always true*). Hence if F states "*Quebec is part of Canada*", the sentence, "*It is irrevocable that F,*" is to be expressed by RSF (read: *it is required that it shall be the case that Quebec is forevermore part of Canada*).

The negation of irrevocability RSF is $AL\neg F$ and is called *contradictability.* Hence if F is defined as in the preceding paragraph, the sentence, "*It is contradictable that Quebec is part of Canada,*" may be translated into ***K*** as $AL\neg F$ (read: *it is admissible that it shall be the case that there is a time t at which Quebec is not part of Canada*). Note that just as the formula ($RSF \rightarrow F$) reflects perfectly the fact that it is irrevocable that Quebec is part of Canada only if Quebec is part of Canada, the contrapositive formula ($\neg F \rightarrow AL\neg F$) reflects perfectly the fact that if Quebec were not part of Canada (at the present time), it would be contradictable that Quebec is part of Canada.

Consider next $RS\neg F$ or "*nonfeasibility.*" The phrase, *"It is not feasible that so-and-so is the case"* is ambiguous: in some cases it means it is not acceptable that so-and-so is the case, while in other cases it simply means that it is not possible that so-and-so is the case. Exactly similar remarks apply to the phrase *"It is irrevocable that so-and-so is not the case";* moreover, I believe it is in accord with standard usage to say that the two phrases are interchangeable. Therefore I propose to translate "*It is nonfeasible that F*" into the formal language ***K*** as the norm $RS\neg F$. For example, if F is "*The fire is lit,*" the sentence "*It is nonfeasible that the fire be lit,*" is to be expressed in the formal language K as $RS\neg F$ (read: *it is required that it shall be the case that the fire is never lit).* On the other hand, since the negation of $RS\neg F$ is ALF, the sentence "*It is feasible that the fire be lit*" is to be expressed as ALF (read: *it is admissible that it shall be the case that sometimes the fire is lit).*

Compulsivity, the next primary norm on our list, is the property of being "compelling" or "compulsory." There are two criteria that are sufficient for concluding "it is compulsive that *F*": one, *F* is a true statement; the other, it is required in some sense that it shall be the case that *F* is sometimes true. The valid formula ($F \rightarrow RLF$) may be interpreted as saying that the first criterion implies the second; accordingly, I translate "It is compulsive that *F*" into ***K*** as *RLF*. For example, if *c* is a state, *x* is a citizen of *c* and *F* is "*x pays taxes to c,*" "It *is compulsive that x pay taxes to c*" is to be expressed as *RLF* (read: *it is required that it shall be the case that at some time x pays taxes to c*). The negation of compulsivity *RLF* is $AS\neg F$ and is called *avoidability*. Hence if *F* is defined as in the previous example, "*It is avoidable that x pays taxes to c*" is to be expressed as $AS\neg F$ (read: *it is admissible that it shall be the case that henceforth x does not pay taxes to c*).

"*It is nontenable that F*," the final primary simple norm on our list, may be taken to mean that "*it is compulsive that* $\neg F$." Consequently, if *F* is "Quebec is part of Canada," I translate "It is nontenable that *F*" into ***K*** as $RL\neg F$ (read: *it is required that it shall be the case that there is a time at which Quebec is not part of Canada*). On the other hand, since the negation of $RL\neg F$ is *ASF*, "*It is tenable that F*" is to be expressed as *ASF* (read: *it is admissible that it shall be the case that forevermore Quebec is part of Canada*). Note that, according to the valid formula ($ASF \rightarrow F$), it is tenable that Quebec *is* part of Canada only if Quebec is part of Canada.

7 The structure of norms

In this section we consider the Boolean algebra constructed out of simple norms. Let *F* be a fixed statement and denote by $\varepsilon(F)$ the smallest Boolean algebra that contains the primary simple norms based on *F*. The elements of $\varepsilon(F)$ are called *elementary norms.* The algebra $\varepsilon(F)$ may be viewed as a finite-dimensional vector space over the integers modulo 2 in which addition is interpreted as the exclusive *or*; i.e. $(p + q) = ((p \;\&\; \neg q) \vee (q \;\&\; \neg p))$. The vector space $\varepsilon(F)$ is a quotient space of the free module generated by the elements in Boole's law of development reduced *modulo* the logical relations among the simple norms. The free module has as a basis the summands in the disjunctive normal form generated from the primary simple norms $\{RSF, RS\neg F, RLF, RL\neg F\}$. There are sixteen of these summands since each of them is a conjunction of four factors in which the *nth* factor is either the *nth* primary simple norm or its negation, $n = 1, 2, 3, 4$. It turns out that eleven of the summands are contradictions; when these are eliminated (and the remainder simplified), five norms remain: *RSF*, *RS¬F, R(LF & L¬F), ASF & AL¬F, AS¬F & ALF*. I conclude: *each elementary norm based on F is expressible uniquely in the* ***disjunctive normal form:***

a(RSF) + b(RS¬F) + c(R(LF & L¬F)) +
d(ASF & AL¬F) + e(AS¬F & ALF)

where the letters a, b, c, d, e are replaceable by 0 or 1.

The summands in the disjunctive normal form, which form a basis for ε*(F)*, are called *atomic norms*. If the *co-norm* of a given norm based on F is defined as the norm obtained from the given norm by replacing *F* by ¬*F*, the atomic norms consist in two pairs of co-norms {*RSF, RS¬F*}, {(*ASF & AL¬F*), (*AS¬F & ALF*)}, together with a fifth norm *R(LF & L¬*F) which is the co-norm of itself. So far we have interpretations for only two of the atomic norms: irrevocability (*RSF*) and nonfeasibility (*RS¬F*). The other three are interpreted as follows: *R(LF & L¬F)* is called *incapacity*, (*ASF & AL¬F*) is called *proper tenability*, and (*AS¬F & ALF*) is called *proper avoidability*.

Definitions

1. It is *properly tenable that F* provided *it is tenable that F but not irrevocable* (i.e. the norm of proper tenability holds).
2. It is *properly avoidable that F* provided *it is feasible that F but not compulsive* (i.e. the norm of proper avoidability holds).

For example, it may be tenable that one owns a house, but not irrevocable; moreover, it may be feasible that one owns a house but not compulsive (i.e. not compelling or compulsory). Another example of proper avoidability is freedom of speech: if speech is free then speech is avoidable, but not nonfeasible.

The atomic norms form an "orthogonal" basis for ε*(F)* in the sense that the conjunction of any two of the atomic norms is 0*. The disjunction of all the atomic norms is equal to 1* and consequently, the disjunction of any subset of the atomic norms is the negation of the disjunction of the remainder. For example,
(*RLF & RL¬F*) ∨ (*ASF & AL¬F*) ∨ (*AS¬*F *& ALF*) = ¬(*RSF* ∨ *RS¬F*).

Since there are five atomic norms, ε*(F)* is a five-dimensional vector space over the integers modulo 2, whence ***the number of elementary norms (excluding the trivial norms 0* and 1*) is thirty.***

Because of the duality between ***and*** and ***or,*** there is also a ***conjunctive normal form*:**

$$\{\boldsymbol{a} \vee (\boldsymbol{AL}\neg\boldsymbol{F})\} \mathbin{\&} \{\boldsymbol{b} \vee (\boldsymbol{ALF})\} \mathbin{\&} \{\boldsymbol{c} \vee (\boldsymbol{ASF} \vee \boldsymbol{AS}\neg\boldsymbol{F})\} \mathbin{\&} \{\boldsymbol{d} \vee (\boldsymbol{ASF} \rightarrow \boldsymbol{RSF})\} \mathbin{\&} \{\boldsymbol{e} \vee (\boldsymbol{ALF} \rightarrow \boldsymbol{RLF})\},$$

where the letters ***a, b, c, d, e*** *are replaceable by 0* or 1*.* The factors in this form, $AL\neg F$, ALF, $(ASF \vee AS\neg F)$, $(ASF \rightarrow RSF)$, $(ALF \rightarrow RLF)$, are all negations of atomic norms. For example, the factor $(AS\neg F \vee ASF)$ is the negation of incapacity and hence called *power* (or "capability"). The factor $(ASF \rightarrow RSF)$, the negation of proper tenability, is called *pro-bondage*; moreover, the co-norm $(ALF \rightarrow RLF)$ is the negation of proper avoidability and is called *pro-restraint.*

Definitions

1. We say it is *conditionally binding that F* whenever the norm of pro-bondage $(ASF \rightarrow RSF)$ with respect to F holds. Since the converse of $(ASF \rightarrow RSF)$ is valid, it is conditionally binding that F provided *it is tenable that F iff it is irrevocable that F* (and contradictable iff nontenable).
2. We say it is *conditionally forbidden that F* whenever pro-restraint $(AS\neg F \rightarrow RS\neg F)$ with respect to F holds. Since the converse also holds, it is conditionally forbidden that F provided *it is avoidable that F iff it is nonfeasible that F* (and compulsive iff feasible).

For example, consider the plight of the hero in the old fairy tale in which once the hero accepts the job as a ferryman he remains stuck in the job. In this case the job of ferryman is conditionally binding since it is tenable iff irrevocable (and contradictable iff nontenable). For an example of pro-restraint consider war. The rationale for a strong defense is the principle that war is avoidable iff war is nonfeasible (and hence feasible iff compulsive). This principle is subject to doubt but, if true, we live in a world in which war is conditionally forbidden. Perhaps a better example of pro-restraint comes from the appetites,—hunger, for example. For some people eating is avoidable only if it is nonfeasible and, therefore, eating is compulsive whenever it is feasible.

As a consequence of the disjunctive normal form each elementary norm in $\varepsilon(F)$ has an associated binary number *(abcde)* formed from the coefficients *a, b, c, d,* and *e* in the disjunctive normal form. Thus (11111) represents the logically true 1*, (00000) (or 0 represents the logically false 0*, (01001) (or (1001))—which is the number of

$(RS\neg F \vee (ALF \ \& \ AS\neg F))$—represents avoidability $AS\neg F$, etc. (As usual, "00000" may be written 0 and in the remaining cases all initial zeroes may be dropped.) The assigning of numbers to the norms makes it easy to take an inventory of all elementary norms in $\varepsilon(F)$. To do this we first break the thirty two norms in $\varepsilon(F)$ (including 0* and 1*) down into three groups. The groups are: one, norms in which a = b = 1; the second, norms in which a = b = 0; the third, norms in which either a = 1 and b = 0 or a = 0 and b = 1. I consider each group in turn.

Four members of the first group have been defined already: *logical truth* 1*, *power* (11011), *pro-bondage* (11101), and *pro-restraint* (11110). Two other norms in this group are the duals (11010) and (11001); the former $(ALF \rightarrow ASF)$ is called *virtual tenability*, the latter $(AL\neg F \rightarrow AS\neg F)$ *virtual avoidability*. (The attribute *virtual* indicates that the norm, or its co-norm, is a conditional norm of the form $(ALF \rightarrow \xi)$.) The two norms that remain, (11000) and (11100) are extremely important. The former $(RSF \vee RS\neg F)$ is called, for obvious reasons, *responsibility*, the latter $(RSF \vee RS\neg F) \vee R(LF \ \& \ L\neg F)$, the conjunction of responsibility and incapacity, is called *constraint*. Constraint has an interpretation as the *principle of* ***noblesse oblige****: (POWER → RESPONSIBILITY).*

Note:: *(RESPONSIBILITY → POWER)*, the converse of the principle of *noblesse oblige*, is easily proved. This principle, which is reminiscent of the Kantian principle—*ought implies can*— implies that ***the norm of constraint*** says, not simply "*ought if can,*" but "*ought iff can.*"

Definitions

1. We say it is *virtually tenable that F* whenever the norm of virtual tenability $(ALF \rightarrow ASF)$ with respect to *F* holds. Since the converse of this norm (as well as the following) is a valid statement, it is *virtually tenable that F* provided *it is tenable that F iff it is feasible that F*
2. We say it is *virtually avoidable that F* whenever virtual avoidability $(AL\neg F \rightarrow AS\neg F)$ with respect to *F* holds and therefore it is *virtually avoidable that F* provided *it is avoidable that F iff it is contradictable that F.*

For example, suppose there is a town in which drinking is tenable whenever feasible, then drinking in this town is virtually tenable. It follows that sobriety (viewed as the negation of drinking) is virtually avoidable in this town.

The members of the next group (for which a = b = 0) are negations of the members of the first group. Their associated binary numbers are easily computed; for the number of the negation of an elementary norm may be found simply by interchanging zeros and ones. For example the negations of the norms 1*, (11011), (11101), and (11110)) are the trivial norm 0* together with the three atomic norms: incapacity (100), proper tenability (10), and proper avoidability (1). Of the four remaining norms, (111), i.e. (*ALF* & *AL*¬*F*), is the negation of the norm of responsibility and hence is called *immunity*. Another (11) is the conjunction of power and immunity and is therefore called *license*. (Appropriately, license is the negation of the norm called "constraint.") Finally, (101) and (110), the negations of virtual tenability and virtual avoidability respectively, are the norms (*RL*¬*F* & *ALF*) and (*RLF* & *AL*¬*F*); the former is called *proper nontenability* and the latter *proper compulsivity.*

Definitions:

(i) It is *properly nontenable that F* provided it is feasible but not tenable that *F*.

(ii) It is *properly compulsive that F* provided it is compulsive but not irrevocable that *F*.

For example, the survival of the human race is feasible (since we are now alive) but ultimately may prove to be nontenable. Moreover, human survival may be compulsive but not irrevocable.

The remaining group is the conjunction of two classes: one, the class of norms of the form (10...); the other, the class of the norms of the form (1...) (where the dots are replaceable by zero or one.) Included in this group are the eight simple norms: *RSF* (10000), *RS*¬*F* (1000), *ASF* (10010), *AS*¬*F* (1001), *RLF* (10110), *RL*¬*F* (1101), *ALF* (10111), and *AL*¬*F* (1111). We may divide the remaining norms into two subgroups such that each subgroup consists of two pairs of co-norms with the property that the members of each pair are negations of members of the other pair. Consider first the subgroup that contains the pair of co-norms (1010) and (10001). For obvious reasons the former is called *virtual proper tenability* and the latter *virtual proper avoidability.* The negations of the elements of this pair are the co-norms (10101) and (1110): the former is called *virtual proper nontenability,* the co-norm is therefore called virtual proper compulsivity. The norms in the remaining subgroup, which contains the pair of co-norms (1011) and (10011), are of more interest. The former (1011) has the form (contradictability & power) and is therefore called *freedom-from,* the latter (10011) has the form (feasibility & power) and is therefore called *freedom-to.* The negation of freedom-from, i.e. (10100), is accordingly called "bondage" and the negation of freedom-to, i.e. (1100), "restraint."

Bondage and restraint are types of constraint and hence imply the principle of *noblesse oblige. Bondage* is of the form

$$(RSF \vee (RLF \,\&\, RL\neg F\,))$$

and *restraint* is its co-norm, $(RS\neg F \vee (RLF \,\&\, RL\neg F\,))$. Bondage has the alternative form, $((ASF \vee AS\neg F\,) \rightarrow RSF)$. If the norm of bondage with respect to F holds we say "*It is **binding** that F.*" If the norm of restraint with respect to F holds we say "*It is **forbidden** that F.*" Like all constraints, bondage and restraint, as forms of duty, are discharged by incapacity. For example, it can be said that the duty to support one's family is irrevocable provided one has the capacity to do it. It can be shown that it is binding that F provided it is not only compulsive that F but conditionally binding that F, i.e. $(RLF \,\&\, (ASF \rightarrow RSF))$.

This completes our survey of elementary norms. *Nonelementary norms* may also be considered. Simple examples of nonelementary norms include $R(LF \rightarrow SF)$ and $A(LF \,\&\, L\neg F\,)$. Nonelementary norms of the form $RS(F \rightarrow G)$ are extremely important; for norms of this type are analogous to "commitments" in standard deontic logic. Another nonelementary norm, which is a generalization of pro-bondage, is the norm $(ASG \rightarrow RSF)$, read: *it is binding that F whenever it is tenable that G.* (Let us symbolize it by $[F/G]$ and call it a *disposition.*) Another important nonelementary norm expresses a principle introduced in Section 4, the principle of constraint. This norm is of the form $((AU \rightarrow \xi) \,\&\, RL\neg\xi)$ and read: *It is admissible that U only if ξ and, moreover, it is nonfeasible that ξ.* Nonelementary norms like the examples given make their appearance in the next section, but their systematic study is deferred to the next chapter.

A few obvious results on elementary norms are as follows:

1) Responsibility implies power since
   $$(RSF \vee RS\neg F\,) \Rightarrow (ASF \vee AS\neg F);$$
   therefore, incapacity implies immunity.
2) Irrevocability implies responsibility since $RSF \Rightarrow (RSF \vee RS\neg F)$.
3) Tenability implies freedom-to and compulsivity,
   and avoidability implies freedom-from and nontenability.
4) Responsibility is constraint in the presence of power.

5) Irrevocability, which implies both power and freedom-to, is bondage in the presence of power.
6) Avoidability is power together with nontenability and tenability is power together with compulsivity.
7) Constraint is the conjunction of pro-bondage and pro-restraint and is the disjunction of bondage and restraint.
8) Power is the disjunction of license and responsibility, while freedom-to is the disjunction of license and irrevocability.
9) The disjunction of freedom-to and freedom-from is power, but the conjunction of freedom-to and freedom-from is license.
10) Incapacity is logically equivalent not only to constraint together with immunity but to bondage together with restraint.
11) The norm of responsibility is valid provided feasibility implies irrevocability and the norm of power is valid provided compulsivity implies tenability.
12) Bondage is compulsivity together with pro-bondage

    i.e. $(RLF \;\&\; (ASF \rightarrow RSF))$

    and restraint is nontenability together with pro-restraint (i.e. $RL\neg F \;\&\; (AS\neg F \rightarrow RS\neg F)$). Moreover *bondage* is the conjunction of compulsivity and constraint and *restraint* is the conjunction of nontenability and constraint.

The following diagram summarizes some of the important relationships among the norms. The arrows represent entailments.

POWER ⟵ *RESPONSIBILITY* ⟶ *CONSTRAINT*

↑ ↑ ↑ ↑

FREEDOM-TO *FREEDOM-FROM* *RESTRAINT* *BONDAGE*

↑ ↑ ↑ ↑

LICENSE ⟶ *IMMUNITY* ⟵ *INCAPACITY*

Often, if a norm is to make sense, the range of application must be restricted; for a norm that applies in one domain may not apply in another. Consider, for example, the law against murder and suppose F states "x kills a human being." Then, if x is an ordinary law-abiding citizen, the norm of nonfeasibility $RS\neg F$ with respect to F applies at all times to x—but this norm does not hold in general since $RS\neg F \Rightarrow \neg F$. Consequently, some norms apply only in ideal worlds. There are at least two ways to get around this restriction: one, use conditional norms (e.g., pro-bondage, pro-restraint, or one of the nonelementary norms); the other, use the directive operator. Examples of the use of the directive operator are given in Section 9.

8 Chisholms's puzzle and other paradoxes

Puzzles are very useful for testing a logical theory—as Russell and others have pointed out. An example of such a puzzle is the "contrary-to-duty" paradox of Chisholm [1963]. It is easy to see that a puzzle of the Chisholm type emerges in any application of standard deontic logic in which there are two statements of conditions that are not identical but whose equivalence is mandated. For let G and H be two such statements and assume:

1) *It is obligatory* that $(G \rightarrow H)$.

2) If G is false, *it is obligatory* that H be false.

Now if we let O stand for "it is obligatory that" and P for $\neg O \neg$; then (according to standard deontic logic) $(PG \rightarrow PH)$ & $(PH \rightarrow G)$ and therefore $(PG \rightarrow G)$. From this it follows that $(OG \rightarrow G)$ & $(\neg G \rightarrow O\neg G)$. I conclude that if for every formula G there is a formula H such that *1*) and *2*) hold, the formula $(OF \leftrightarrow F)$ is valid. This, of course, is the crux on which Mally's system of deontic logic came to grief.

Now let us consider Chisholm's original example as a test of the present system. Suppose a certain man has a duty to go the assistance of his neighbors and, moreover, the man acts contrary to his duty and does not go. Consequently, assuming it ought to be that before coming to their assistance he tells them he is coming, it follows that the man not only has a duty to tell his neighbors he is coming to their assistance, but, since he is not going, he has a counter duty not to tell them. In standard deontic logic this is an outright contradiction.

In order to see whether a similar contradiction occurs in the present framework, suppose a normative system $\boldsymbol{N}$ for governing behavior in the village is given and suppose $\boldsymbol{N}$ has one (and only one) proper axiom (over **IMP**):

$$(*)\ S(F \leftrightarrow G)$$

Moreover, let the following two premises be assumed:

$$\neg F \quad \text{and} \quad RLF.$$

The intended interpretations of F and G are as follows: F is the statement that the man is prepared (at time t) to go to the assistance of his neighbors and G is the statement that he alerts them that he is coming. $\neg F$ is read: *The man is not prepared (at time t) to go to the assistance of his neighbors; RLF* is read: *It is compulsive that the man be prepared to go to the assistance of his neighbors*. The proper axiom $S(F \leftrightarrow G)$ may be read:

It shall be the case that for all time t, F iff G!

It follows from (*) and (10) that

(i) $RS(F \rightarrow G)$ and (ii) $RS(\neg F \rightarrow \neg G)$.

Hence, applying (7) and (11)), we may infer *RLG* from the premise *RLF*. Since (according to (8)) $\neg F$ implies $RL\neg F$, we may in a similar fashion infer $RL\neg G$ from the premise $\neg F$. No contradiction follows from the premises—although if RSF (which contradicts $\neg F$) were a premise we *would* obtain a contradiction. The closest we come to a contradiction is that as a consequence of conflicting compulsions, the incapacities (RLF & $RL\neg F$) and (RLG & $RL\neg G$) hold.

In the face of these conflicting compulsions, can the system still tell the man what to do? The answer lies in the result (ii) $RS(\neg F \rightarrow \neg G)$ (read: *it is required that it shall be the case that for all time, if the man does not prepare to go to the assistance of his neighbors, he does not inform them that he is coming*). Note that this result is not marred by the fact that (ii) does not hold in general but only in those deontically ideal worlds that conform to the normative system ***N***. Similarly, it is easy to see that *this result is not vitiated by a so-called "paradox of commitment,"* (even though, admittedly, if *RSF were* true—and (i) held—a contradiction *would* follow, *even if neither* $\neg F$ *nor* (*) *were assumed.*) Therefore Chisholm's puzzle has within the present system a noncontradictory formulation that not only avoids a "paradox of commitment" but supplies an unequivocable answer to the question of what the man is to do in this "contrary-to-duty" situation. I conclude that the *functorial framework* ***K*** affords a straight-forward solution to the "contrary-to-duty" paradox of Chisholm.

Another puzzle for deontic logic is the *paradox of the knower* (cf. Åqvist [1967]). This paradox stems from the assumption that it is provable that if a statement is known to be true, the statement itself is true and, therefore, *if it ought to be known that a statement is true, it ought be that the statement itself is true.* Deontic theories that are static cannot handle this paradox. But the functorial framework ***K*** is not a static one and as a consequence there is no paradox of the knower in the present system. Consider irrevocability, for example. It cannot be irrevocable that it is *known* that so-and-so is married, unless it irrevocable that so-and-so *is* married. Similarly, there is nothing paradoxical about inferring that it is tenable that *x* murders *y* from the fact that it is tenable that it is known that *x* murders *y*. Similar considerations apply to the other deontic notions of ***K*** such as compulsivity and feasibility.

There are, of course, other paradoxes to be considered. Most of the others can be easily disposed of. For example, unless I am very much mistaken, the "good Samaritan" paradox yields to that "paradigm of philosophy," Russell's theory of descriptions (Cp. Frank Ramsey [1950]). I return to these matters in the next chapter.

9 On the logic of one-agent norms

Given an agent x and a statement F, the class of *one-agent norms* (with respect to (x, F)), is built out of a set of three formulae: $[x, F]$, $[x, \neg F]$, and $([x, F] \vee [x, \neg F])$. Let us call this set of three formulas the *ur-set with respect to* (x, F) (or simply *ur-set* when x and F are tacitly understood). ("*ur*" is a Swedish prefix with the connotation of primitive or "original.") A rough description of the class of *one-agent norms* with respect to an agent x and a statement F is that it is the class of norms constructed out of the elementary norms with respect to one or another of the elements in the *ur*-set. A *basic* one-agent norm with respect to (x, F) is an elementary norm with respect to one of the elements of the *ur*-set.

Let us call $([x, F] \vee [x, \neg F])$ *engagement;* then a simple example of a basic one-agent norm is responsibility with respect to engagement, i.e. $RS([x, F] \vee [x, \neg F]) \vee RS\neg([x, F] \vee [x, \neg F])$. Two other examples of basic one-agent norms include the responsibilities based on the formulas $[x, F]$ and $[x, \neg F]$ respectively, i.e. $(RS[x, F] \vee RS\neg[x, F])$ and $(RS[x, \neg F] \vee RS\neg[x, \neg F])$. Let us call the three responsibilities based respectively on one of the elements of the *ur*-set, the *basic responsibilities* (with respect to (x, F)). In a similar way three *basic immunities*, three *basic powers,* three *basic incapacities,* three *basic bondages*, three *basic restraints*, etc. may be constructed out of the elements of the *ur*-set. Obviously, other, more complex, one-agent norms can be constructed by means of elementary logical operations on the basic one-agent norms. For example, the conjunction of the three basic responsibilities is a one-agent norm called *total responsibility.* Total responsibility is of the form

(i) $(RS[x, F] \vee RS[x, \neg F] \vee RS\neg([x, F] \vee [x, \neg F]))$.

Total responsibility can also be put into one of two alternative forms:

(ii) $(AL([x, F] \vee [x, \neg F]) \rightarrow (RS[x, F] \vee RS[x, \neg F]))$,

(iii) $(AL\neg[x, F] \,\&\, AL\neg[x, \neg F]) \rightarrow RS\neg([x, F] \vee [x, \neg F])$.

As an illustration, suppose Mary, who is a teller at a bank, has total responsibility for seeing to it that each check presented to her is either accepted or rejected. Now suppose Mary is presented with a check y, and let Mary be denoted by x and the statement "y is cashed" by F; then according to (iii) whenever it is not only contradictable that Mary cashes the check but contradictable that Mary rejects the check, it is nonfeasible that she either accepts or rejects y. (There are many ways such a nonfeasibility could occur, e.g. Mary might have a heart attack or the bank might be low on cash.)

I turn now to consider further types of one-agent norms. The conjunction of the three basic powers is called *total power* and is of the form $(AS[x, F] \vee AS[x, \neg F] \vee AS\neg([x, F] \vee [x, \neg F])$. The conjunction of the three basic immunities is called *total immunity* and is of the form $(AL[x, F]\ \&\ AL[x, \neg F]\ \&\ AL\neg([x, F] \vee [x, \neg F]))$. Moreover, *total license* is the conjunction of total power and total immunity:

$$(AS[x, F] \vee AS[x, \neg F] \vee AS\neg([x, F] \vee [x, \neg F]))\ \&$$
$$AL[x, F]\ \&\ AL[x, \neg F]\ \&\ AL\neg([x, F] \vee [x, \neg F]).$$

Finally, the conjunction of the three basic bondages is called *total bondage* and the conjunction of the three basic restraints is called *total restraint.*

The negation of total immunity is called *limited responsibility* and is of the form $(RS\neg[x, F] \vee RS\neg[x, \neg F] \vee RS([x, F] \vee [x, \neg F]))$. Similarly, the negation of total responsibility is called *limited immunity* and is of the form $AL\neg[x, F]\ \&\ AL\neg[x, \neg F]\ \&\ AL([x, F] \vee [x, \neg F])$. The negation of responsibility with respect to engagement is called *immunity against engagement* and is of the form

$$AL([x, F] \vee [x, \neg F])\ \&\ AL\neg([x, F] \vee [x, \neg F]).$$

It is easy to see that total immunity implies immunity against engagement and immunity against engagement implies limited immunity; whence, total responsibility implies responsibility with respect to engagement, which in turn implies limited responsibility. Finally, note that total responsibility implies total power.

Since limited responsibility is the negation of total immunity, which we defined as the conjunction of the three basic immunities, limited responsibility is the *disjunction* of the three basic responsibilities. But total responsibility, which we defined as the conjunction of the three basic responsibilities, may be reduced to the conjunction of two responsibilities, namely those based on $[x, F]$ and $[x, \neg F]$ respectively. Consequently, limited immunity, which we defined as the negation of total responsibility, is the disjunction of the (two) corresponding immunities.

As an application of the theory, consider the "right" we call freedom of speech. In Section 7 it was indicated that freedom of speech for the citizen satisfies the norm of proper avoidability. Consider now freedom of speech as it applies to the state. Let *c* be the state and let *F* be the statement that *x speaks* where *x* is an arbitrary citizen; then the norm of free speech (as applied to the state) is *total restraint*, which may be viewed as the conjunction of the two pro-restraints $(AS\neg[c, F] \rightarrow RS\neg[c, F])$ and $(AS\neg[c, \neg F] \rightarrow RS\neg[c, \neg F])$ together with the nontenability $RL\neg([c, F] \vee [c, \neg F])$. The norm of total restraint says, in part, that interference by the state with respect to speech is avoidable only if it is nonfeasible. Since all states seem to have an almost insatiable appetite for interference with free speech, this norm suggests that such an appetite can only be thwarted by creating institutions that ensure that the state has the power to curb this appetite. In a deontically ideal world, in which state interference with speech is avoidable, the law against abridging free speech can be reduced to the nonfeasibility with respect to engagement: $RS\neg([c, F] \vee [c, \neg F])$.

The class of *atomic* one-agent norms may be defined as the set of all conjunctions of the form $(X \& Y \& Z)$ where X is an atomic norm with respect to $[x, F]$, Y is an atomic norm with respect to $[x, \neg F]$, and Z is an atomic norm with respect to $([x, F] \text{ v } [x, \neg F])$. There are potentially (5 x 5 x 5) or 125 atomic one-agent norms, but only twenty four remain after contradictions are winnowed out. This means that the number of norms in the Boolean algebra based on the atomic one-agent norms is two-raised-to-the-twenty-fourth-power. If we call these norms *elementary one-agent norms* there are over sixteen million elementary one-agent norms. Consequently, as one might expect on *a priori* grounds, the logic of directives is much more complicated than the logic of fiats.

Lindahl [1977], who bases his work on Kanger, gives examples of one-agent norms, which he calls "legal positions." Lindahl's examples are meant to illustrate the seven basic legal positions $T_1, T_2, \ldots, T_7$ out of which, according to Lindahl, all one-agent norms may be constructed. Lindhal considers T_3 separately, but for the rest he assumes there is an agent called John who owns a property called Whiteacre and, furthermore, there is an adjacent property called Blackacre. The examples are expressed in terms of the following six statements:

F_1. The main building on Whiteacre is painted white.
F_2. A fence is erected between Whiteacre and Blackacre.
F_3. A factory is built on Blackacre close to the border of Whiteacre.

F_4. The electrical installations on Whiteacre are in accord with existing security provisions.
F_5. The main building on Blackacre is painted white.
F_6. The requisite fire-fighting equipment does not exist at Whiteacre.

Let us abbreviate "John" by "*J*." Then one may suppose that John's behavior is guided by the following norms:

1) Total license with respect to (J, F_1)
2) A license with respect to $[J, F_2]$ together with a nonfeasibility with respect to $[J, \neg F_2]$, i.e.

$$\{(AS[J, F_2] \vee AS\neg[J, F_2]) \,\&\, AL[J, F_2] \,\&\, AL\neg[J, F_2]\} \,\&\, RS\neg[J, \neg F_2]$$

3) A license with respect to $[J, \neg F_3]$ together with a nonfeasibility $RS\neg[J, F_3]$ with respect to $[J, F_3]$
4) An irrevocability with respect to $[J, F_4]$, i.e. $RS[J, F_4]$
5) A nonfeasibility with respect to engagement $([J, F_5] \vee [J, \neg F_5])$, i.e. $RS(\neg[J, F_5] \,\&\, \neg[J, \neg F_5])$
6) A nonfeasibility with respect to $[J, F_6]$

The examples of T_3 that Lindahl gives are like the example of the teller who has the responsibility to cash checks. I propose to take another example: that of an editor who has the responsibility to accept or reject papers. If F is the sentence *"A paper y is accepted"* and x is an editor, it may be assumed that two of the norms governing the behavior of the editor are *the norm of responsibility with respect to engagement*

$$RS([x, F] \vee [x, \neg F]) \vee RS\neg([x, F] \vee [x, \neg F])$$

and the norm of limited immunity

$$AL\neg[x, F] \,\&\, AL\neg[x, \neg F] \,\&\, AL([x, F] \vee [x, \neg F]).$$

The conjunction of these two norms, which can be written as

$$AL\neg[x, F] \,\&\, AL\neg[x, \neg F] \,\&\, RS([x, F] \vee [x, \neg F]),$$

is the counterpart of Lindahl's T_3. This norm says: *it is contradictable that the editor accepts the paper and contradictable that the editor rejects the paper, but it is irrevocable that the editor either accepts the paper or rejects the paper.* This irrevocability assumes a deontically ideal world in which it is feasible for the editor to perform his duties (i.e. accept or reject *y*).

10 A comparison with the Kanger-Lindahl theory

It may be asked: why use *the functorial framework* instead of the Kanger-Lindahl theory? I propose to answer this question by comparing the two theories on the basis of the following criteria: *fineness of discrimination, inevitability, breadth, depth, ease of getting results, felicity of notation, and richness of language.* I consider each criterion in order.

It is easily seen that everything that can be expressed in the Kanger-Lindahl theory can be expressed in *K;* hence the functorial framework is relatively adequate. The reverse is not true, however, since there are twenty-four atomic one-agent norms and only seven basic one-agent legal positions. For example, the Kanger-Lindahl theory has no concept of *incapacity* and therefore cannot express conflicting compulsions without running into contradiction. Other glaring omissions include examples of bondage and constraint.

For an example of *bondage* consider the case in which it is binding on an agent, John, to mow the lawn at Blackacre. Let *F* state that *the lawn at Blackacre is mowed*; then the relevant norm (of bondage) is $RL[J, F] \ \& \ (AS[J, F] \rightarrow RS[J, F])$, i.e. *it is not only compulsive, but **(if tenable)** it is irrevocable that John sees to it that the lawn is mowed.* (Note that if it is irrevocable that John sees to it that the lawn is mowed, it is *true* that the lawn is mowed.) There are two (mutually exclusive) conditions which contradict this norm: one, it is avoidable that John sees to it that the lawn is mowed, i.e. $\neg RL[J, F]$ (which implies $\neg AS[J, F]$) or two, it is contradictable *although tenable* (i.e. $AL\neg[J, F] \ \& \ AS[J, F]$) that John sees to it that the lawn is mowed. These are mutually exclusive since avoidability implies nontenability.

Now consider a contrary-to-duty case in which it is binding on the agent (John) that he mow the lawn and yet the lawn is not mowed (i.e. $\neg F$). We can immediately infer not only that it is not only compulsive ($RL[J, F]$) that John see to it that the lawn is mowed, but (since $\neg F \Rightarrow \neg AS[J, F]$) it is nontenable ($RL\neg[J, F]$) that John see to it that the lawn is mowed. In other words, John suffers from (the norm of) *incapacity* ($RL[J, F] \ \& \ RL\neg[J, F]$). This, I believe, is typical of *contrary-to-duty* cases. I conclude that the *functorial framework for practical logic **K*** is superior to the Kanger-Lindahl theory in fineness of discrimination.

This brings us to the next criterion: *inevitability.* I say a theory is "inevitable" if it comes from a single central conception rather than being introduced *ad hoc*. While I admire Kanger's logical skill, I do not find that his semantical analysis makes his theory inevitable. The difficulty with *any* semantical approach to modal logic is that the selection of the "alternative" relation is open-ended. In particular, Kanger's semantic analysis fails to make it clear whether *Shall F* is to be distinguished from !*F*, on one hand, and *Ought F,* on the other. Is "Shall *F*" a sentence in the imperative mood or a declarative sentence? He himself says that the concept *shall* is "vague" and admits "different specifications in different contexts" (cf. Kanger & Kanger [1966, p. 88]). Since the semantic analysis of shall in Kanger [1972] does not distinguish *shall* from *ought,* his theory is subject to all the drawbacks of standard deontic logic.

On the other hand, I submit that a *functorial theory* is inescapable. Adjoint functors lie at the foundation of most, if not all, far-reaching formal theories (cf. Lawvere [1969]). Moreover, once it is realized that normative systems involve more than one category, it is inevitable that one examine the possibility of functors between them. Furthermore, the existence of the *let* functor, once admitted, immediately suggests that there is a right adjoint. As we have already stated, the *let* functor yields the syntactical structure of the logic of imperatives, while its right adjoint *requirement* gives that syntax normative import. I conclude that the present theory is superior to the Kanger-Lindahl theory in inevitability.

Consider next *breadth.* It is easy to see that ***K*** has a broader coverage than the Kanger-Lindahl theory; for it not only yields a deontic logic and a theory of norms capable of supplanting the Kanger-Lindahl theory but an imperative logic. I therefore turn to the next criterion in our list, *depth.*

Though I think we all owe a debt to Kanger for his broad ranging analysis, his pioneering theory of "rights" suffers from some serious failings. One of these is a failure to make a distinction between *requirement* and a*dmissibility*. Without such a distinction it is difficult if not impossible to make sense of the meaning of imperatives. Another failure of the Kanger-Lindhal theory is that the relationships between "rights" are not very well developed. In the functorial theory ***K***, on the other hand, the system of elementary norms forms a recognizable structure and theorems showing the logical relations between the various norms are easily proved. I conclude that ***K*** is superior to the Kanger-Lindahl theory in depth.

Consider next *ease in getting results*. In ***K*** a calculus based on the properties of the operators *L, S, A, R* and [,] is easily established. Once this calculus is established most theorems can be proved by simple algebraic manipulation. Even where the relationships to be deduced cannot be so established, the deductions are easy. To take a simple example, consider a deduction parallel to a deduction in Lindahl [1977, p. 95] where he shows that $(Shall\ (Do(p, F) \vee Do(p, \neg F))$ yields $[MayDo(p, F) \leftrightarrow May\neg Do(p, \neg F)] \& [MayDo(p, \neg F) \leftrightarrow May\neg Do(p, F)]$. The parallel statement in the imperative theory is that $RS([p, F] \vee [p, \neg F])$ implies $(AL[p, F] \leftrightarrow AL\neg[p, \neg F])$ & $(AL[p, \neg F] \leftrightarrow AL\neg[p, F])$. Lindahl's deduction consists in a long series of formulas without explanation. In the present theory, on the other hand, the deduction follows almost immediately from the two valid formulas:

(i) $(RSF \& ALG) \rightarrow AL(F \& G)$ and (ii) $[p, F] \rightarrow \neg[p, \neg F]$.

(The first of these is a consequence of (10).) The deduction goes as follows: Since (ii) implies not only that $AL[p, F] \rightarrow AL\neg[p, \neg F]$ but $AL[p, \neg F] \rightarrow AL\neg[p, F]$, we need only show that the converses follow from the hypothesis. Assume $RS([p, F] \vee [p, \neg F])$; then according to (i), $AL\neg[p, \neg F]$ yields $AL(([p, F] \vee [p, \neg F]) \& \neg[p, \neg F])$, whence $AL\neg[p, \neg F] \rightarrow AL[p, F]$. Moreover, interchanging F with $\neg F$, we infer $AL\neg[p, F] \rightarrow AL[p, \neg F]$. This completes the proof.

Felicity of notation, the next item on the list, is important in any theory. According to Russell [1922]: "...***a good notation has a subtlety and suggestiveness which at times makes it seem almost like a live teacher. Notational irregularities are often the first sign of philosophical errors, and a perfect notation would be a substitute for thought.***" Felicity of notation is not a notable feature of Kanger's theory; moreover, the Kanger notation is anything but concise. Strange as it may seem, this does not make for ease in reading. On the other hand, if the present theory has any merit, it is owing to the subtlety and suggestiveness of the notation in category theory.

Finally, I come to the matter of *richness of language*. Lindahl, in referring to the seven basic one-agent norms, has to call them by the code-names T_1, T_2,..., T_7. These norms need to be named, or at least have verbal descriptions. In the present framework, on the other hand, each elementary norm has a name which, to a greater or lesser extent, gives an intuitive feel to its purpose, function, and applicability. It is true that Kanger talks about claim, power, immunity, and freedom, but these are two-agent legal positions. The analysis of two-agent norms is beyond the scope of this work.

To sum up: The Kanger-Lindahl theory has one feature to

recommend it: it has a simple structure. But the price it pays for this feature is high. Perhaps the worst failure of this theory is that it is subject to *all* the paradoxes of standard deontic logic—simply substitute *shall* for *ought.* Moreover, there are many important norms for which it has no formulation. On the grounds of all the criteria we have been examining, the *functorial framework for practical logic* ***K*** is, I submit, superior to the Kanger-Lindahl theory.

11 Final remarks

As we indicated earlier in this work, John Stuart Mill, in his *A SYSTEM OF LOGIC* [1898 ed.] suggests that what is important in practical logic is the relation between two realms: one, the realm of imperatives; the other, the realm of fact. (Mill calls one realm "art" and the other "science.") Mill's idea is that though practical problems have their origin in the realm of imperatives, those problems that have solutions may be transformed into problems that are statable in the realm of fact. The present work may be viewed as an attempt to work out the logical details of this idea.

I have tried to show that the functorial framework for practical logic is superior in many respects to its rivals. A salient aspect of this framework is the central role it assigns to imperative logic. Without imperative logic, one is unable to treat properly that part of the subject matter of moral philosophy which is in the form of imperatives (or optatives), e.g. maxims, advice, appeals, and instructions. Moreover, as I suggested in the ***Introductory***, the use of imperative logic makes moot the question of whether or not ethical judgments (as usually conceived) are true or false, and therefore removes an epistemological obstacle to establishing proposed solutions to problems in the field.

One feature of the present theory is the one that motivated Mill to stress the relationship between the practical realm of imperatives and the theoretical realm of scientific fact; namely, that the use of functors between the two realms should make it possible to reduce problems expressible in imperatives to problems to which the methods of science can be applied. The first step in the use of functors in this way is to build up, for each practical realm, e.g. ethics or politics, a normative system based on proper axioms over *K*. To see how this might be done the reader may consult ***Part Three***.

As already stated (see ***Introductory***), the next chapter, to which we now turn, goes over the same ground as the present chapter but does so in greater depth, and, in addition, treats some material not in the present chapter.

Chapter Ten

Imperative Logic

(**Note**: After reading the concluding remarks (Section 6), this chapter may be skipped by those not interested in technical results.)

1 Introduction

The contents of this chapter are: first, a generalization of the theory of admissibility and requirement whose main features were discussed in the previous chapter; secondly, a more complete listing of results than is found in the previous chapter, together with their proofs (mostly, a reprise of results found in Johanson [1988]); thirdly, an exploration of dyadic norms; fourthly, a discussion of paradoxes not treated in Chapter Nine; finally, a brief discussion of the formal semantics of admissibility and requirement.

As explained in the previous chapter, imperative logic may be viewed as a simple application of category theory and we take advantage of this fact in this chapter. The category theorist will notice that we do not use the definitions of categorical terms like "functor" and "adjoint" in their full generality. This is because—as we explained in Chapter Nine—our attention in this work is confined to logic and lattice theory, and hence we are able to tailor the definitions to fit the subject.

2 The formal structure

In Chapter Nine, we outlined a theory $\boldsymbol{K}$ virtually identical to the system explored here. The essential difference is that in this chapter *the law of excluded middle is not assumed.* Nonetheless some results that depend on the law of excluded middle are included in this chapter, *but these are marked with an asterisk.* Those formulas from intuitionistic calculus that we make use of are listed at the end of this section. Finally, it may be noted that, as already anticipated in Chapter Nine, statements of the form "(x => y and y => z)" will be shortened to "(x => y => z)."

Again, as in the previous chapter, we take advantage of the fact that the underlying structure of any logic may be regarded as a lattice ordered by entailment. Such a lattice is an example of a *Heyting algebra* $\boldsymbol{H}$. In the terminology of category theory, a Heyting algebra is a *cartesian-closed category* (with an initial element 0*) in which given any pair of objects (unordered) there is at most one *arrow* => between them (cf. Freyd [1972], Johnstone [1977, p. 137], Goldblatt [1979, p. 183]). What this means in lattice theoretical terms is that $\boldsymbol{H}$ is a partly ordered set together with a *reflexive, transitive, and antisymmetric relation* => with the following properties:

(a) $\boldsymbol{H}$ has a least element 0*
(b) For every pair of elements (x, y) in $\boldsymbol{H}$ there is a least upper bound $(x \vee y)$ and a greatest lower bound $(x \,\&\, y)$
(c) For every pair of elements (x, y) in $\boldsymbol{H}$ there is an *exponential object* $(x \rightarrow y)$ in $\boldsymbol{H}$ having the property that for every triple x, y, z in $\boldsymbol{H}$, $(x \,\&\, y)$ => z if and only if x => $(y \rightarrow z)$

In a logic, the relation => is interpreted as *entailment*, 0* is the *equivalence class of invalid sentences,* and the exponential object defines the familiar connective "$\rightarrow$" called the *conditional.* The condition (c) essentially says that *the logical laws of exportation and importation are valid.*

Since for every pair of elements (x, y), $(x \,\&\, y)$ => y it follows immediately from (c) that for every x and y, x => $(y \rightarrow y)$ and hence for every x, $(x \rightarrow x)$ is *the greatest* (or *terminal*) element. (I say "the" because for every x and y, $(x \rightarrow x)$ => $(y \rightarrow y)$ and $(y \rightarrow y)$ => $(x \rightarrow x)$, whence $(x \rightarrow x) = (y \rightarrow y)$. In particular, for every x, $(x \rightarrow x) = (0^* \rightarrow 0^*)$. Let us denote $(x \rightarrow x)$ by 1*.

The rule of inference (*modus ponens*) follows from (c) by replacing

x by $(y \to z)$ and noting that the relation $\Rightarrow$ is reflexive. Moreover, the operation of *negation* $\neg$ (or "complementation") is obtained in H by defining $\neg x$ as the exponential object $(x \to 0^*)$ for each x in $\boldsymbol{H}$. The contrapositive rule—if $(x \Rightarrow y)$, then $(\neg y \Rightarrow \neg x)$—follows from the result, easily proved (using the properties of greatest lower bounds together with *modus ponens*) that negation is order reversing; moreover, a simple use of *modus ponens* and exportation yields $x \Rightarrow \neg\neg x$. Since in this algebra x cannot be deduced from $\neg\neg x$, a Heyting algebra is sometimes called a *Brouwerian* (or *pseudo-complemented*) *lattice*. Thus, rather than having the structure of classical logic, a Heyting algebra has *the algebraic structure of intuitionistic sentential calculus*.

The symbols used here are the same as those in Chapter Nine. As in that chapter linguistic expressions are often treated autonymously, i.e. as names of themselves. As before, in $\boldsymbol{K}$ there are two lattices: one, **PROP** (a logic of statements); the other, **IMP** (a logic of imperatives). The only essential change is the one already indicated: the lattices **PROP** and **IMP** are Heyting algebras, but not necessarily Boolean. In most other respects the conventions and definitions of Chapter Nine remain in force; in particular, it is assumed that **PROP** is a complete lattice and, moreover, it is assumed that the definitions $L = !(\exists t)$ and $S = !(\forall t)$ remain in force, as well as the definitions of the simple norms, avoidability, irrevocability, compulsivity, tenability, etc.

As before the main action in $\boldsymbol{K}$ occurs between the two logics as shown in the following diagram:

$$\mathbf{PROP}\ \underset{R}{\overset{L}{\rightleftarrows}}\ \mathbf{IMP}\ \underset{S}{\overset{A}{\rightleftarrows}}\ \mathbf{PROP}$$

As explained in the previous chapter, the mapping R is called *requirement* and the mapping A is called *admissibility*. The mapping L is called *let* and the mapping S is called *sustain*. Recall that the application of the mapping R to an imperative U yields the expression RU (read: *It is required that U*); the application of the mapping A to U yields the expression AU (read: *it is admissible that U*); moreover, the application of L to an element F in **PROP** yields an imperative LF (read: *it* ***shall*** *be the case that at some time t, F is true)*; finally, the application of S to an element F in **PROP** yields an imperative SF (read: *it* ***shall*** *be true for all time t that F is the case*).

Note: *Note that if F does not contain t free,* $!F = !(\exists t)F = LF$.

We may summarize the underlying structure from which to derive the formal syntax of the functors *R, L, A,* and *S* by the following list of statements: (2.01-2.06):

(2.01) **PROP** *and* **IMP** *are Heyting algebras in which* ($0^* \neq 1^*$)

(2.02) *A and R are associate functors, i.e.* $A = \neg R \neg$ *and* $R = \neg A \neg$

It follows from D9 in Chapter Nine that a logical formula in **IMP** and **PROP** is *valid* provided it is equal to 1*.

(2.03) *The operator A is a functor from* **IMP** *to* **PROP** *that preserves validity*

(2.04) (*A, S*) *and* (*L, R*) *are* ***adjoint*** *pairs*

Note: We noted in Chapter Nine that as if by *pre-established harmony*, *L* is *left* adjoint and *R* is *right* adjoint. We also noted that if *L* is given, and *R* is defined as the right adjoint of *L*, then if *R* exists it is unique up to logical equivalence.

(2.05) *The operators L and S are functors from* **PROP** *to* **IMP** *that preserve validity*

Note:. Not all lattice functors preserve 1*. As an example consider the functor *j*: **PROP** ⟶ **PROP** defined as follows: let *F* be a fixed statement and set *jp* = *(p & F)*; then *jp* is never valid unless *F* = 1*. On the other hand, if a lattice functor *f*, defined on a lattice that has a terminal element 1*, is *onto*, then *f*1* = 1*. Note that unless *F* is a valid formula, the functor *j* is not onto.

The following formulae are valid:

(2.06) ($RU \vee A\neg U$), ($AU \vee R\neg U$)

The first formula says "Either it is required that *U* or it is admissible that ¬*U*; the second says "Either it is admissible that *U* or it is required that ¬*U*." *These formulae are provable if we assume the law of excluded middle.*

It was asserted in the previous chapter that *requirement is analogous to truth.* I based this assertion on the following string of results, which are simple consequences of the assumption that *R* is *right adjoint* to *L*.

(2.10)

(a) *For every imperative U,*

$$L1^* \Rightarrow U \text{ iff } 1^* \Rightarrow RU,$$

whence (since *L* preserves validity)

(b) *U is valid iff RU is valid*

It follows that

(c) *The functor R preserves validity,* i.e. $R1^* = 1^*$

Moreover, if U and V are imperatives; then we have the corollary:

(d) *U entails V iff $R(U \rightarrow V)$ is valid*

We have already shown that $1^* = \neg 0^*$. It remains to show that the initial object 0^* is the complement of the terminal object 1^*.

(2.11) $0^* = \neg 1^*$

Proof: $\neg 1^* = (1^* \rightarrow 0^*) \Rightarrow (1^* \,\&\, (1^* \rightarrow 0^*)) \Rightarrow 0^*$. But $0^* \Rightarrow \neg 1^*$ and hence $0^* = \neg 1^*$. ♦

In the following we let F be a statement.

The theorem

(2.20) $(SF \rightarrow LF)$

is obvious since $((\forall t)F \rightarrow (\exists t)F)$ is valid.

By adjointness, we have as a corollary:

(2.21) $(RSF \rightarrow F)$

Another theorem is

(2.22) $S\neg F = \neg LF$

This is obvious since $(\forall t)\neg F = \neg(\exists t)F$. Corollaries (using first (2.20) and then the contrapositive) are

(2.23) $(\neg LF \rightarrow L\neg F)$

and

(2.24) $(LF \rightarrow \neg S\neg F)$,

whence

(2.25) $(SF \rightarrow \neg S\neg F)$, $(S\neg F \rightarrow \neg SF)$

and

(2.26) $\neg(SF \,\&\, S\neg F)$

Using the law of excluded middle, we have

*(2.27) $LF = \neg S\neg F$, $SF = \neg L\neg F$

In the remainder of this chapter we assume F and G are statements and U and V are imperatives.

The theorem that admissibility commutes with disjunction and requirement commutes with conjunction was already noted in Chapter Nine. That is to say:

(2.30) $A(U \vee V) = (AU \vee AV)$, $L(F \vee G) = (LF \vee LG)$,

(2.31) $R(U \,\&\, V) = (RU \,\&\, RV)$, *and* $S(F \,\&\, G) = (SF \,\&\, SG)$

These theorems follow from the fact that A is left adjoint to S and R

is right adjoint to L. Moreover, as already noted in Chapter Nine, (by *modus ponens*) we have as a corollary:

(2.32) $R(U \to V) \to (RU \to RV)$, $R(U \vee V) \to (AU \vee RV)$,

$$S(F \to G) \to (SF \to SG),\ S(F \vee G) \to (LF \vee SG),$$

$$RS(F \to G) \to (RSF \to RSG)$$

As another consequence of adjointness, $LF \Rightarrow LF$ and $RU \Rightarrow RU$ yield:

(2.33) $(F \to RLF)$ and $(LRU \to U)$

Hence, we also have

$$LF \Rightarrow LRLF \Rightarrow LF \quad \text{and} \quad RU \Rightarrow RLRU \Rightarrow RU,$$

whence

(2.34) $L = LRL$ and $R = RLR$.

Moreover, it follows from (2.33) and the law of the excluded middle that the following formula is valid:

*(2.35) $(RLF \vee RL\neg F)$,

i.e. *either it is nonavoidable that F or it is nontenable that F.*

This result should be carefully distinguished from $(RLF \vee R\neg LF)$ (i.e. the statement, "It is either nonavoidable that F or it is nonfeasible that F") which is not a theorem. On the other hand, given the law of the excluded middle, $R(LF \vee \neg LF)$ is provable; that is to say: the imperative, *Let F or Sustain ¬F*, is logically valid. It should also be pointed out that although $R(LF \& LG) = (RLF \& RLG)$, the formula $RL(F \& G) = (RLF \& RLG)$ is not a theorem, whence it can happen that conditions described by F and G are both nonavoidable but the condition described by their join is avoidable.

The laws of noncontradiction,

(2.40) $\neg(U \& \neg U)$ and $\neg(F \& \neg F)$,

are easy consequences of $(U \& \neg U) \Rightarrow 0^*$ and $(F \& \neg F) \Rightarrow 0^*$, which are mere expressions of *modus ponens.* For instance, $U \& (U \to 0^*) \Rightarrow 0^*$, whence the contrapositive (together with the identities, $\neg U = (U \to 0^*)$ and $1^* = \neg 0^*$) yields $1^* \Rightarrow \neg(U \& \neg U)$.

Note:: This theorem says that $(U \& \neg U) = 0^*$ and $(F \& \neg F) = 0^*$.

In Chapter Nine we made use of the following form of the law of contradiction for imperatives:

(2.41) $\neg(RU \& R\neg U)$

Proof: This result follows from the fact that requirement, like truth, commutes with conjunction and that admissibility preserves validity:

$$\neg(RU \& R\neg U) = \neg R(U \& \neg U) = \neg R0^* = A1^* = 1^* \blacklozenge$$

An immediate corollary—as pointed out in Chapter Nine—is the logical equivalent:

(2.42) $(RU \rightarrow AU)$

Other logical equivalents include

(2.43) $(R\neg U \rightarrow \neg RU)$

and

(2.44) $(\neg AU \rightarrow \neg RU)$

Note: If R commuted with "$\vee$" and $(U \vee \neg U))$ were valid, **IMP** would be complete in the sense that $(RU \vee R\neg U)$ (read: either it is required that U or it is required that $\neg U$); moreover, this result would imply that $(AU \rightarrow RU)$, i.e. whatever is admissible is required.

Our proof that the theorem

(2.45) $(R\neg U = \neg AU)$

is valid consists in two parts. First observe that $(R\neg U \rightarrow \neg AU)$ follows from $(AU \rightarrow \neg R\neg U)$; then infer $(\neg AU \rightarrow R\neg U)$ from (2.06).

Dual to (2.33), we have

(2.50) $(U \rightarrow SAU)$ *and* $(ASF \rightarrow F)$;

consequently $AU \Rightarrow ASAU \Rightarrow AU$ and $SF \Rightarrow SASF \Rightarrow SF$, whence

(2.51) $A = ASA$ and $S = SAS$

Note: The results $(F \rightarrow RLF)$ and $(ASF \rightarrow F)$ may at first sight appear to be problematic. The first simply says that if F holds then it is nonavoidable that F and the second says that it is tenable that F only if F already holds.

Now, because of adjointness, we have

(2.52) $1^* \Rightarrow SF$ *iff* $A1^* \Rightarrow F$

Since admissibility preserves validity, it follows that an imperative SF is valid (at time t) iff the statement F is valid (at time t). Thus RSF is an axiom in a normative system only if F is also an axiom. Since A and L are order-preserving we have

(2.60) $L(F \& G) \rightarrow (LF \& LG)$

and

(2.61) $A(U \& V) \rightarrow (AU \& AV)$

A further result is

*(2.62) $L(F \rightarrow G) \rightarrow (SF \rightarrow LG))$, $A(U \rightarrow V) \rightarrow (RU \rightarrow AV)$,

from which we obtain the following, using the relation

$$(\alpha \rightarrow (\beta \rightarrow (\alpha \& \beta))):$$

*(2.63) $(LF \& SG) \rightarrow L(F \& G)$, $(AU \& RV) \rightarrow A(U \& V)$

*(2.64) $S(F \vee G) \rightarrow (SF \vee LG)$, $R(U \vee V) \rightarrow (RU \vee AV)$

Further results include:

*(2.65) $L(F \to G) \to (SF \to LG)$, $A(U \to V) \to (RU \to AV)$

*(2.66) $S(F \to G) \to (LF \to LG)$, $R(U \to V) \to (AU \to AV)$

Moreover, applying (2.32), we obtain the important result,

*(2.67) $(RS(F \to G)\ \&\ RLF) \to RLG$, $(RL(F \to G)\ \&\ RSF) \to RLG$

Note, however, in contrast to (2.26) and (2.42) neither $\neg(LF\ \&\ L\neg G)$ nor $\neg(AU\ \&\ A\neg U)$ are theorems. In the preceding we have made free use of formulas (in both **IMP** and **PROP**) which are familiar from intuitionistic calculus such as the following:

$(\alpha\ \&\ (\neg\alpha \vee \beta)) \to \beta$

$(\alpha \to \beta) \to (\neg\beta \to \neg\alpha)$, $\quad (\alpha \to \neg\beta) \to (\beta \to \neg\alpha)$

$\neg(\alpha\ \&\ \beta) \leftrightarrow (\alpha \to \neg\beta)$

$(\alpha \to \beta) \to \neg(\alpha\ \&\ \neg\beta)$

$(\alpha \vee \beta) \to (\neg\alpha \to \beta)$

$\neg(\alpha \vee \beta) \leftrightarrow (\neg\alpha\ \&\ \neg\beta)$

$\neg(\exists\alpha)\Phi = (\forall\alpha)\neg\Phi$

$\alpha \to (\alpha \vee \beta)$

The last formula in this list is the logical basis for the "paradox of Ross" in deontic logic. Ross's paradox and other related paradoxes are discussed below in Section 4.

3 Dyadic logic

The theory of elementary norms is explored in the previous chapter. In this section we explore norms defined in terms of two statements F and G. In particular we explore generalizations of pro-bondage and responsibility. Recall that pro-bondage is of the form $(ASF \to RSF)$ and its generalization $(ASG \to RSF)$ (abbreviated $[F/G]$) is called *disposition.* On the other hand, "*round-bracket disposition*" (abbreviated (F/G)), which is of the form $(ALG \to RSF)$, is a generalization of responsibility $(RSF \vee RS\neg F)$. *Disposition* is read: "*it is binding that F on condition that it is tenable that G". and round bracket disposition is read: "*it is binding that F on condition it is feasible that G*." This section is devoted to the *dyadic operators* (/) and [/].

Notes

1. I sometimes refer to the disposition $[F/G]$ as "square-bracket disposition."

2. Note that a necessary condition for $[F/G]$ is that it is tenable that G only if F is true. A sufficient condition for $[F/G]$, on the other hand, is that it is

nontenable that *G*. Note, however, that there is a commitment to *F* only if it is tenable that *G*. A lack of power releases one from a commitment in bondage; in a similar way, lack of tenability releases one from the commitment implicit in [*F/G*]. Furthermore, if (*F/G*) is true, not only does ($G \to F$) hold, but a necessary condition for it to be feasible that G is that F be true.

3. The mappings (/) and [/] are functors defined on the category (***Op*** x **PROP**) where ***Op*** is the opposite category of **PROP**, i.e. the same category as **PROP** except that the arrows go in the opposite direction. Thus if $P \Rightarrow G$ and $F \Rightarrow Q$, then $(ASG \to RSF) \Rightarrow (ASP \to RSQ)$ and $(ALG \to RSF) \Rightarrow (ALP \to RSQ)$. In particular if *H* is a fixed statement in **PROP**, (/*H*) and [/*H*] are functors from **PROP** to **PROP** that preserve validity.

The operator (/) satisfies a set of axioms introduced by von Wright [1971]. (These axioms are a variation of an original set which turned out to have certain undesirable features.) Von Wright introduces an operator *O* such that the result of applying *O* to an ordered pair of statements (*F, G*) yields *O*(*F/G*) (read: *one ought to see to it that F given G*). The axioms of von Wright are as follows: (Following Hansson [1969], I add an "innocuous" axiom (3.0).)

(3.0) $O(1^*/F) \;\&\; O(F/0^*)$
(3.1) $\neg O(0^*/1^*)$
(3.2) $O((F/H) \;\&\; (G/H)) \leftrightarrow O((F \;\&\; G)/H)$
(3.3) $O(F/(G \vee H)) \to (O(F/G) \;\&\; O(F/H))$
(3.4) $(O(F/G) \;\&\; O(F/H)) \to O(F/(G \vee H))$

All of these axioms are satisfied if *O(F/G)* is interpreted as (*F/G*). And similarly—if we substitute [*F/G*] for *O(F/G)*—the operator [/] satisfies axioms (3.0) through (3.3). That is to say, the following theorems are valid:

(3.0a) $(1^*/F) \;\&\; (F/0^*)$
(3.1a) $\neg(0^*/1^*)$
(3.2a) $((F/H) \;\&\; (G/H)) \leftrightarrow ((F \;\&\; G)/H)$
(3.3a) $(F/(G \vee H)) \to ((F/G) \;\&\; (F/H))$
(3.4a) $((F/G) \;\&\; (F/H)) \to (F/(G \vee H))$
(3.0b) $[1^*/F] \;\&\; [F/0^*]$
(3.1b) $\neg[0^*/1^*]$
(3.2b) $([F/H] \;\&\; [G/H]) \leftrightarrow [(F \;\&\; G)/H]$
(3.3b) $[F/(G \vee H)] \to ([F/G] \;\&\; [F/H])$

In any system satisfying axioms (3.0) through (3.3) the formula

(3.5) $O(F/G) \to O(F/(G \,\&\, H))$

is valid. Hence the following are also valid:

(3.5a) $(F/G) \to (F/(G \,\&\, H))$,

(3.5b) $[F/G] \to [F/(G \,\&\, H)]$,

Results of this kind are considered counter-intuitive. In none of Hansson's [1969] systems of dyadic logic is (3.5) valid, and as a justification, Hansson presents the following example, which he attributes to Powers: suppose John Doe makes Suzy Mae pregnant and suppose this obligates John Doe to marry Suzy Mae. According to (3.5), if John Doe impregnates Suzy Mae and then kills her, he is still obligated to marry her. This is of course absurd. Hansson takes this to mean that the axioms are "incorrect," but there are other options.

Ordinarily one would say that the existence of a counter-example means that one of the premisses—in this case the statement "If John Doe makes Suzy Mae pregnant, he is obligated to marry her"—is false. Surely it is absurd to make such a blanket statement regardless of circumstances. Suppose, for example, Suzy Mae is already married. One would think that the correct statement is that "If x makes y pregnant and x *can* marry y, then x is obligated to marry y." Even such a true believer as John Searle [1964] in his famous "How to derive ought from is" recognizes the need for a *ceteris paribus* clause. To make this clear, let F be "x marries y," G be "x impregnates y," H be "x can marry y," and J be "x murders y." Then instead of $O(F/G)$—which, even on Kantian principles, is certainly false (recall "ought implies can")—let us assume $O(F/(G \,\&\, H))$. One may surely assume $(J \to \neg H)$ (as an axiom), whence $(G \,\&\, H) \,\&\, J = 0^*$ and $O(F/(G \,\&\, H) \,\&\, J)$ is true by axiom (3.0); consequently $O(F/(G \,\&\, H)) \to O(F/(G \,\&\, H) \,\&\, J)$ holds.

Another apparently counter-intuitive aspect of these formulas is that they allow for "conflicting duties." I will not comment further on problems of "obligation," as found in standard deontic logic, but restrict my remarks to the present theory. Accordingly, instead of conflicting duties let us speak of "conflicting dispositions." In the case of the round-bracket dispositions, let us suppose both (F/G) and $(\neg F/G)$ both hold. Then it follows that $(0^*/G)$ (i.e. $(ALG \to 0^*)$ holds), whence $\neg ALG$, i.e. $RS\neg G$, holds. It follows that a disposition toward both F and $\neg F$ of the round-bracket kind can only occur under circumstances G that are nonfeasible and contrary to fact. A similar result occurs in the case of conflicting dispositions of the "square-bracket" kind.

Invalid are the following formulas (analogs of Rescher's axioms for dyadic logic (cf. Hansson [1969])):

$$[F/(G \mathbin{\&} H)] \to [(G \to F)/H],$$
$$[(F \vee G)/H] \mathbin{\&} \neg[F/H] \to [G/(F \to H)]$$

Finally, I note that the following formulas are valid:

(3.6) $([F/H] \mathbin{\&} [G/1^*]) \to [(F \mathbin{\&} G)/H]$
(3.7) $([F/(G \mathbin{\&} H)] \mathbin{\&} [G/H]) \to [F/H]$
(3.8) $([(F \vee G)/H] \mathbin{\&} [G/G]) \to [G/(H \mathbin{\&} \neg F)]$
(3.9) $([F/G] \mathbin{\&} [\neg H/1^*]) \to [F/(G \vee H)]$
(3.10) $([F/H] \mathbin{\&} RS(F \to G)) \to [G/H]$.
(3.11) $(F/G) \leftrightarrow (\neg G/\neg F)$.
(3.12) $(F/G) \to [F/G]$

4 A budget of paradoxes

Some paradoxes such as Chisholm's contrary-to-duty paradox and Åqvist's paradox of the knower were already discussed in Chapter Nine. In this section, I discuss the "paradox of the good Samaritan" as well as Ross's paradox. Both of these are in one way or another concerned with the problem of "commitment." I begin with the paradox of the good Samaritan.

Can one condition commit us to another? This question is important in the case of the so-called "good-Samaritan paradox," a paradox that goes like this. Suppose there are two individuals Bob and Jerry with the following properties: Bob owes five hundred dollars to Jerry and Bob will murder Jerry (and no one else) a week from now. One may conclude that Bob is obligated to pay five hundred dollars to the man he will murder next week. It is apparently the case that the statement that Bob pays five hundred dollars to the man he will murder next week logically entails that Bob murders someone next week. Hence, it follows—according to standard deontic logic—not only is Bob obligated to pay the man but he is obligated to commit murder. I claim, however, that this conclusion does not follow if the statement "Bob is obligated to pay five hundred dollars to the man he will murder next week" is properly interpreted within formal logic.

In formal systems, such as *PRINCIPIA MATHEMATICA*, denoting phrases such as "the man Bob will murder next week" are treated as "incomplete symbols." This means that in order to forestall logical blunders any statement in which such an offending phrase appears must be revamped in such a way as to eliminate it. In *PRINCIPIA MATHEMATICA* (Whitehead & Russell [1910]) this is done by means of Russell's theory of descriptions. I claim that Russell's theory may be applied profitably to the case at hand.

Let us write Px for "Bob pays five hundred dollars to x," and write Qx for "Bob will murder x next week"; then "Bob pays five hundred dollars to the one he will murder next week" becomes *P*(*the x such that Qx*) which, according to the theory of descriptions, may be translated into the formal language as:

$$(\exists y)((\forall x)(Qx \leftrightarrow (x = y)) \,\&\, Py).$$

Consider now the translation of *OP*(*the x such that Qx*) where *O* is the obligation operator discussed in the previous chapter. According to the same rule of translation this is of the form "$(\exists y)((\forall x)(Qx \leftrightarrow (x = y)) \,\&\, OPy)$," which may be read: *There is a y such that for all x, Bob murders x iff x is y and it is obligatory that Bob pays five hundred dollars to y.* From this you cannot conclude it is obligatory that Bob murders y. (It is true that $(\exists y)(Qy \leftrightarrow (y = y))$ but this is not a logical equivalence and hence we cannot conclude OQy from $O(y = y)$.) This, in my opinion, effectively disposes of the paradox of the good Samaritan.

There are classic attempts to define the concept of "commitment" in terms of the obligation operator *O*. Two formulas vie for the translation of the phrase "F commits one to G": one is $O(F \rightarrow G)$ the other is $(F \rightarrow O(G))$. Since $(F \rightarrow (F \vee G))$ and the related formula $(\neg F \rightarrow (F \rightarrow G))$ are valid, it follows that $(O(F) \rightarrow O(F \vee G))$, $(\neg F \rightarrow (F \rightarrow O(G)))$, and $(O(\neg F) \rightarrow O(F \rightarrow G))$ are also valid formulas. The last three formulae are considered strange and paradoxical in deontic logic. The first of these, the "paradox of Ross," goes back to Ross [1941]. It implies, for example, "If it is obligatory that you pay your taxes then it is obligatory either that you pay your taxes or you stand on your head." There is no way to escape a consequence of this type in any theory that depends on a functor such as *O* (or on functors such as *RS* and *RL*). The last two formulae are called "paradoxes of commitment." The first of these is interpreted as saying "Whatever is not the case, commits one to anything," and the second has the interpretation, "If anything "forbidden" is done (or is the case), one has an arbitrary commitment" (cf. Føllesdal & Hilpinen [1971]). Chisholm showed that not only are these results strange, but the theory leads in certain cases to a contradiction (see Chapter Nine).

The concept of one condition committing one to another should perhaps be abandoned. It cannot be expressed either by a disposition $[F/G]$ or round-bracket disposition (F/G). $[F/G]$ as "commitment" leads to the "paradox" that if G is false or nontenable then G "commits" one to F—showing that doing anything false commits one to everything. Another alternative is the nonelementary norm $(RLF \& (ASG \rightarrow RSF))$, a generalization of bondage, but accepting this as a form of commitment would have the consequence that anything nontenable commits one to the status quo. We do not need a concept of commitment, however, to solve paradoxes like the contrary-to-duty paradox of Chisholm or the paradox of the knower of Åqvist (see Chapter Nine).

Ross's paradox is usually thought to be the root cause of the paradoxes of commitment. The parallel in the present theory is a formula of the type $(RLF \rightarrow RL(F \vee G))$; for example, if F states "This letter is mailed" and G states "This letter is burned" then from the requirement that it shall at some time be the case that this letter is mailed, one may derive the requirement that it shall at some time be the case that *either this letter is mailed or this letter is burned.* Philosophers in the field of practical logic have been needlessly disturbed by results of this kind. The mistake here, common to philosophers and the man or woman on the street, is the failure to take each instruction in any realm with a grain of salt. An instructive example is found in a fable invented by the eminent science fiction writer and popular thinker H.G.Wells. Wells [1924] in *THE TRUTH ABOUT PYECRAFT* tells a story about an uncommonly fat man named Pyecraft who, desperate for weight reduction, followed an ancient Hindustani recipe for "Loss of Weight." The instructions worked only too well: poor Pyecraft was found floating in his apartment up near the ceiling, *weightless—but as fat as ever.*

To follow single instructions without question is the essence of casuistry, and I suggest that casuistry is the road to hell. To take a controversial example, blindly following the *First Amendment to the United States Constitution* is counter productive; for (according to the Supreme Court) this amendment stands in the way of reasonable control of the access to office by those whose only virtue is that they have lots of money. Kant, among others, was mistaken when he thought that commands like "Do not lie!" can contain the whole of wisdom about the matters with which they deal. *No one statement*—except possibly this one—*can be totally relied on to guide conduct on any matter.*

The mistake of many deontic logicians is to take an *atomistic* approach to the realm of practice. I recommend a *wholistic* approach. As Quine [1992] points out, no scientific statement stands or falls by itself in isolation from a total theory; similarly, no imperative can reasonably be accepted as valid without reference to the totality of human wisdom. Confronted with an instruction of the form "Burn this letter or mail it!" discretion is the better part of valor. If the original prescription was to burn the letter it *may* make sense to burn it, but I would recommend an inquiry first. The *Nuremberg convention* prohibits following instructions blindly. Computers and robots are built to follow orders, but human beings—unless they have been turned into robots—have the God-given capacity to make choices based on inquiry and on the whole range of human knowledge and wisdom. Single instructions, on important matters, even with a wide-ranging knowledge of the context and reasons for the instructions, have to be approached with suspicion, for nothing is more common in human affairs than error and folly. (I refer the reader back to Chapter Two where the matter of the possibility—or rather the *impossibility*—of a scientific ethic, complete with a set of instructions on how to deal with each and every important practical problem, is discussed.)

5 Modal semantics

Note: In this section we assume the law of excluded middle.

This section is devoted to an exploration of the meaning of the functors *R* (and *A)* through a semantic analysis of a normative system defined in terms of functors *R* and *S*. A *normative system N*, ***as defined here***, is a first-order modal logic which, in addition to the usual logical apparatus, contains the operators *RS* and *AS* (where $A = \neg R \neg$). More precisely, *N* is obtained from the basic logic **PROP**, considered as a first-order predicate logic with equality, by adding the following rule: If *F* is a formula in **PROP** then *RSF* and *ASF* are formulas of *N*. Here "*RS*" and "*AS*" are modal operators.

Definition: A *Kanger* [1957] *frame* is a structure <W, D, B, V> defined as follows:

(i) W is a nonempty set (of "possible worlds");
(ii) D is a set of "individuals,"
(iii) B is a binary relation of *accessibility* on W;
(iv) V is a function that assigns to each world *w*, a *valuation* V(*w*), i.e. a function that assigns to each n-ary predicate letter a subset of the set of all finite sequences of length *n* where the elements of the sequence are individuals in D.

Notes

1. One could require that the set D vary as the world *w* varies, but for the sake of simplicity this is not done (cf. Turner [1984]).

2. The terminology used here is not Kanger's in every respect.

3. In this analysis a static model is assumed for the sake of simplicity. At the expense of complicating the theory the time dimension could be introduced, but from a formal standpoint this appears to be unnecessary.

Now let g be an assignment function that assigns a member of D to each individual variable, let Φ be an *n*-ary predicate letter, and write [g, *w*, *P*] for "*g satisfies the formula P in world w.*" Satisfaction is defined recursively as follows:

(i) [g, *w*, $\Phi(a_1, \ldots, a_n)$] iff (g(a_1), g(a_2), ..., g(a_n)) is a member of V(*w*)(Φ)

(ii) [g, *w*, (x = y)] iff g(x) = g(y)

(iii) [g, *w*, ¬*F*] iff it is false that [g, w, *F*]

(iv) [g, *w*, (*F* & *G*)] iff [g, *w*, *F*] and [g, *w*, *G*]

(v) [g, *w*, (∀x)*F*] iff [g*, *w*, *F*] for every assignment function g* that agrees with g except possibly at *x*

(vi) [g, *w*, *RSF*] iff [g, *w**, *F*] for all *w** such that *w**B*w*

(vii) [g, *w*, *ASF*] iff [g, *w*, *F*] and [g, *w**, *F*] for some *w** such that *w**B*w* and *w** is not *w*

Definition:. A formula *P* is *valid* provided [g, w, *P*] for every assignment g, and every world *w*. *P entails Q* (written *P* => *Q*) whenever the formula (*P* → *Q*) (i.e. ¬(*P* & ¬*Q*)) is valid.

Note that items (vi) and (vii) of the definition of satisfaction imply that *RS* and *AS* are functors that preserve validity. One may also prove (using (vii)) that *ASF* => *F*, whence it follows (from the contrapositive) that *F* => *RLF*. (Here we use *A* = ¬*R*¬ and *S* = ¬*L*¬.)

We are now in a position to characterize the accessibility relation B. Based on the fact that (*L*, *R*) is an adjoint pair, it may be proved that

(5.1) *RSF* => *ALF*, *RSF* => *F*, and *F* => *RSALF*

(see below). The formula (*F* → *RSALF*) is the analogue in *N* of the so-called *Brouwersche axiom* (cf. Lewis and Langford [1959, p. 497]); "Brouwer" is therefore an appropriate characterization of the modal logic of the present kind (cp. Snyder [1971]). (When expressed in terms of the necessity operator, often written "L", the Brouwersche axiom is equivalent to "*what is possibly necessary, is so.*") According to modal semantics (cf. Snyder [1971], Kanger [1972]), the results corresponding to (5.1) are valid in a Kanger frame provided the accessibility relation is *symmetric and reflexive with domain* D. Accordingly, let us hypothesize that B *is the relation that holds between two worlds w and w* provided either they are the same world or one is an ideal alternative of the other; then B is a symmetric and reflexive relation with domain D.*

Such an accessibility relation has implications for the operator "*AS*" as well as "*RS.*" For example, it follows from (vi) and (vii) and the reflexivity of B that $(RSF \rightarrow ASF)$.

As an illustration consider the following example. Assume W is the set of all possible cities each with its set of traffic laws (expressed as formulas of *N*) and let B be a subset of (W x W). Assume further that there is a consistent set of traffic laws that apply to all cities in W. A city *w* is called *ideal* if all the traffic laws are scrupulously obeyed in *w*. Finally, assume that B is a relation such that the pair (w, w^*) is in B iff either $w^* = w$ or one of the pair (w, w^*) is an ideal version of the other. Then B is a reflexive and symmetric relation whose domain is W.

Now suppose there is a traffic light *z* controlling the traffic flow through a certain intersection in a certain city *w*. Assume a vehicle approaches the traffic light and there is a traffic law that may be stated as follows: *It is tenable that a vehicle approaching traffic light z does not stop at z only if z is green.* Now let Φ and Γ be singulary predicate letters and assign $\Phi(z)$ the interpretation "A vehicle approaching traffic light *z* does not stop at *z*" and assign $\Gamma(z)$ the interpretation "*z* is green." Then the emphasized sentence has a translation into *N* as the formula $(AS\Phi(z) \rightarrow \Gamma(z))$. This formula is certainly not valid in the sense that for every world *w* and every assignment g, $[g, w, (AS\Phi(z) \rightarrow \Gamma(z))]$; however, it is easy to see, as follows, that *the formula is valid in every ideal city*. Since traffic laws are obeyed in ideal cities, the city w^* is ideal only if $[g, w^*, (\Phi(z) \rightarrow \Gamma(z))]$ for all assignments g; that is to say, $V(w^*)(\Phi)$ is contained in $V(w^*)(\Gamma)$. Moreover, since $(AS\Phi(z) \rightarrow \Phi(z))$ *is* valid, $[g, w, (AS\Phi(z) \rightarrow \Phi(z))]$ is true for every g and *w*. It follows that $[g, w^*, (AS\Phi(z) \rightarrow \Gamma(z))]$ is true for all ideal cities w^*.

As should be expected, the formal semantics for the two modal operators *RS* and *RL* are quite different. We have seen that the semantics of the operator *RS* is of the *Brouwer* type (i.e. the alternative relation satisfies existence, reflexivity, and symmetry); for *RL*, on the other hand, it can be shown that the alternative relation satisfies only transitivity (cf. Snyder [1971], Kanger [1972]). I leave it to others to say how illuminating this is.

6 Concluding remarks

We have shown that the imperative domain has a logical structure similar to but not identical to ordinary logic. The semantics in the imperative domain are, however, entirely different. For although requirement resembles truth, it does not have all the properties of truth: for example, if U and V are imperatives, $(U \vee V)$ may be *required* while neither U nor V is *required.* (On the other hand, if $(U \vee V)$ is admissible, then either U is admissible or V is admissible.) The main point of the logic, besides its intellectual interest, is that it relates imperative logic to ordinary logic and makes it possible to solve problems cast in the language of imperatives by recasting them as problems in the language of declarative sentences. This process serves us in at least three ways: one, it avoids the use of predicates like "right," "good," and "obligatory" as applied to actions or things, predicates whose meaning are perennially a subject of philosophical dispute—we shave these away by Occam's razor; two, it affords a relatively simple and understandable way of avoiding standard deontic logic with its many paradoxes. Finally, what is more important, it works!

Note

In Chapter Nine it was noted that, *without an explicit axiom*, one can *prove* that admissibility A preserves validity *provided* ***K*** *is a complete system..* The proof is indirect. Assume $R0^*$, i.e. that $R0^*$ is a valid formula. It then follows that $1^* \Rightarrow R0^*$, whence, by adjointness, $L1^* \Rightarrow 0^*$. Since $L1^* = 1^*$, we conclude $1^* \Rightarrow 0^*$. But since the converse always holds, this implies $0^* = 1^*$, contrary to D5, proving $\neg R0^*$. The conclusion follows on putting $0^* = \neg 1^*$ and $A = \neg R \neg$.

Part Three

A DEFENSE OF A PRACTICAL PHILOSOPHY

Chapter Eleven

Arguments for Ethical and Political Principles: A Formal Analysis

1 Introduction

Karl Marx is often quoted (e.g. in Ayer & O'Grady [1992]) to the effect that philosophers have only interpreted the world in various ways, but *the real task is to change it.* That is to say, according to Marx, philosophers have been concerned to study models *of* the world—as it exists now—while what the world requires is a model *for* a newer and improved world. A normative system (as understood in this work) fits Marx's specifications to a T since, unlike a theory in physics or biology, a normative system is not simply a model OF reality as it is today; on the contrary, a normative system is, as it were, a *blue print for changing or reconstructing the world.* Accordingly, it is clear that, as noted in Chapter Two (see Note 1), normative systems may be called *paradigms.*

As already stated in Chapter Two, Hume [1739] makes a related point. In a familiar passage, he writes:

> In every system of morality, which I have hitherto met with, I have always remark'd, that the author proceeds for some time in the ordinary way of reasoning,...when of a sudden I am surpriz'd to find, that instead of the usual copulations of propositions, *is,* and *is not*, I meet with no proposition that is not connected with an *ought*, or an *ought not* [emphasis in original].

Hume's point seems to me well taken. Despite all the controversy concerning "ought" and "is" in recent times, it remains the case that a normative system cannot be built without special normative assumptions or definitions (cf. Johanson [1973]). In other passages Hume seems to recognize the need for assumptions based on empirical evidence as well. Recall the passage (already quoted in Chapter Two): *It is full time that they [i.e. philosophers]...reject every system of ethics, however subtle and ingenious, which is not founded on fact and observation.* The need for empirical axioms in the present work is clear; for without an empirical base, the method of Mill cannot possibly work. A normative system like those in this work—which contain both empirical and normative axioms—is able to accommodate itself not only to the crude demands of nature, but to the kinder, gentler world of human aspirations.

In Chapter Nine the following is stated: The purpose of the theory ***K*** is to serve as a foundation for normative systems. Informally, a normative system is a system of norms, rules, or imperatives such as one finds in law, religion or personal ethics. Formally, by a *normative system **N**,* I understand a logical system, axiomatized over ***K*** with the following property: ***N*** contains, as subsystems, two theories (formal languages): one, a first-order theory axiomatized over **PROP**; the other, a theory (not necessarily first-order) axiomatized over **IMP**.

Recall that the axioms of a logical system come in two varieties; logical axioms, on one hand, and proper axioms, on the other. The *logical* axioms of a normative system are the axioms of the framework *K*; the *proper* axioms are those axioms that are over and above the logical axioms. It is the proper axioms of a normative system which enable us to accommodate the demands of those like Hume and Marx who ask for both normative and empirical axioms. The proper axioms are derived from two sources: one, the *design principles* of the system; the other, *special rules of inference* (see below), which are over and above the strictly logical rules of inference. Design principles reflect the aims of the designer or designers of the system. The rationality of the choice of the particular design principles in the systems studied here is discussed in ***Part One***, especially Chapters Two, Three, Four, and Six.

In each of the normative systems (or paradigms) studied in this work there is a single design principle Δ. In the case of the paradigm of (societal) functionality, Δ is the principle of functionality (see Section 2) and in the case of the growth paradigm, Δ is the principle of growth (see Section 3). Moreover, there are two rules of inference over and above the strictly logical rules of inference. These rules are

(1) *If F is an empirical generalization expressed in the language of* ***K*** *that is regarded as established (under modern conditions) in the relevant discipline (such as political science or psychology) then infer F*

(2) *Let* ***B*** *be the* ***binding operator*** —i.e., the operator ***B*** such that for each statement *F*, ***B****F* is the norm (*RLF* & {*ASF* $\rightarrow$*RSF*}) read "It is binding that *F*"—*then if F is a statement such that* Δ => *F*, *infer* ***B****F*

Rule (1) enables us to introduce empirical data into the system; rule (2) enables us to introduce normative assumptions. Moreover, proper axioms of either the empirical or normative variety may be added at will. For example, in traffic theory one would add traffic laws expressed in terms either of requirement or admissibility (see Section 4 in Chapter Nine).

The theory of normative systems can be applied in all practical realms, ranging from computer programming to aesthetics, but has its most important application in ethics and politics. As already stated (see ***INTRODUCTORY***), the present chapter sets up, in a formal way, the two normative systems (or paradigms)—one for ethics and the other for politics—that were introduced in ***Part One*** and backs up some of the arguments in ***Part One*** by formal and logical demonstration. Section 2 sets up a normative system for politics and, within that system, argues for some of the positions taken in Chapters Four and Five. Section 3 gives a similar treatment to the normative system for ethics which is the subject of Chapters One, Six, and Seven. Section 4 argues that the two normative systems are complementary to each other. At the end of the chapter are some remarks that bring this work to a close.

2 The paradigm of (societal) functionality

In this section we study the logical structure of a normative system ***D*** formulated within the language ***K***. It is assumed that there is a fixed society and the purpose of the normative system ***D*** is to fix the rules that regulate the behavior of the people in that society. ***D*** has only one design principle—the functionality principle—denoted by *E*. The intended interpretation of *E* is the statement "The people live in a society that is functional." (The letter "*E*" stands for "equilibrium": cp. Spencer [1879, p.198], [1892, p.233].) The two special rules of inference may therefore be expressed as follows:

(2.1) *If F is an empirical generalization expressed in the language of **K** that is regarded as established (under modern conditions) in the relevant discipline (such as political science) then infer F*

(2.2) *If F is a statement and E => F, then infer **B**F*

Rule (2.2) says that if E entails F then *it is binding that F.* Rule (2.1) has the interpretation that if F satisfies the antecedent of the rule, *then F is a proper axiom of **D**.*

Note: Except when an agent is specified, imperatives in the language ***D*** apply to the people (collectively) in society. For example, the command *SE*, which says "*It shall be the case that from this point onward the people are living in a society that is functional*" must be understood as an imperative that is binding on the people in the fixed society.

Now since $E \Rightarrow E$ is trivially true in ***K***, we may obtain from rule (2.2) the principle:

$$\boldsymbol{B}E$$

which says that *it is binding that the people live in a functional society.* Since according to Chapter Nine, this norm may be interpreted informally as *can implies ought,* one can say *power "commits" society to equilibrium.* More generally, if one can establish an empirical result of the form $(E \rightarrow F)$ or $(E \rightarrow \neg F)$, then, with the help of rules of inference (2.1) and (2.2), one can establish either the norm ***B****F* (*it is binding that F*) or $\boldsymbol{B}\neg F$ (*it is forbidden that F*: see Chapter Nine).

For example, let F be a statement of the principle of the consent of the governed. In Chapter Four, we argue that a state in which the principle of the (unmanipulated) consent of the governed is violated is not properly functional. Assuming that this conclusion is in accord with modern political science, one may conclude that $\neg F$ entails $\neg E$, whence according to rule (2.2), ***B****F* holds. Consequently, since in Chapter Four we define a democracy as any society *bound* by the principles of functionality and the consent of the governed, it follows that ***D*** ("*D*" for democratic) ***is a normative system for a democracy***.

In certain cases, however, neither $(E \rightarrow F)$ nor $(E \rightarrow \neg F)$ can be established. In such a case, if one wants to establish either ***B****F* or its co-norm, an additional axiom is needed. In the absence of such an axiom *constitutional laws of impotence* apply. We have already argued that laws are not to be multiplied beyond necessity and therefore, when it cannot be proved that functionality requires the imposition of either F or $\neg F$, the principle of functionality requires *total restraint* (see

Chapter Nine) on the part of the people. There are numerous examples, including monitoring the press and speech, establishing religion, monitoring consensual sexual activity between adults, prohibiting the use by adults of alcohol and other substances, and interfering, in early stages, with a woman's choice to end—or not to end—her pregnancy.

For an example, consider abortion. Suppose F states that in the early stages it is admissible that it shall be the case that a woman aborts her pregnancy; then there is no reason to believe either $(E \rightarrow F)$ or $(E \rightarrow \neg F)$; for no matter what the people do about abortion there is bound to be a lot of trouble; moreover, there is no knock-down argument to show that society functions better either way. In any case, it is difficult, if not impossible, in an advanced country to obtain consent of the governed on this matter. The ruling principle is that laws are not to be multiplied beyond necessity and therefore government is impotent in this case. One could envision a constitutional provision—like the constitutional amendment to the *United States Constitution* outlawing the interference by Congress with free speech—prohibiting the legislature from interfering with abortion. (Since there is no such provision in the *United States Constitution*, the *Supreme Court of the United States* relies on a "right-of-privacy" based on the 14th or 9th amendment.) The relevant norm here is therefore total restraint on the part of the people. In Chapter Nine we saw (in the case of speech) that total restraint reduces to nonfeasibility when interference by the people (collectively) is avoidable. Let us assume that *interference with abortion by the state is avoidable;* then if we let c be the state, the norm of total restraint with respect to abortion reduces to $RS\neg([c, F] \vee [c, \neg F])$, which says *it is nonfeasible for the state to interfere with abortion "rights."*

In other cases, it can be argued that for the sake of functionality the government needs to be involved. For example, consider first the case of *equal suffrage*. Let F be the statement that for each pair of individuals (x, y) of sound mind and having reached the age of suffrage it is the case that the electoral power of x is equal to the electoral power y. F is meant to apply to all pairs of citizens who meet the requirements of age and mental competence; say, reaching eighteen years of age, and not being so mentally impaired as to be unable to register a preference. If F does not hold with a sufficient degree of probability, it is easy to see that elections become relatively useless as a measure of how the people think and hence the principle of the consent of the governed is violated. Let $F(p)$ be the statement that F is true with probability not less than p; then one may suppose that there is a number p such that $(E \rightarrow F(p))$ and therefore the norm of bondage $\boldsymbol{B}F(p)$ holds. If a norm of this type holds, one can say that *equal suffrage to the degree p is binding on a democratic society*. If consent is sufficiently manipulated, however, it is possible that a society can appear to be functional even if p is much less than one, but this is dangerous since once a sufficiently large segment of the electorate realizes it is cheated it could throw society into turmoil.

Finally, let us consider entitlements. To say you own something (a property), or have an entitlement, is to say you have a claim in the form of a feasibility or power with respect to something. Entitlement comes about in many ways: conquest, inheritance, exchange, gift, tradition, physical possession, etc. The function of the law is largely to allocate the scope and limits of liberties, responsibilities, and entitlements. Some entitlements are indistinguishable from liberties. For example, am I entitled to swear in public? Though the detailed working out of the scope and limits of entitlement is something that is the function of legislators, some remarks are in order.

With respect to entitlements, government functions either to preserve entitlements or to redistribute them among the people. The entitlements in question may be material, like money or oil wells, or may be nonmaterial, like power, culture, or health. In the case of money (and other material resources) before it can be redistributed, the government must—like Robin Hood—take it from those who have it, using force if necessary. The taxation power of the state is limited by the danger of revolt, but if a tax is well chosen, the individual may not have the power to stop a tax.

Originally government intervened in entitlements only to preserve them. "Justice" may be as Thrasymachus says "the interest of the stronger," but in a liberal democracy the people become the stronger and the government tends to redistribute wealth. Let us end our examination of entitlements with one example of a resource that democracies attempt to distribute equally: health care. In this case the state must not only attempt is to ensure that people get the care they need, but also to ensure that people in sufficient numbers are motivated to enter the health care field. One way to do this is to have a single payer, and a collective bargaining system that enables the health care worker to earn a wage that is competitive with what people of comparable skills can earn elsewhere. The norm to be satisfied can be expressed as follows. Let H be the statement that the people (collectively) make universal health care available. A comparative study of health systems throughout the world would show that $(\neg H \rightarrow \neg E)$; for experience shows that societies without universal health care suffer needlessly and cannot be considered truly functional. As in other cases of this kind one may conclude ***BH;*** in other words, universal health care is binding on society.

As a final example, consider the case of slavery. In the past, fueled by economic interests, the norm of freedom for all people was not recognized despite the ringing phrase in *The Declaration of Independence*: ***All men are created equal.*** Nowadays it is accepted that slavery cannot be squared with the principle of the consent of the governed. In Aristotle's time a state that permitted slavery was to a certain extent functional, but this is no longer the case and Americans fought a terrible war to prove it. The past can be viewed with much less passion than the present, and norms that were once controversial no longer are so. If we let F be that slavery is the case, it is now obvious that $(F \rightarrow \neg E)$—and assuming this is empirically established—we may infer the *restraint* $\boldsymbol{B}\neg F$ (i.e. $\{RL\neg F\ \&\ (AS\neg F \rightarrow RS\neg F)\}$) Since apparently *slavery is avoidable in the modern world*, we may shorten this norm to $RS\neg F$, which states that slavery is nonfeasible.

We mentioned in Chapter Nine the means-ends principle, i.e. the principle that the end does not justify the means. A means to attaining pleasure may be admissible, but not if the means has an inadmissible consequence. For example, consider smoking. It is a well-known consequence of smoking that it causes lung cancer; moreover, if this consequence is to be avoidable, one can prove that a public policy that does not discourage smoking as a means to satisfaction is inadmissible. Suppose F is the statement, "*People smoke in public buildings*," and G is the statement that "*Many people suffer from lung cancer*;" then we may represent the relationship of the freedom to smoke in public buildings to lung cancer by

(*) $((\{ASF \vee AS\neg F\}\ \&\ ALF) \rightarrow RLG)$,

i.e., *people are free to smoke in public buildings only if it is compulsive that many people suffer from lung cancer.* Hence, ***if it is to be avoidable that many people suffer from lung cancer*** (i.e. $AS\neg G$) ***it follows*** (from the contrapositive of (*)) ***that the norm of restraint*** $(\{ASF \vee AS\neg F\} \rightarrow RS\neg F)$ ***holds***, whence we obtain the conclusion that it is forbidden to smoke in public buildings. Moreover, assuming not only that it *is* avoidable that many people suffer from lung cancer (i.e. $AS\neg G$) but that *the people have power with respect to smoking in* *public buildings* (i.e. $\{ASF \vee AS\neg F\}$), we can get the stronger conclusion that $RS\neg F$: *it is required that it shall be the case that from now on people avoid smoking in public buildings*. Therefore (given some rather natural assumptions) the system ***D*** yields a formal proof that it is inadmissible that people from time to time smoke in public buildings.

2.1 Social norms as consequences of functionality

There are some norms that hold only in a "perfect society," i.e. in a society that is in perfect equilibrium *E*. These are apt to be absolute norms in the form of an irrevocability *RSF*. (In a "perfect society," we may assume *F* would be true.) We are not interested in these here since what we want to know is what to do when society is not in equilibrium. Consequently, the norms we are interested in here are not absolute; instead we are interested in conditional norms of the form ***BF*** or ***B***$\neg F$. For example, if *F* is the statement, "The people are gentle," the norm of *gentleness* is of the form ***BF*** and may be read: *It is binding that the people are gentle*. The question to be answered here therefore is how one can prove norms of the form ***BF*** (or ***B***$\neg F$).

Now let *G* (instead of *F*) stand for the statement: "The people are gentle." The statements of the sages, from Laotse to Russell, regarding gentleness amount to the claim that $(\neg G \rightarrow \neg E)$, i.e. people are free from gentleness only in nonfunctional societies. It seems very likely that this claim can be supported by political science and historical research. Assuming this is the case, we may (according to Rule (2.1)) infer $(\neg G \rightarrow \neg E)$ as a proper axiom and therefore (by *modus ponens*) $E \Rightarrow G$; whence we conclude that the norm of gentleness is valid.

The arguments for the other norms are similar. For example, if we write *W* for "The people have awareness" (where by "awareness" we mean "curiosity guided by rationality and tempered by sensitivity and gentleness"), then the arguments in the previous section tend to show that one may establish by empirical observation that $(\neg W \rightarrow \neg E)$, i.e. freedom from awareness (in the population) yields disequilibrium. Hence we may conclude, in the usual manner, that $E \Rightarrow W$ and therefore the bondage ***BW*** holds. Similarly, if we write *"X"* for "The people are responsible" and *"I"* for "The people have integrity" then we may conclude both ***BX*** and ***BI*** since, given the relevant empirical axioms, one can show that those societies in which either responsibility or integrity is not binding on individuals are not fully functional.

In the preceding argument we have assumed that awareness W is compatible with gentleness G; for otherwise freedom from gentleness will by itself produce disequilibrium. Similar remarks apply *mutatis mutandis* to the other norms. In particular, it follows that responsibility and integrity are not to be pursued at all costs; in particular, they must not be pursued in such a way as to cause unnecessary harm to others. If this is correct, then it follows that Kant is wrong in his contention that the norm of integrity is absolute. For even Kant admits that there are certain cases in which to tell the truth (or to be completely honest) is sure to result in the destruction of the innocent. The assumption that gentleness takes precedence over integrity (see Chapter Five) amounts to the assumption that if gentleness is tenable, then gentleness is irrevocable even at the expense of integrity. For example, suppose there is a situation in which it is tenable that gentleness is incompatible with integrity, i.e. $AS(G \rightarrow \neg I)$. Then it follows that $(RSG \rightarrow AS\neg I)$ (see *2.62, Chapter Ten), whence if gentleness is irrevocable, integrity is avoidable! In such cases people have an incapacity $R(LI \,\&\, L\neg I)$ with respect to integrity.

I come next to the norm of *health*. It may seem strange to say that it is binding on one to be healthy since it hardly seems a moral norm and health is somewhat out of our control. In answer to this there are two remarks to be made:

(1) Being healthy requires the moral courage to be hard on oneself; for it is difficult to eat right, to not succumb to alcohol and drugs, to get the right amount of sleep, to exercise sufficiently, and to stay out of unnecessary trouble;

(2) The norm of health says that you must (i.e. you are *required* to) be healthy *only insofar as being healthy is within your control.*

The norm of health may be established in a manner similar to the preceding norms. A similar remark applies to the norm of altruism. If we write H for "The people are healthy" and J for "The people are altruistic" I do not believe it is hard to establish that $(E \rightarrow H)$ and $(E \rightarrow J)$ on the basis of the discussion in the preceding section. The central problem with respect to altruism is to decide when it is we have the capacity to be altruistic. For example, suppose that in a particular instance H and J are incompatible. If H takes precedence over J, it seems to follow that one is "off the hook"—so to speak—when one's own life is threatened.

In a manner now familiar, we may also conclude that the norm of trustworthiness $\boldsymbol{B}Y$, where Y stands for "The people are trustworthy," is valid. This is the case since it is easily shown by empirical observation and logical analysis that the defense, stability, and welfare of any society generally demands loyalty and obeying the law. For similar reasons, the norm of cooperativeness is entailed, as a conditional norm, by the principles of social construction, particularly the principles of stability and welfare. If we write C for "People cooperate" the norm of cooperativeness may be written $\boldsymbol{B}C$. Certainly, there are valid empirical generalizations about society that ensure that $\neg C$ entails $\neg E$.

Piety is a conditional norm since respect, politeness, and telling the truth have their limits. Hence if we write Z for "People are pious," the principle of piety may be written $\boldsymbol{B}Z$. As in previous cases, proof of this principle follows from the conclusion, $(\neg Z \rightarrow \neg E)$, inferred on the basis of empirical observation and logical analysis.

Finally, the norm of industry follows—in the same manner as the other norms—from the principle of functionality together with the relevant empirical generalization concerning the consequences to society of a lack of industry on the part of the people, which may be assumed to be derivable from empirical observation and logical analysis.

2.2 From norms to rules

We need in addition to norms, individual rules of conduct like those in the original Decalogue. For example, what is wanted are rules like "Thou shalt not kill!" As a test case let us see if we can derive this command—or something like it—from the system.

The rule "Thou shalt not kill!" can be shown to follow from the norm of gentleness, but only in cases in which it is within the capacity of a person to be gentle. More precisely, I intend to show that if G is "The people are gentle" and V is "The people avoid killing," the command SV (i.e. *It **shall** be the case that from now on the people avoid killing*!) follows from the norm $\boldsymbol{B}G$ together with the assumption ASG (i.e. *It is tenable that the people are gentle*).

I assume that part of what it means to be gentle is that one does not kill. Hence, I assume (as an axiom) $(G \rightarrow V)$. From this axiom one may conclude $RS(G \rightarrow V)$ since R and S are functors that preserve validity. Moreover, bondage $\boldsymbol{B}G$ implies the norm of *pro-bondage* $(ASG \rightarrow RSG)$, and therefore the *disposition* $(ASG \rightarrow RSV)$ is an easy consequence. (Here one may apply (10) from Chapter Nine or, as an alternative, (3.10) from Chapter Ten.) Finally, from the assumption *ASG*, we conclude *RSV*, whence, according to (2.11b) in Chapter Ten, the desired result *SV* follows.

3 The paradigm of growth

In this section we make a brief formal study of the paradigm of growth Γ, whose informal presentation is given in Chapter Six. Γ, like the democratic paradigm ***D***, is a normative system axiomatized over the formal language ***K***. A normative system is based on one or more design principles. Moreover, the design principles of any normative system embody, at least in part, the aims of those for whom the system is designed. In this instance we want the aim or aims to be widely, if not universally, shared. I assume that for everyone a life in which there is insufficient reason for remaining alive is intolerable. In order to make a reasoned choice for principles on which to base a life, the aim chosen here is the aim of the attainment of importance, where *importance* is defined as that which, *outside of fear of death*, confers a reason for being and makes sense out of life. Let us call the design principle that embodies the chosen aim the *principle of individual growth*. The statement of this principle *Q* has two parts:

Q1: *The precursors to growth are fulfilled in one's life.*

Q2: *The aim of attaining the maximum of importance in life is also fulfilled.*

In addition to the design principle *Q*, Γ (like ***D***) has two special rules of inference:

(3.1) *If F is an empirical generalization that is regarded as established in the relevant discipline (such as psychology) then F is an axiom of Γ*

(3.2) *If F is a statement and Q => F, then* ***BF***

Rule (3.2) says that if *Q* entails *F* then *it is binding that F*.

Now since *Q* => *Q* is trivially true in ***K***, we may obtain from the rule of inference (3.2) the principle:

BQ

which says that *it is binding that the growth principle holds*. Since according to Chapter Nine, this norm may be interpreted informally as *can implies "ought,"* one can say: *power "commits" one to satisfying the growth principle*. More generally, if one can establish an empirical result of the form ($Q \rightarrow F$) or ($Q \rightarrow \neg F$), then with the help of rules of inference (3.1) and (3.2) one can establish either the norm ***BF*** or its co-norm.

If ξ is a characteristic (or tendency) such as sensitivity, let $[\xi]$ be the statement that one acts consistently in a manner characterized by ξ. Then $S[\xi]$ is the imperative: *It* ***shall*** *be the case that from now on one acts consistently in a manner characterized by* ξ. For example, if ξ is "self-reliance" then $S[\xi]$ states: *It* ***shall*** *be the case that from now on one acts consistently in a manner characterized by self-reliance!* The *norms of higher growth* are defined as the norms of the form $\boldsymbol{B}[\xi]$ where ξ is one of the five higher tendencies: *sensitivity, sapience, self-reliance, dedication, and transcendence.* The norm $\boldsymbol{B}[\xi]$ states that it is binding that one acts in a manner characterized by ξ. It follows that if we want to show that it is binding that one acts in a manner characterized by ξ, we "need only" show $Q \Rightarrow [\xi]$. According to rule (3.1), *this may be done by showing that according to the best available empirical evidence it is true that* $(\neg[\xi] \rightarrow \neg Q)$.

For example, let ξ stand for sensitivity. Then $(\neg[\xi] \rightarrow \neg Q)$ says that in the absence of aesthetic appreciation one is lacking in personal growth; that is to say, one does not attain, to the degree possible, the maximum of importance or significance in life. A failure to make the most of life owing to a lack of aesthetic appreciation can happen in two different ways: one, the precursors to growth have not been fulfilled; the other, the precursors *have* been fulfilled and *yet* one suffers from a deficient sense of beauty. While scientific data may be demanded to back me up, it appears to me a foregone conclusion that people (otherwise well provided for) who are full of aesthetic appreciation are more likely than others to find life to be full of significance. I conclude we are justified in applying rules (3.1) and (3.2) and inferring that the sensitivity norm is valid.

The other principles (or norms) of growth in Chapter Six may be argued for in the same manner. That is to say, if ξ is one of the higher tendencies, autonomy, self-reliance, dedication, or transcendence, the method of proof in each case is to show $Q \Rightarrow [\xi]$ (or $\neg[\xi] \Rightarrow \neg Q$).

It remains to give a formal analysis of the argument in Chapter Six that a fully developed person must obey the imperative to be moral. What is to be shown is that given that T stands for transcendence and M stands for the property of being moral, the imperative $S[M]$ (which says, roughly, it is imperative to be moral) is a consequence of the following two conditions: one, $Q \Rightarrow [T]$ (i.e., maximal importance in life cannot be achieved without transcendence); the other, *ASQ* (i.e., less than maximal importance in life is avoidable). According to the argument in Chapter Six, not only does transcendence imply self-transcendence, but self-transcendence implies morality; hence we conclude [*T*] => [*M*] and therefore Q => [*M*]. But the latter implies not only *the principle of morality*,

(i) (*RL*[*M*] & (*AS*[*M*] → RS[*M*]))
(which says "it is binding that one be moral"),

but, given *ASQ*, implies

(ii) *AS*[*M*] (i.e., to be moral is tenable).

Finally, from (i) and (ii) we infer *RS*[*M*], from which the desired conclusion *S*[*M*] (which says *"it **shall** be the case that from now on one acts consistently in a moral manner!"*) is a simple consequence.

4 Are the two normative systems compatible?

As stated in Chapter Two, we need to show that the paradigms are *complementary* to each other. This is not difficult to do. First of all, it is easy to see that the two paradigms are based on complementary principles. This is the case since the measure of how far a society satisfies the principle of functionality (on which the societal paradigm is based) is the degree to which it fosters the satisfaction of the principle of individual growth, i.e. the degree to which it not only promotes the health, happiness, and welfare of the individual citizens (basically, the precursors to growth) but promotes the maximal attainment of importance—or what is the same thing, the satisfaction of the higher needs. Secondly, it is not hard to see that the rules of behavior derived from each paradigm are compatible; that is to say, the set of rules derived from one are compatible with the set derived from the other. For example, since moral principles—like gentleness—are implicit in the higher needs of human beings, the norm of gentleness, which in Chapter Five was derived from the principle of (societal) functionality, could just as well have been derived from the principle of growth.

Finally, let us consider whether the two systems differ from one another in that one is "deontological" and the other "teleological." Recall that an ethical theory is said to be "teleological" if it bases ethical judgments on a certain end (or collection of ends) and an ethical theory is said to be "deontological" if it bases ethical judgments not on ends but on duties. Kant's ethics is considered the epitome of a deontological ethic, but we saw in Chapter Two that Kant's arguments are based at least in part on ends. We noted in the same chapter that ethical systems may be based on different types of requirements. We defined "deontological requirements" as "moral rules (or "duties") imposed by society" and "teleological requirements" as "ends chosen by the individual." It may be noted that the social ethic in Chapter Four, derived from the paradigm of (societal) functionality, is based on moral rules imposed by society and hence is, in a sense, deontological. On the other hand, the personal ethic in Chapter Six, derived from the paradigm of growth, is based on the ends we called "the higher needs" and hence is clearly teleological. Since the two ethics are complementary and hardly in conflict, it follows that whether an ethic is looked at from its deontological aspect or its teleological aspect is perhaps not a matter of great import. Nonetheless, I think the effort to make ethics autonomous is based on a mistake since it is not reasonable to suppose that ethical principles are not, at bottom, based on their use either to society or to the individual. In any case, for my part, I should not be sorry to see ethical principles disappear like the dinosaurs *if they were no longer of any human use.*

If this sounds like utilitarianism, so be it. Labels don't matter. Utilitarianism has as least the virtue that it recognizes, like Protagoras and Christ, that ethical principles were made for human use rather than human beings being made to serve ethical principles. (Recall Mark [2.23]: *The Sabbath was made for man, and not man for the Sabbath.*) The thrust of what I have to say in this work is that if we wish to make sense of practical philosophy, it is time to base it firmly on the logic of human requirement, rather than on semi-mystical intuitions about the meanings of ethical words and phrases (like "good," "bad," "ought," and "ought not") or on a transcendental theology. But this is to anticipate the remarks in the final section.

5 A remark on logical form in religion

The most distinctive utterances in religion are in imperative form—*Thou shalt have no gods before me; Be fruitful and multiply; Love thy neighbor as thyself,* etc.—and yet philosophers of religion, so far as I am aware, devote little study to this form. Instead of devoting their attention to the realm of requirement, they devote their attention to the realm of truth. Religions tend to have a list of *requirements* thought to be imposed by the *deity.* (It is of course assumed that a requirement imposed by the deity is *ipso facto* valid.) The source of the list is usually one or more *men*, who tend to show their holiness by *signs and wonders.* Jesus of Nazareth, sometimes irascible, whom the faithful believe is God himself, considered the necessity for signs and wonders an abomination. (He says (cf. Matt. 12.39): *An evil and adulterous generation seeketh after a sign...*) Nonetheless, it is written that Jesus gave many signs and wonders: walking on water, turning water into wine and most impressively what Jesus (ibid.) called "the sign of the prophet Jonas": *For as Jonas was three days in the whale's belly; so shall the Son of Man be three days and three nights in the heart of the earth.* There are theologians like St. Thomas Aquinas, however, who attempt to back up scripture (considered to be the word of God) by reason, but the conclusions to be reached are preordained and their arguments are apt to be disingenuous; moreover, as we already said, they (the theologians) fail to distinquish between requirement and truth.

Of course, some of "the requirements of God," presuppose some truths; for example, the existence of higher powers. It is requirements that presuppose truths of this kind that primarily distingish traditional religious requirements from secular ones. The arguments of theologians focus, therefore, on the existence, uniqueness, and "goodness" of God, on one hand, and the existence and immortality of the soul, on the other; and hence their concern is mostly with metaphysics, rather than with either epistemology or practical inquiry. For my part, I think "true religion" lies not in the major concerns of theologians but with mysticism, social "justice," and, as Aurobindo puts it, "the growth of the soul."

6 Epilogue

In this work, I have not in general striven for novelty; for it has not been my intention to invent new systems of ethics and politics, nor even to find new arguments for the old ones; rather it has been my intention to develop a new framework into which the heritage of moralists from Socrates to Whitehead and Dewey can find its proper place. I do not abandon the old framework, based on traditional logic and on the traditional language of ethics and politics (like "good," "right" and "ought"), for trivial reasons. Like Hume [1739, p. 269], one would like to "*live, and talk, and act like other people in the common affairs of life.*" But common ways of speaking interfere in significant ways not only with human understanding but, more importantly, with human growth. Moreover, just as the old associative psychology could not advance until its practitioners abandoned concepts such as "ideas" and "sensations" in favor of operationally defined stimuli and responses, ethics and politics cannot advance unless philosophers abandon the vague concepts and unproductive methods that have hitherto characterized moral philosophy and adopt a new way of thinking.

Over 150 years ago, John Stuart Mill expressed great hope for practical philosophy. In the last paragraph of his *A SYSTEM OF LOGIC*, after holding out the prospect for the development of what he called the logic of practice, he writes:

> Notwithstanding the extreme generality of the principles of method which I have laid down,..., I have indulged the hope that to some of those on whom the task will devolve on bringing those most important of all sciences [i.e. the "moral sciences"] into a more satisfactory state these observations may be useful... Should this hope be realized, what is probably destined to be the great intellectual achievement of the next two or three generations of European thinkers will have been to some degree forwarded.

The present book was written out of the conviction that the modern version of practical logic which it contains has the potential to fulfill the hopes of Mill of bringing the moral sciences "into a more satisfactory state." The extent to which this book lives up to these hopes is not for me to say.

Note: In Section 5, I did not mean to imply that Christian theologians try to back up *all* religious dogma by reason. St. Thomas Aquinas in *SUMMA CONTRA GENTILES* distinguishes between the parts of the Christian faith that can be proved by reason from those that cannot.

Bibliography

Alterman, E. 1992. *Sound and Fury.* New York: Harper Collins

Anscombe, G. 1958. "Modern moral philosophy." *Philosophy* 33,1-19

Apter, M. 1989. *Reversal Theory.* London, New York: Routledge

Åqvist, L. 1967. "Good Samaritans, contrary-to-duty imperatives, and epistemic obligations." *Nous* 1 361-379

Aurobindo 1950. *The Human Cycle.* New York: Sri Aurobindo Library

Ayer, A. 1936. *Language, Truth and Logic.* London: Gollanz

Ayer, A. & O'Grady, J. 1992. *A Dictionary of Philosophical Quotations.* Oxford: Blackwell

Baumol, W. and Blinder, A. 1985. *Economics.* New York: Harcourt Brace Jovanovich

Bentham, J. 1932 ed. in Ogden, C. K. ed. *Bentham's Theory of Fictions.* London: Kegan Paul, Trench, Trubner & Co.

________. 1789. *The Principles of Moral and Legislation.* 1948 ed. New York: Hafner

Bergson, H. 1935. *The Two Sources of Morality and Religion.* New York: Holt

Birkhoff, G. 1948. *Lattice Theory.* New York: American Mathematical Society

Blackwell, K. 1985. *The Spinozistic Ethics of Bertrand Russell.* London: Allen & Unwin

Blake, W. 1946 ed. *Complete Poetry of Donne and Blake.* Modern Library New York: Random House

Brown, M. & Carmo, J. (eds.) 1996. *Deontic Logic, Agency and Normative Systems.* London: Springer

Bucke, R. 1901. *Cosmic Consciousness.* (reprint) 1969 New York: Dutton

Burke, E. 1790. *Reflections on the French Revolution.* 1910 ed. London: Dent

Burt, E. A. 1955. *The Compassionate Buddha.* New York: Mentor

Campbell, J. 1988. *The Power of Myth.* New York: Doubleday

Carnap, R. 1935. *Philosophy and Logical Syntax.* London: Routledge & Kegan Paul

Castañeda, H. 1983. *Thinking and Doing.* Dordrecht: Reidel

Cecil, R. 1985. *Textbook of Medicine.* Philadelphia: Saunders

Chisholm, R. 1963. "Contrary-to-duty imperatives and deontic logic." *Analysis* 26 33-36

________. 1964. "The ethics of requirement." *American Philosophical Quarterly* 1 147-153

Chomsky, N. 1987. *The Chomsky Reader.* New York: Pantheon

________. 1997. *The Prosperous Few and the Restless Many.* Berkeley: Odonian Press

Clarke, R. 1976. *The Life of Bertrand Russell.* New York: Knopf

Clarke, T. 1996. "Mechanisms of corporate rule." in Mander, J. & Goldsmith, E. [1996]

Clifford, W. 1886. *Lectures and Essays.* Second edition. London: Macmillan

Clinton, H. 1996. *It Takes a Village*. New York: Simon & Schuster

Coughlin, M. 1983. "Mathematics rooted in mystery." *The Mathematical Intelligencer* vol.5, 1 48-51

Cousins, N. 1979. *Anatomy of an Illness.* New York: Norton

Cowell, A. 1999. "Annan fears backlash over global crisis." *The New York Times* Feb. 1, p.A10

Current, R. 1996. "Lincoln" in *The Encyclopaedia Brittanica.* Chicago: Encyclopaedia Brittanica, Inc.

Daly, H. 1993. "The perils of free trade." *Scientific American,* Nov., 50-57

Darwin, C. 1872. *The Origin of Species.* (Sixth Edition) in Eliot, C. ed. 1909. Harvard Classics New York: Collier

________. 1874. *The Descent of Man.* Second Edition. New York: Hurst

Dewey, J. 1898. in Boydston, ed. 1967. *The Early Works of John Dewey.* Carbondale: Southern Illinois

________. 1902. *The Child and the Curriculum.* Chicago: University Press

________. 1920. *Reconstruction in Philosophy.* New York: Holt

________. 1922. *Human Nature and Conduct.* New York: Holt

________. 1938. *The Theory of Inquiry.* New York: Holt, Rinehart, & Winston

________. 1939. *Theory of Valuation.* Chicago: University Press

Durning, A. and Brough, H. 1992. *Taking Stock.* Worldwatch Paper 103, Worldwatch Institute

Eilenberg, S. & MacLane, S. 1945. "A general theory of natural equivalences." *Transactions AMS* 58, pp. 231-294

Einstein, A. 1930. "What I believe" in Fadiman, C. ed. *Living Philosophies.* New York: Forum

Eisenberg, D. 1987. *Encounters with Qi.* New York: Penguin

Emerson, R. 1841a. "The over-soul" ***and***

________. 1841b. "Self-reliance"

in Eliot, C. (ed.) 1909. *Essays & English Traits.* New York: Collier

________. 1860. "Power" in *Works of Ralph Waldo Emerson.* (19 —) New York: Black

Engels, F. 1878. *Anti-Dühring.* Burns, E. (tr.) 1996. New York: International Pub.

Feldman, M. (co-writer) 1980. *IN GOD WE TRU$T.* (filmplay)

Fichte, J. 1794. *Wissenschaftslehre*. Heath & Lachs (tr.) Cambridge: University Press

Freud, S. 1930. *Das Unbehagen in der Kultur.* Joan Riviere (tr.) New York: Cape & Smith

Freyd, F. 1972. "Aspects of topoi." *Bull. Austral. Math. Soc.* Vol. 7, 1-76, 467-480

Fromm, E. 1941. *Escape From Freedom.* New York: Holt, Rinehart, & Winston

________. 1973. *The Anatomy of Human Destructiveness.* New York: Holt, Rinehart, & Winston

Galbraith, K. 1967. *The New Industrial State*. Boston: Houghton Mifflin

Gill, B. 1989. "The faces of Joseph Campbell." *The New York Review of Books* Sept. 28, pp. 16-19

Gödel, K. 1931. "Über formal unentscheidbare Sätze der Principia Mathematica und Verwandter System I." *Monatschefte für Matematik und Physik* 38, 173-198

Goldblatt, R. 1979. *Topoi.* Amsterdam: North-Holland

Goldstein, K. 1939. *The Organism.* reprint 1995. New York: Zone Books

Goodall, J. 1971. *In the Shadow of Man.* London: Collins

Goodman, A. & Gilman, A. 1941. *The Pharmacological Basis of Therapeutics.* New York: Macmillan

Greider, W. 1987. *Secrets of the Temple.* New York: Simon & Schuster

________. 1992. *Who Will Tell the People?* New York: Simon & Schuster

________. 1997a. "When optimism meets overcapacity." (Op. Ed.) *The New York Times*, Oct. 1

________. 1997b. *One World, Ready or Not.* New York: Simon & Schuster

Hägerstrom, A. 1911 *Om Moraliska Föreställingars Sanning.* Uppsala

Hall, A. 1929. "Agriculture: A General Survey." *Encyclopaedia Brittanica* (14th ed.) London: EB

Hamblin, C. 1987. *Imperatives.* Oxford: Blackwell

Hampshire, S. 1993. "Liberalism: The New Twist." (a review of Rawls, J. [1993]) *New York Review of Books* Aug. 12, 43- 47

Hansson, B. 1969. "An analysis of some deontic logics." *Nous* 373-398. Reprinted in Hilpinen [1971]

Hare, R. 1952. *The Language of Morals.* Oxford: Clarendon

Harlow, H. 1953. "Mice, monkeys, men & motives." *Psychological Review* 60, 23-32

Hartman, R. S. 1967. *The Structure of Value.* Urbana: South Illinois University Press

Hebb, D. 1953. "On motivation and thought." *Contributions a L'etude des Sciences de L'Homme* 2, 41-47

________. 1955. "Drives and the CNS (conceptual nervous system)." *Psychological Review*. 62, 243-254

Both of the above articles by Hebb are reprinted in

________. 1982. *The Conceptual Nervous System.* London: Pergamon

________. & Donderi, D. 1987. *A Textbook of Psychology.* Hillsdale, N.J.: Erlbaum

Hegel, G. 1837 "Introduction to the philosophy of history." trans. Sibree, J. 1857. in Loewenberg, J. ed. 1929 *Hegel Selections* New York: Scribner

Heidegger, M. 1949. *Sein und Zeit.* Tübingen: Neomarius

Heilbroner, R. 1980. *The Worldly Philosophers.* New York: Simon & Schuster

________. 1990. Quoted in Uchitelle, L. "A Challenge to Scientific Economics." *The New York Times* Jan. 23, pp. A19, 21

Hilpinen, R. (ed.) 1971. *Deontic Logic: Introductory and Systematic Readings.* Dordrecht: Reidel

________. (ed.) 1981. *New Studies in Deontic Logic.* Dordrecht: Reidel

Hofstadter, A. & McKinsey, J. 1939. "On the logic of imperatives." *Philosophy of Science* vol. 6, 446-457

Howard, P. 1995. *The Death of Common Sense.* New York: Random House

Humboldt, W. 1854. *The Spheres and Duties of Government.* London: Chapman

Hume, D. 1739. *A Treatise of Human Nature* London: Noon

________. 1751. *An Inquiry Concerning the Principles of Morals.* (1957 ed.) Indianapolis: Bobbs-Merill

Huxley, A. 1932. *Brave New World.* [1953 ed.] New York: Bantam

James, W. 1890. *The Principles of Psychology.* New York: Holt

________. 1895. "Is life worth living?" *International Journal of Ethics.* Oct. reprinted in

________. 1897. *Will to Believe.* New York: Longmans, Green, & Co.

________. 1900. *Talks to Teachers.* New York: Holt

________. 1911. *Memories and Studies.* New York: Longmans, Green, and Co.

Jefferson, T. 1787a. Letter to Edward Carrington (January 16)

________. 1787b. Letter to James Madison (January 30)

________. 1787c. Letter to William S. Smith (November 13)

________. 1810. Letter to John Langdon

in Padover, S. 1956. *A Jefferson Profile.* New York: John Day

Johanson, A. 1973. "A proof of Hume's separation thesis, based on a formal system for descriptive and normative statements." *Theory and Decision* 3, 339-350

________. 1988. "Imperative logic as based on a Galois connection." *Theoria.* 54 part 1 1-24

________. 1996. "The logic of normative systems." in Brown, M. & Carmo, J. (eds.) [1996]

________. & Puligandla, R. 1972. "Ethics, human adaptation, and the social bond." *Journal of Thought.* vol. 7, 1. 7- 18

Johnson, S. 1755. *Dictionary.* McAdam, E. & Milne. G. (eds.) 1965. *Selections.* New York: Modern Library

Kan, D. 1958. "Adjoint functors." *Trans. AMS.* 87, pp. 294-329

Kanger, S. 1957. *New Foundations for Ethical Theory.* Stockolm. Reprinted in R. Hilpinen [1971] pp. 36-58

________. 1963. "Rättighetsbegreppet." in "Sju filosofistica studier tillägnade Anders Wedberg" *Philosophical Studies* #9, University of Stockholm

________. & Kanger, H. 1966. "Rights and parliamentarianism." *Theoria* 32, 6, 85-115

________. 1972. "Law and logic." *Theoria* 38, 3, 105-132

________. 1977. "Några synpunktor på begreppet inflytande." *Filisofiska Smulor tillägnada Konrad Marc-Wogau* Uppsala Universitetet

________. 1985. "On the realization of human rights." preprint

Kant, I. 1785. *Groundwork of the Metaphysics of Morals.* Paton, H. J. (tr.) 1956. London: Hutchinson

________. 1788. *The Critique of Practical Reason.* Abbott, T (tr.). 1952. Chicago: Benton

Kaufmann, W. 1968. *Nietzsche* 3rd ed. New York: Vintage

Keynes, J. 1933. "National self-sufficiency." in *The Collected Writings of John Maynard Keynes,* v. 21, London: Macmillan

Kierkegaard, S. 1843. *Euten-Eiler* (Either-Or). trans. vol. I, 1941. Swenson & Swenson, vol. II, 1944. Lowrie Princeton: University Press

Kohler, W. 1938. *The Place of Value in a World of Facts.* New York: Liverwright

Kohn, A. 1999. "In pursuit of affluence, at a high price." *The New York Times* Feb. 2, p.D7

Kripke, S. 1963. "A semantic analysis of modal logic I." *Zeitschrift für Mathematische Logic und Grundlagen der Mathematik* vol 9, 67-96

Krishna, Gopi. 1970. *Kundalini.* London: Stuart and Watkins

Krotpotkin, P. 1902. *Mutual Aid.* London: Heineman

Kuhn, T. 1970. *The Structure of Scientific Revolutions.* Chicago: Chicago University Press

Lawvere, W. 1969. "Adjointness in foundations." *Dialectica* vol. 23, No 3/4, 281-295

Lazerson, A. 1975. *Psychology Today.* Third Edition. New York: Random House

Lewis, C. & Langford, C. 1959. *Symbolic Logic.* New York: Dover

Lind, M. 1996. *Up from Conservatism.* New York: Free Press

Lindahl, L. 1977. *Position and Change.* Dordrecht: Reidel

Lorenz, K. 1966. *On Aggression.* New York: Harcourt, Brace, & World

MacIntyre, A. 1981. *After Virtue.* South Bend: Notre Dame

MacLane, S. 1971. *Categories for the Working Mathematician.* New York: Springer

Mailer, N. 1998. review of "A Man in Full" by Tom Wolfe. *The New York Review of Books* (XLV, 20) Dec. 17

Mander, J. & Goldsmith, E. 1996. *The Case Against the Global Economy.* San Francisco: Sierra Club

Markham, E. 1899. "The Man With the Hoe." in Untermeyer, L. (ed) 1942. *A Treasury of Great Poems* New York: Simon & Schuster

Marx, K. 1873. *Das Kapital.* Untermann, E. (ed.) 1906. New York: Kerr

________. 1977 ed. *Selected Writings.* David McLellan, ed. Oxford: Oxford University Press

________. &. Engels, F. 1848. *The Communist Manifesto.* (English Edition 1888.) Beer, S. ed. 1955. New York: Appleton-Century-Crofts

Maslow, A. 1954. *Motivation and Personality.* New York: Harper & Row

________. 1968. *Toward a Psychology of Being.* Second edition. New York: Van Nostrand

________. 1970a. (Second Edition) *Motivation and Personality.* New York: Harper & Row

________. 1970b. *Religion, Values, and Peak Experiences.* New York: Viking

________. 1971 *The Farther Reaches of Human Nature.* Harmondsworth: Penguin

Maugham, S. 1915. *Of Human Bondage.* Salem: Ayer

May, R. 1953. *Man's Search for Himself.* New York: Norton

________. 1969. *Love and Will.* New York: Norton

McKenney, J. 1955. "Rhythmic cycles of skepticism." *Educational Theory* Vol. V, No. 4, 235-241

________. 1958. "Concerning Russell's analysis of value judgments." *Journal of Philosophy* vol. LV, 9, 382-389

McLarty, C. 1993. *Elementary Categories, Elementary Toposes.* Oxford: Clarendon

Mendelson, E. 1987. *Introduction to Mathematical Logic.* Monterey: Wadsworth & Brooks

Meyer, J. 1988. "A different approach to deontic logic: deontic logic as a variant of dynamic logic." *Notre Dame Journal of Formal Logic* 29 (1) 109-136

Mieder, Kingsbury, & Harder 1992. *A Dictionary of American Proverbs.* Oxford: University Press

Mill, J. 1859. *On Liberty.* Shields, C. (ed.) 1956 Indianapolis: Bobbs-Merill

________. 1871. *Utilitarianism.* London: Longmans, Green, & Co.

________. 1873. *Autobiography.* in Eliot, C. (ed.) 1909. Harvard Classics, vol. 25 New York: Collier

________. 1898 ed. *A System of Logic.* Eighth ed. London: Longmans, Green, & Co.

Miller, M. 1996. "Free the media." *The nation* V.262, N.3, June 3

________. 1997. "The crushing power of big publishing." *The Nation* V. 264, N.10, Mar. 17

Monk, R. 1990. *Wittgenstein, the Duty of Genius.* New York: Penguin

Monro, D. 1967. "Bentham, Jeremy" in *Encyclopedia of Philosophy* vol 1 New York: Macmillan

More, T. 1516. *Utopia.* Rhys, E. (ed.) 1910. London: Dent

Morris, C. 1964. *Signification and Significance.* Cambridge (USA): MIT

Morris, D. 1996. "Free trade: the great destroyer." in Mander & Goldsmith

(eds.) [1996]
Moore, G. 1903. *Principia Ethica.* Cambridge: University Press
Munro, H. 1886. translation of *De Rerum Natura.* Reprinted in Oates, W.(ed.) 1940. *The Stoic and Epicurean Philosophers.* New York: Random House
Munroe, M. E. 1963. *The Language of Mathematics.* Ann Arbor: University of Michigan Press
Murray, H. *et al.* 1938. *Explorations in Personality.* New York: Oxford University Press
Myers, D 1987. *Social Psychology.* New York: McGraw-Hill
________. 1989. *Psychology.* Second Edition New York: Worth
Nietzsche, F. 1885. *Thus Spake Zarathustra.* Common, T. (trans.) 190? New York: Boni & Liveright
________. 1886. *Beyond Good and Evil.* Zimmern, W. (trans.) 190?. New York: Boni & Liveright
________. 1889. *Twlight of the Idols.* Hollindale, R. (trans.) 1968. Harmondsworth: Penguin
________. 1974 ed. *The Gay Science.* Kaufmann (trans.) New York: Vintage
Nozick, R. 1974. *Anarchy, State, and Utopia.* New York: Basic Books
Odets, C. 1945. "None but the lonely heart." in Gassner, J. & Nichols, D. eds. 1946. *Best Film Plays of 1945.* New York: Crown
Olson, R. 1967. "Deontological ethics." in *The Encyclopedia of Philosophy* vol 2. London: Macmillan
Paine, T. 1791. *Rights of Man.* 1951 ed. New York: Dutton
Pauling, L. 1986. *How to Live Longer and Feel Better.* New York: Freeman
Perry, R. 1926. *A General Theory of Value.* New York: Longmans Green & Co.
Perutz, M. 1992. "The fifth freedom." *New York Review of Books* Oct. 8, 3-7
Poincaré, H. 1913. "Science and Method." in Halstead, G.(tr.) 1946. *Foundations of Science.* (reprint) Lancaster: Science Press
Pojman, L.1995. *Ethics.* Second Edition Belmont: Wadsworth
Popper, K. 1950. *The Open Society and its Enemies.* Princeton: Princeton University Press
________. 1976. *Unended Quest.* Lasalle: Open Court
________. 1981. *Objective Knowledge.* Oxford: Clarendon
Pörn, I. 1970. *The Logic of Power.* Oxford: Blackwell
Price, L. 1954. *Dialogues of Alfred North Whitehead.* Boston: Little, Brown, & Co.
Ramsey, F. 1950. *The Foundations of Mathematics.* New York: Humanities
Rawls, J. 1971. *A Theory of Justice.* Cambridge (USA): Belnap
________. 1993. *Political Liberalism.* New York: Columbia University Press
Rice, P. 1955. *On the Knowledge of Good and Evil.* New York: Random House
Riesman, D. 1965. *The Lonely Crowd.* New Haven: Yale University Press
Rifkin, J. 1992. *Beyond Beef.* New York: Dutton
________. 1995. *The End of Work.* New York: Putnam

Rogers, C. 1961. *On Becoming a Person.* Boston: Houghton Mifflin
Ross, A. 1941. "Imperatives and logic." *Theoria* 7, 53-71.
Ross, W. 1930. *The Right and the Good.* Oxford: Clarendon
Roy, D. 1950. *Among the Great.* Bombay: Jaico
Russell, B. 1903. "A freeman's worship." reprinted in Russell [1917]
________. 1910. *Philosophical Essays.* London: Longmans, Green, & Co.
________. 1916a. *Justice in Wartime.* Chicago: Open Court
________. 1916b. *Principles of Social Reconstruction.* London: Allen & Unwin
________. 1917. *Mysticism and Logic.* London: Allen & Unwin
________. 1918. *Proposed Roads to Freedom.* London: Allen & Unwin
________. 1922. Introduction to Wittgensteins's *Tractatus Logico-Philosophicus.* London: Kegan Paul *et. al.*
________. 1925. *What I Believe.* New York: Dutton
________. 1926. Education and the Good Life. New York: Boni & Liveright
________. 1927. *Philoso*phy. New York: Norton
________. 1930. *The Conquest of Happiness.* New York: Horace Liveright
________. 1932. "In praise of idleness." *Review of Reviews* v.82, Oct. 48-54 reprinted in
________. 1935. *In Praise of Idleness.* New York: Norton
________. 1938. *Power: A New Social Analysis.* London: Allen & Unwin
________. 1940. *An Inquiry into Meaning and Truth.* London: Allen & Unwin
________. 1942. (p. 418) in *Invitation to Learning.* Cairns, Tate, van Doren (eds.) New York: New Home Library
________. 1944. "Reply to critics." in *The Philosophy of Bertrand Russell.* Schilpp, P. (ed.) Evanston: Northwestern
________. 1945 *A History of Western Philosophy.* New York: Simon & Schuster
________. 1949. *Authority and the Individual.* London: Allen & Unwin
________. 1954. *Human Society in Ethics and Politics.* London: Allen & Unwin
________. 1956. *Portraits from Memory.* New York: Simon & Schuster
________. 1962. *Has Man a Future?* New York: Simon & Schuster
________. 1967. *The Autobiography of Bertrand Russell.* vol. I New York: Simon & Schuster
________. 1969. *The Autobiography of Bertrand Russell.* vol. III New York: Simon & Schuster
________.1972. *The Collected Stories of Bertrand Russell.* New York: Simon & Schuster
________. 1979. "On keeping a wide horizon." *Russell* 33-34 pp. 5-11
________. 1988. "The philosophy of pacifism." in *The Collected Papers of Bertrand Russell* vol. 13 London: Unwin & Hyman
Sacks, O. 1992. "The last hippie." *New York Review of Books* Mar.26, 53-62

Sale, P. 1980. *Human Scale.* New York: Coward, McCann & Geoghegan

Santayana, G. 1913. *Winds of Doctrine.* London: Dent

Sartre, J. 1943. *L'être et Neant.* Paris: Gallimard

________. 1946. *L'existentialism est un Humanisme.* Paris: Nagel

Schopenhauer, A. 1901 ed. *The Wisdom of Life and Other Essays.* Saunders, B. & Bax, E. trans. New York: Dunne

________. 1928 ed. *The Philosophy of Schopenhauer.* Edman, I. ed. New York: Random House

Schumacher, E. 1973. *Small is Beautiful.* New York: Harper

Searle, J. 1964. "How to derive ought from is." *Phil. Rev.* 73. pp. 43-58

Sen, Amartya 1997. "Tagore and his India." *The New York Review of Books* June 26, p. 62

Shelley, P. 1811. Notes on "*Queen Mab*,"

________. 1817. "Hymn to Intellectual Beauty" and

________. 1821. "A Defense of Poetry"

all in

Shelley, M. (ed.) 1889. *Shelley's Poems and Essays.* London: Ward, Lock & Co.

Sidgwick, H. 1901. *Methods of Ethics.* London: Macmillan

Sinclair, T. 1962. "Introduction" in Aristotle. *The Politics.* Harmondsworth: Penguin

Singer, P. 1992. "Bandit and friends" *New York Review of Books* April 9, 9-13

________. 1995. *How Are We to Live?* Amherst: Prometheus

Skinner, B. 1953. *Science and Human Behavior.* New York: Macmillan

Smart, J. 1967. "Utilitarianism." in *The Encyclopedia of Philosophy* vol 8. London: Macmillan

Smith, A. 1776, 1778. *Wealth of Nations.* Everyman ed. in two vols. (date uncertain) London: Dent

Snyder, P. 1971. *Modal Logic.* New York: Van Nostrand Reinhold.

Soros, G. 1998. "Toward a global open society." *The Atlantic Monthly.* Jan., 20-22, 24, 32

Stein, B. 1974. *Size, Efficiency, and Community Enterprise.* Cambridge: Center for Community Economic Development

Spencer, H. 1879. *The Data of Ethics.* New York: Am. Pub. Corp.

________. 1892. *Social Statics* (*Abridged*). New York: Appleton

________. 1897. *The Principles of Ethics. Vol. 1.* (reprint: 1987) Indianapolis: Liberty Fund

________.1900. *First Principles.* Fifth Edition (reprint) New York: Burt

Spinoza, B. 1677. *Ethic.* (translated by W. H. White) in Wild, J. (ed.) 1930. *Spinoza Selections.* New York: Scribner

Stefansson, V. 1939. Essay in Fadiman, C. (ed.) *I Believe.* New York: Simon & Schuster

Stephen, L. 1882. *The Science of Ethics.* London: Smith, Elder

Stevens, W. 1993. "River life through U.S. broadly degraded." *The New York*

Times Jan. 26

Stillman, J. 1993. Letter to the New York Times. *New York Times* National edition, Mar. 19, p. A14

Stolnitz, J. 1967. "Beauty." in *Encyclopedia of Philosophy* New York: Macmillan

Taylor, P. 1975. *Principles of Ethics.* Encino: Dickenson

Teresa, St. 1577. *The Interior Castle.* Peers, E. (trans) 1961. Garden City: Image Books

Thomas, L. 1979. *The Medusa and the Snail.* New York: Viking

Thomason, R. 1981. "Deontic logic as founded on tense logic. in Hilpinen, R. (ed.) [1981]

Thoreau, H. 1849. "Civil disobedience" in 1906 ed. *The Writings of Henry David Thoreau.* Boston: Houghton Mifflin

Thurow, L. 1995. "Why their world might crumble." *The New York Times Magazine* Nov. 19, 78- 79

________. 1996. *The Future of Capitalism.* New York: Penguin

________. 1999. "The boom that wasn't." (op. ed.) *The New York Times*, Jan. 18

Tocqueville, A. de 1862. *Democracy in America.* trans. Reeve, H. 1945 ed. New York: Knopf

Turnbull, C. 1972. *The Mountain People.* New York:Simon & Schuster

Twain, M. 1901. "To the person sitting in darkness." in Smith, J. ed. 1962. *Mark Twain on the Damned Human Race.* New York: Hill & Wang

________.1974 ed. *Letters from the Earth.* New York: Harper & Row

Walljasper, J. 1998. "What Works? Denmark!" *The Nation*, Jan. 26

Ward, G. 1993. "Loved to death." *The New York Times Book Review* March 28

Washington, G. 1796. "Neutrality and natural growth." in Hefner, R. (ed.) 1952. *A Documentary History of the United States* New York: Mentor

Watts, A. 1970. *The Meaning of Happiness.* New York: Harper & Row

Wells, H.G. 1891. "The Rediscovery of the Unique." *Fortnightly Review*, July

________. 1906. *In the days of the Comet.* London: Macmillan

________. 1908. *First and Last Things.* London & New York: Putnam

________. 1913. *The Discovery of the Future.* New York: Huebsch

________. 1920. *The Outline of History.* Garden City: Garden City

________. 1928. *The Open Conspiracy.* London: Gollanz

________. 1931. *The Work, Wealth, and Happiness of Mankind.* Garden City: Doubleday

________. 1935. *What Are We to Do with Our Lives?* London: Watts

Whitehead, A. 1911. *Introduction to Mathematics.* London: Williams & Norgate

________. 1927. *Symbolism, Its Meaning and Effect.* Cambridge: University Press

________. 1929a. *The Aims of Education.* New York: Macmillan

________. 1929b. *Process and Reality.* New York: Macmillan

________. 1933. *Adventures of Ideas.* New York: Macmillan

________. 1938. *Modes of Thought.* New York: Macmillan

________. 1941. "Mathematics and the good." reprinted (pp. 666-681) in

________. 1951. *The Philosophy of Alfred North Whitehead.* Library of Living Philosophers, vol. III New York: Tudor

________. & Russell, B. 1910. *Principia Mathematica.* Cambridge: Cambridge University Press

Whitman, W. 1921ed. *Leaves of Grass.* New York: Modern Library

Wilson, E. 1992. *The Diversity of Life.* Cambridge, Mass: Belnap

________. 1993. "Is humanity suicidal?" *The New York Times Magazine.* May 30, 24-29

Wilson, W. 1996. *When Work Disappears.* New York: Knopf

Wittgenstein, L. 1922. *Tractatus Logico-Philosophicus.* London: Kegan Paul, Trench, Trubner & Co.

________. 1953. *Philosophical Investigations.* London: Blackwell

Wright, G. von 1951. "Deontic logic." *Mind.* 60, 1-15

________. 1981. "On the logic of norms and actions." in Hilpinen, R. (ed.) [1981].

Yutang, L. 1942. *The Wisdom of China and India.* New York: Random House

INDEX